Mathematical Approaches to Modeling, Optimally Designing, and Controlling Electric Machine

Mathematical Approaches to Modeling, Optimally Designing, and Controlling Electric Machine

Editors

Vladimir Prakht
Mohamed N. Ibrahim
Aleksey S. Anuchin

MDPI • Basel • Beijing • Wuhan • Barcelona • Belgrade • Manchester • Tokyo • Cluj • Tianjin

Editors

Vladimir Prakht
Ural Federal University
Russia

Mohamed N. Ibrahim
Ghent University
Belgium

Aleksey S. Anuchin
Moscow Power Engineering Institute
Russia

Editorial Office
MDPI
St. Alban-Anlage 66
4052 Basel, Switzerland

This is a reprint of articles from the Special Issue published online in the open access journal *Mathematics* (ISSN 2227-7390) (available at: https://www.mdpi.com/journal/mathematics/special_issues/Modeling_Optimally_Designing_Controlling_Electric_Machine).

For citation purposes, cite each article independently as indicated on the article page online and as indicated below:

LastName, A.A.; LastName, B.B.; LastName, C.C. Article Title. *Journal Name* **Year**, *Volume Number*, Page Range.

ISBN 978-3-0365-2644-7 (Hbk)
ISBN 978-3-0365-2645-4 (PDF)

Contents

About the Editors

Vladimir Prakht received a graduate degree in engineering and a Ph.D. degree from the Department of Electrical Engineering, Ural Federal University, Yekaterinburg, Russia, in 2004 and 2007, respectively. He submitted his Ph.D. dissertation on optimal control and mathematical modeling induction heating systems in 2006. He is an Associate Professor with the Department of Electrical Engineering, Ural Federal University. His research interests include mathematical modeling and optimal design of energy efficient electric motors and generators.

Mohamed N. Ibrahim received a B.Sc. degree in electrical power and machines engineering from Kafrelshiekh University, Egypt, in 2008, an M.Sc. degree in electrical power and machines engineering from Tanta University, Egypt, in 2012, and a Ph.D. degree in electromechanical engineering from Ghent University, Belgium, in 2017. In 2008, he became a Teaching Assistant in the Electrical Engineering Department, Kafrelshiekh University. He is currently a Post-Doctoral Researcher with the Department of Electromechanical, Systems and Metal Engineering, Ghent University, Belgium. He is an Assistant Professor (on leave) with the Department of Electrical Engineering, Kafrelshiekh University. His major research interest includes design and control of electrical machines and drives for industrial and sustainable energy applications. Dr. Ibrahim was several times a recipient of the Kafrelshiekh University Award for his international scientific publications. He serves as a Reviewer in several journals and conferences including *IEEE Trans. on Industrial Electronics, Industry Applications, Magnetics*, etc. He serves as a Guest Editor in *Energies, Mathematics,* and *Electronics* journals.

Aleksey S. Anuchin received their B.Sc., M.Sc., Ph.D., and Dr.Eng.Sc. degrees from Moscow Power Engineering Institute, Moscow, Russia, in 1999, 2001, 2004, and 2018, respectively. He delivers lectures on "control systems of electric drives," "real-time software design," "electric drives," and "science research writing" at Moscow Power Engineering Institute. He has been in a head position with the Electric Drives Department for the last eight years. He has more than 20 years of experience covering control systems of electric drives, hybrid powertrains, and real-time communications. He is the author of three textbooks on the design of real-time software for the microcontroller of the C28 family and Cortex-M4F, and control system of electric drives (in Russian). He has authored or coauthored more than 100 conference and journal papers.

Preface to "Mathematical Approaches to Modeling, Optimally Designing, and Controlling Electric Machine"

An electric machine is the main core of electric drives in industrial, transportation, and domestic applications, as well as in traditional and renewable energy generation systems. A pre-experimental evaluation of the electric machine performance for a given application is always based on a mathematical model. The mathematical model accuracy and methodology vary depending on the application requirements. In addition, the methods of optimal design of electric machines significantly facilitate reaching these requirements. Most of the requirements for electrical machine design are in contradiction to each other (reduction in volume or mass, increase in efficiency and power density, etc.). Therefore, finding the optimal design that will achieve all of them can be a massive task due to a large number of varied parameters whose effects on the machine performance and quality of the design are strongly coupled. Therefore, the optimal design methodology of electric machines is always necessary. The reliable and efficient operation of the electrical machine is impossible without precise control. Therefore, the control strategies, state observers, and their mathematical models which help to check the approaches of optimal and efficient control, are important. The main topics of this Special Issue include, but are not limited to:

- Analytical models (electromagnetic, thermal, etc.) of electric machines;

- Numerical models (finite element method, boundary element method, equivalent circuits, etc.) of electric machines;

- Multi-physics models of electric machines;

- Lifetime modeling of electric machines;

- Losses modeling of electric machines;

- Optimal design methodologies of electric machines;

- Optimization techniques for fast and efficient optimal design of electric machines;

- Optimal control techniques of electric.

Vladimir Prakht, Mohamed N. Ibrahim, Aleksey S. Anuchin
Editors

 *mathematics*

Article

Optimal Design of a High-Speed Flux Reversal Motor with Bonded Rare-Earth Permanent Magnets

Vladimir Prakht, Vladimir Dmitrievskii and Vadim Kazakbaev *

Department of Electrical Engineering, Ural Federal University, 620002 Yekaterinburg, Russia;
va.prakht@urfu.ru (V.P.); vladimir.dmitrievsky@urfu.ru (V.D.)
* Correspondence: vadim.kazakbaev@urfu.ru; Tel.: +7-343-375-4507

Abstract: Single-phase flux reversal motors (FRMs) with sintered rare-earth permanent magnets on the stator for low-cost high-speed applications have a reliable rotor and a good specific power. However, to reduce eddy current loss, the sintered rare-earth magnets on the stator have to be segmented into several pieces and their cost increases with the number of magnet segments. An alternative to the sintered magnets can be bonded magnets, in which eddy current loss is almost absent. The remanence of bonded magnets is lower than that of sintered magnets, and they are prone to demagnetization. However, the cost of low-power motors with bonded magnets can be lower because of the simpler manufacturing technology and the lower material cost. This paper discusses various aspects of the optimal design of FRM with bonded magnets, applying the Nelder–Mead method. An objective function for optimizing an FRM with bonded magnets is designed to ensure the required efficiency, reduce torque oscillations, and prevent the bonded magnets from demagnetizing. As a result, it is shown that the FRM with bonded magnets has approximately the same efficiency as the FRM with sintered magnets. In addition, the peak-to-peak torque ripple is minimized and the minimal instantaneous torque is maximized.

Keywords: demagnetization; electric machine; flux reversal machine; high-speed electrical machine; high-speed electrical motor; Nelder–Mead method; optimal design

Citation: Prakht, V.; Dmitrievskii, V.; Kazakbaev, V. Optimal Design of a High-Speed Flux Reversal Motor with Bonded Rare-Earth Permanent Magnets. *Mathematics* **2021**, *9*, 256. https://doi.org/10.3390/math9030256

Academic Editor: Mario Versaci
Received: 12 December 2020
Accepted: 25 January 2021
Published: 28 January 2021

Publisher's Note: MDPI stays neutral with regard to jurisdictional claims in published maps and institutional affiliations.

1. Introduction

Single-phase variable-speed motors are widely used in low-cost applications such as vacuum cleaners [1], blowers [2], power tools [3], pumps [4], compressors [5], and fans [6]. The main advantage of single-phase motors is the lower cost of the semiconductor inverter that has fewer transistors than in the case of a three-phase motor.

In turbochargers, vacuum cleaners, blowers, and other high-speed applications, a retaining ring on the rotor of such a brushless motor is used to provide strength against centrifugal forces and to increase its reliability. However, this also increases the cost of the rotor and its complexity. In addition, applying a retaining ring also increases the equivalent air gap, i.e., the gap between the rotor magnets and the stator teeth. This reduces the specific torque and efficiency [7].

Therefore, in low-cost high-speed applications, it can be preferable to use motors with a simple and reliable toothed rotor and with magnets on the stator, such as flux switching motors (FSMs) [8,9], flux reversal motors (FRMs) [10,11], or hybrid switched reluctance motors (HSRMs) [12,13]. Applying a structurally simple toothed rotor made of a steel lamination reduces the cost of rotor manufacturing in comparison with a rotor with magnets on its surface and with a retaining ring [14] and also increases the reliability of the motor. However, to reduce eddy current loss, the sintered rare-earth magnets on the stator have to be segmented into several pieces [10,11,15]. The cost of the magnets increases with the number of magnet segmentations [16,17].

An alternative to rare-earth sintered magnets can be bonded magnets, in which eddy current loss is almost absent. The remanence of bonded magnets is lower than that of sintered rare-earth magnets, but the cost of low-power motors with bonded magnets is lower compared to motors with sintered rare-earth magnets because of the simpler manufacturing technology and also the lower material cost [18–20]. For this reason, the use of motors with bonded magnets is also prospective in low-cost applications. Therefore, the aim of this work is to optimize a single-phase FRM with bonded magnets to use in low-cost high-speed applications.

The optimization of motors with bonded magnets on the rotor is considered in [21–23]. An evolutionary approach is applied in [21]. A hand-made optimization is performed in [22]. A quasi-Newtonian technique is used in [23]. However, optimization of a single-phase FRM with bonded magnets has not been considered yet.

When designing electrical machines, multi-criteria optimization methods are extensively used [24–26]. Such methods can create a Pareto front that includes solution points in which no objective can be improved without degrading others. Then, from all points of the obtained Pareto front, one point is manually selected, the characteristics of which most satisfy the solution of a given technical problem. Typically, multi-criteria methods require a substantial number of function calls. For example, in [24], about 3000 calls are required.

A one-criterion method can also be applied to the problem of optimization of electric machines. In this case, the objective function is constructed with a number of multipliers, each of which is an expression of a separate criterion. The importance of each multiplier can be adjusted when constructing such a function. Therefore, if, in the case of the Pareto approach, the final design is chosen among the points of a Pareto front after the optimization process to satisfy the practical importance of each of the optimization criteria, in the case of a one-criterion method, it is necessary to determine the importance before the start of the optimization process. However, the use of one-criterion methods allows a significant reduction in computational time. For example, in [10], the Nelder–Mead method was applied and only 115 function calls were required, which is 26 times less than in [24], when a multi-criteria method was applied.

It can be concluded that although the Nelder–Mead method does not provide confidence in finding the global optimums according to the Pareto criterion, it significantly reduces the computational time and can provide good practical results.

In this work, a multi-criteria product-type optimization function was built and the one-criterion Nelder–Mead method was applied to reduce the calculation time.

The Nelder–Mead method has previously been successfully applied to optimize various types of electrical machines using various objective functions [10,11,14,27,28]. In each of these works, a special objective function was constructed which depends on the specifics of an electrical machine and application. In [28], optimization of a synchronous reluctance motor was considered by applying this method. A motor efficiency much higher than the IE5 level ("ultra premium efficiency") according to the IEC (International Electrotechnical Commission) 60034–30-2 standard was reached. Torque ripple and underload efficiency also were improved. In [11], optimization of a single-phase high-speed FRM with sintered rare-earth magnets was carried out to reduce torque oscillations and power loss when operating with a fan load profile.

However, the optimization criterion developed for an FRM with sintered rare-earth magnets [11] cannot be applied to an FRM with bonded magnets, since in this case, there is a risk of demagnetization of bonded magnets. This paper discusses various aspects of the optimal design of FRMs with bonded magnets applying the Nelder–Mead method. An objective function for optimizing the FRM with bonded magnets is designed to ensure at least 84% efficiency, reduce torque oscillations, and prevent the bonded magnets from demagnetizing.

2. Construction of an Objective Function for Single-Phase Flux Reversal Motor with Bonded Permanent Magnet

In [11], the optimization of a 754-W, 18-krpm single-phase FRM with segmented sintered rare-earth magnets (NdFeB) is considered for fan load (the torque is proportional to the square of the rotational speed). Table 1 determines 2 considered loading modes of the motor range.

Table 1. Loading points of the flux reversal motor (FRM).

Mode Number	Torque, N·m	Rotational Speed, rpm	Rotational Speed, %	Mechanical Power, W
1	0.256	14,400	80	386
2	0.4	18,000	100	754

Considering that the portion of time taken by a particular loading mode is supposed to be approximately inversely proportional to active (real) power, the arithmetic average efficiency in the two modes ($<\eta>$) was chosen in [11] as one of the optimization objectives of the FRM with sintered rare-earth magnets. The efficiency of the FRM with sintered rare-earth magnets is 85.7% in mode 1 and 84.5% in mode 2. The average efficiency is $<\eta>$ = 85.1% in this case.

In addition to the efficiency, other objectives for optimizing the FRM with sintered rare-earth magnets described in [11] are:

1. Maximizing the minimum instantaneous value of the torque waveform and making it positive if it is possible.
2. Reducing the peak-to-peak value of torque ripple (*PPTR*).

To achieve both of these objectives using the Nelder–Mead method, the optimization criterion was formulated as $A = <PPTR + 2 \cdot AMinTD>$, where *AMinTD* is the average to minimum torque difference; "$< >$" is the sign of the arithmetic mean over both loading modes considered. Therefore, the following optimization function was applied in [11] to the FRM with sintered rare-earth magnets:

$$F = A \cdots B = < PPTR + 2 \cdot AMinTD > (1 - < \eta >), \tag{1}$$

Modern sintered rare-earth magnets not only have high remanence but are also highly coercive magnets. In Gaussian units, the coercivity of such magnets can be 2 or more times higher than their remanence. For this reason, the risk of demagnetization was not taken into account in the objective function when optimizing the FRM with sintered rare-earth magnets in [11]. The modern bonded magnets have significantly lower remanence than sintered rare-earth magnets, which leads to an increase in the required magnetomotive force of the winding (MMF) and, therefore, to the risk of demagnetization. In addition, the coercivity of bonded magnets is much lower than that of sintered rare-earth magnets. Therefore, when designing FRMs with bonded magnets, it is necessary to consider the risk of demagnetization. Bonded magnets of 3210 grade (B_r = 7 kG, H_{ci} = 9.5 kOe) [29] were chosen for the FRM.

The main objective in this study was also to minimize the torque oscillations (term *A* in [1]) of the FRM with bonded magnets. In addition, the average efficiency of the motor must be at least 84% (i.e., approximately the same value as in [11]) and the risk of demagnetizing the bonded magnets must be avoided. It was assumed that the demagnetizing magnetic field must not exceed the coercivity by more than 0.3% of the total volume of the magnets.

The unconstrainted one-criterion Nelder–Mead method is applied in this study to optimize the FRM design. The Nelder–Mead method belongs to unconstrained optimization methods. Therefore, there is no need to set the permissible range of parameters before the optimization. It is necessary to determine only the initial design and the objective function. The noise of the objective functions is caused, firstly, by rounding errors of the finite element method and, secondly, by differences in the mesh that is rebuilt with each different call of the objective function.

It is possible to compose an objective function that takes the value *A* if $<\eta>$ is more than 84% and the volume of the demagnetized magnets is less than 0.3% and goes to infinity otherwise. However, first, the initial set of design parameters may not satisfy the formulated constraints, and the optimization direction cannot be determined. Second, whenever these conditions are not satisfied, the size of the simplex constructed by the Nelder–Mead algorithm decreases, and this will slow down the search for the optimum.

In this work, instead of processing the constraints precisely, an objective function with soft constraints is adopted; that is, penalty factors increasing rapidly if the constrained characteristics of the machine go beyond the specified limits are added to the optimization function [30]. In the considered case, the second factor will increase exponentially if the average efficiency becomes less than 0.84, and the third factor will increase exponentially if the volume of demagnetized magnets becomes greater than 0.3%:

$$F = A(1 + e^{\delta_1(0.84-<\eta>)})(1 + e^{\delta_2(\max(S)-0.003)}) \tag{2}$$

where η is the motor efficiency; S is the volume part of the demagnetized magnets; max is the maximum value among the loading modes considered. In addition to the first factor A, expression (2) contains two other factors. The second factor grows rapidly at $<\eta>$ <0.84 and rapidly tends to unity at $<\eta>$> 0.84. The third factor grows rapidly at $\max(S)$ >0.003 and rapidly tends to unity at $\max(S)$ <0.003. Multipliers $\delta_1 = 200$ and $\delta_2 = 3000$ set the slope of F. Both exponents equal to 1 at $<\eta> = 0.84$ and $\max(S) = 0.003$. The exponent in the first multiplier becomes equal to e when $<\eta>$ increases by 0.5%. The exponent in the second multiplier becomes equal to e when $\max(S)$ increases by 0.00033, which is 11% of its reference value 0.003.

3. Initial Design and Parameters Varied During Optimization

Figure 1 demonstrates the main geometric parameters of the FRM with bonded magnets. As in the case of the FRM with sintered magnets, there are two poles on the surface of each stator tooth; there is the same direction of magnetization of adjacent magnetic poles on adjacent stator teeth. In contrast to sintered magnets, the segmentation of the magnets is not required in the case of bonded magnets. Therefore, the pole pair placed on each tooth can be formed by, for example, one two-pole magnet, which simplifies the production of the rotor assembly and reduces the cost. Here, 35PN440 electrical steel with a thickness of 0.35 mm [31] was selected to determine the properties of the stator and rotor laminations.

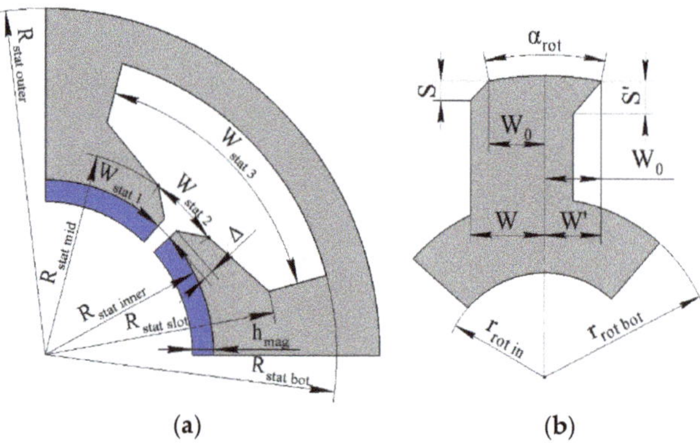

Figure 1. Design parameters of the FRM: (a) stator; (b) rotor.

Table 2 shows the design parameters varied during the optimization of the FRM with bonded magnets: the number of turns per phase; outer stator radius $R_{stat,inner}$; stator slot dimensions $R_{stat,inner}$, $R_{stat,bottom}$ and $W_{stator1}$; rotor dimensions α_{rot}, w/w_0, w'/w_0, and shifting angle of the supply voltage. Furthermore, $R_{stat,middle} = 0.5\cdots(R_{stat,slot} + R_{stat,inner})$.

Table 2. Parameters that changed during the optimization.

Parameter	Before	After
Number of turns per phase	80–10	111
$R_{stat,inner}$, mm	18 + 1	19.2
$R_{stat,slot}$, mm	22 + 1	23.8
$R_{stat,bottom}$, mm	26 + 1	28.3
$W_{stator1}$, degrees	12.6 + 1.8	12.5
α_{rot}, degrees	27–4.5	22.9
w/w_0	1.5–0.2	1.78
W'/w_0	0.8 + 0.2	0.72
Voltage shift, electrical radians	0.015 + 0.01	0.033

In the beginning, the Nelder–Mead algorithm constructs the initial simplex and calculates the value of the objective function in each point of this simplex. In the MATLAB function "fminseach", to define points of the simplex, each parameter of the initial design, in turn, increases by 5%. Such an approach has some disadvantages. First, changing some parameters by 5% can lead to a very significant change in the objective function, but not for others. Changing various parameters by 5% can affect the change in the objective function to a very different extent.

Table 3 shows the parameters that did not vary during the optimization.

Table 3. Parameters that were fixed during the optimization.

Parameter	Value
Supply voltage, V	320
Stator stack length L, mm	30
Stator outer radius, R_{stat_outer}, mm	32
Stator slot width W_{stat3}, degrees	72.9
Stator slot width W_{stat2}, degrees	36.5
Δ, mm	0.007
Magnet thickness, mm	2
Air gap, mm	0.5
Magnet's remanence, T	0.65
s, mm	2
S', mm	3
$R_{rot\,bot}$, mm	7
R_{inr}, mm	3

For example, an increase in $R_{stat,bottom}$ by only 5% causes a significant decrease in the stator yoke thickness and, therefore, can lead to a significant change in saturation level, while an increase in α_{rot} by 5% does not lead to a significant change in the performance. Of course, it is possible to try to select the parameters in such a way so as to avoid this problem. For example, the thickness of the yoke can be used instead of $R_{stat,bottom}$. However, it is more convenient to set the increment for each parameter separately. Second, with an increment in a certain parameter when constructing the next point of the simplex, the objective function may accidentally decrease. Then, having a more optimal design, it is not reasonable to increase the next parameter of the initial design. Therefore, in this study, augmentation of the next parameter is performed for the best design from the already constructed points of the simplex. Thus, in this work, the simplest optimization is carried out already at the stage of constructing a simplex. To construct a simplex, an initial design is first calculated. Then, at each step, the best of the calculated points is selected, and the next parameter is changed. The procedure continues until the number of points of the simplex is one more than the number of parameters.

Therefore, the procedure of building the simplex is as follows:

1. The array x of the initial design parameters and the array d of their increments are given.

2. $F_{min} = F(x)$. $x_{min} = x$. Simplex = {x}.
3. For i from 1 to n (where n is the number of parameters):

    a. $x = x_{min}$.
    b. $x(i) = x(i) + d(i)$.
    c.   Simplex = Simplex $\cap$ {x}.
    d.   If $F(x) < F_{min}$, then $x_{min} = x$, $F_{min} = f(x)$.

After the simplex was built, the Nelder–Mead algorithm, as it is described in [32], was applied. The reflection, expansion, and contraction coefficients are 1, 2, and $\frac{1}{2}$, respectively.

4. Optimization Results of FRM with Bonded Permanent Magnets and Discussion

The Nelder–Mead algorithm, described in [32], was used in designing a new FRM with bonded magnets. The number of optimization parameters was nine. The mathematical model described in [33] was used to evaluate the objectives <η> and max(S) included in the optimization function (2).

Figure 2 shows the FRM designs and the flux density magnitude plots before and after the optimization. An asymmetrical rotor was chosen for the FRM since it was shown in [11,14] that such a rotor design provides positive values of the torque waveform during the entire period, in contrast to a symmetrical rotor.

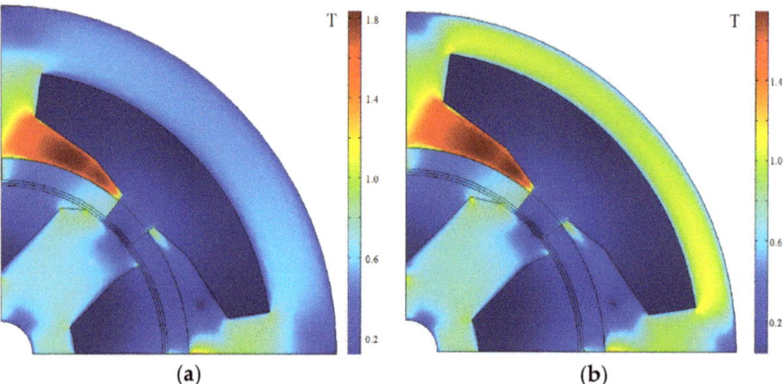

Figure 2. Calculation area arrangement and flux density magnitude plot (T) in the FRM (**a**) before optimization and (**b**) after.

Table 2 shows the variable design parameters of the FRM before and after the optimization. Next to the initial value of the variable parameter, its increment at constructing the initial simplex is given. As seen in Table 2 and Figure 2, the maximum flux density decreased as a result of optimization. In addition, the thickness of the yoke, which was underutilized in the initial design, was reduced, which increased the flux density in the yoke. It made it possible to increase the value of $R_{stat,inner}$. Figure 3 shows the dependence of the torque on the rotor angular position before and after optimization for two loading modes of the FRM with bonded magnets. Figures 4 and 5 demonstrate the waveforms of the current and voltage depending on the rotor angular position before and after optimization for two loading modes of the FRM with bonded magnets. *PPTR* and *AMinTD* changed only slightly. This is because the initial design was obtained by scaling the design of the optimized FRM with sintered rare-earth magnets [11].

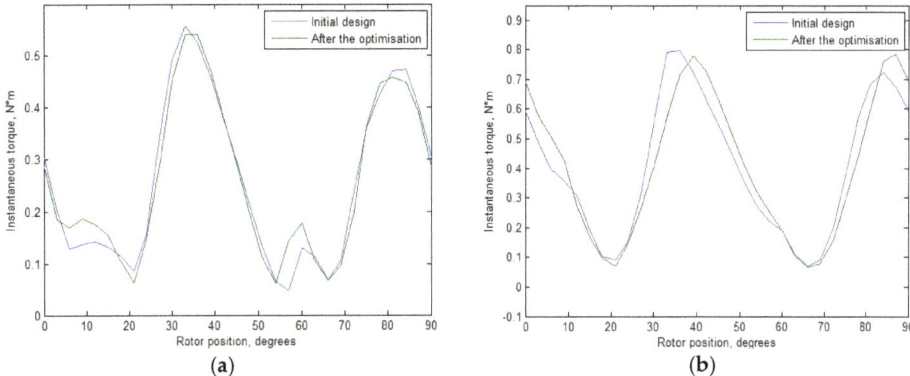

Figure 3. Torque waveforms versus the rotor angular position before and after the optimization: (**a**) mode 1; (**b**) mode 2.

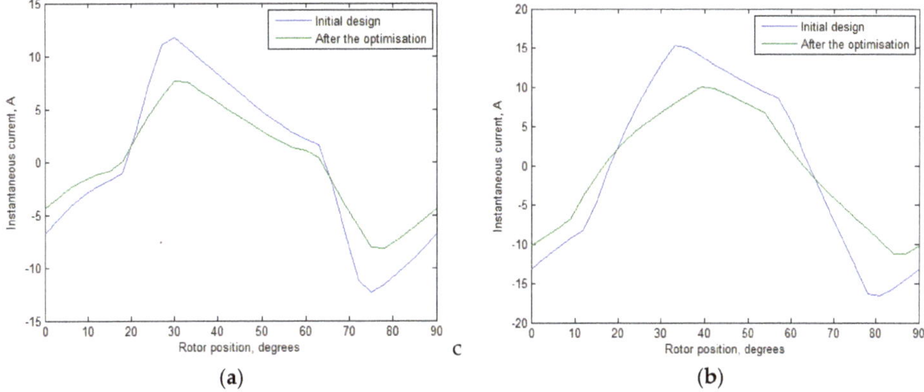

Figure 4. Current waveforms versus the rotor angular position before and after the optimization: (**a**) mode 1; (**b**) mode 2.

Figure 5 shows the dependence of the voltage on the rotor position. The voltage waveform is rectangular. The discretization that leads to the non-rectangularity of this voltage waveform on the plots is clearly visible. The motor torque is controlled by adjusting the voltage duty cycle. The more torque (mechanical power) that is required, the longer the duty cycle. After optimization, the duty cycle increased in both considered loading modes, taking the optimal values. This is achieved, in particular, by increasing the number of turns. With increasing the number of turns, the current decreased, as can be seen in Figure 4.

Figure 6b shows the dependence of the objective function on the number of the function calls. The very large values of the objective function at the beginning of the optimization process are not shown on the plot for better visibility of small changes in the objective function.

Figure 6a shows this dependence at the stage of constructing the simplex. The value of the objective function corresponding to the initial design is 6.84. Already at the stage of constructing the simplex, it was possible to find an improved design with an objective function value of 1.77. This improvement was achieved by fulfilling the soft constraints on the motor efficiency and the volume of the demagnetized magnets. Figure 7a shows the change in average efficiency during optimization. Figure 7b shows the variation in the motor torque oscillations A during optimization. Figure 8a shows the change in the volume percentage of demagnetized magnets during optimization. Figure 8b shows the comparison of the motor losses at two considered operating points before and after optimization. Table 4 shows the characteristics of the FRM before and after optimization.

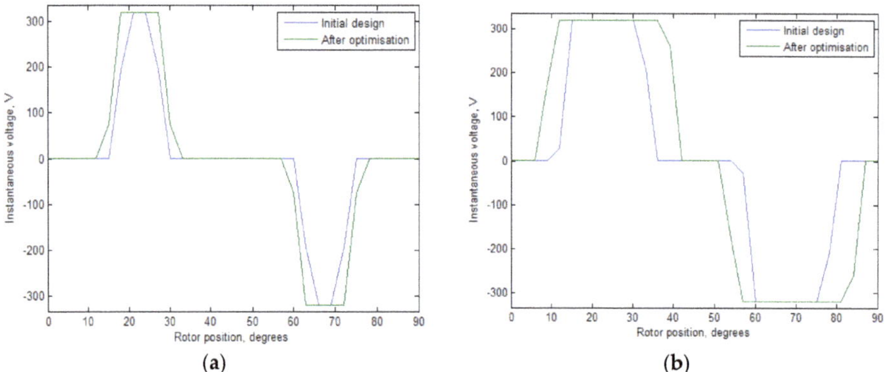

(a)   (b)

Figure 5. Voltage waveforms versus the rotor angular position before and after optimization: (**a**) mode 1; (**b**) mode 2.

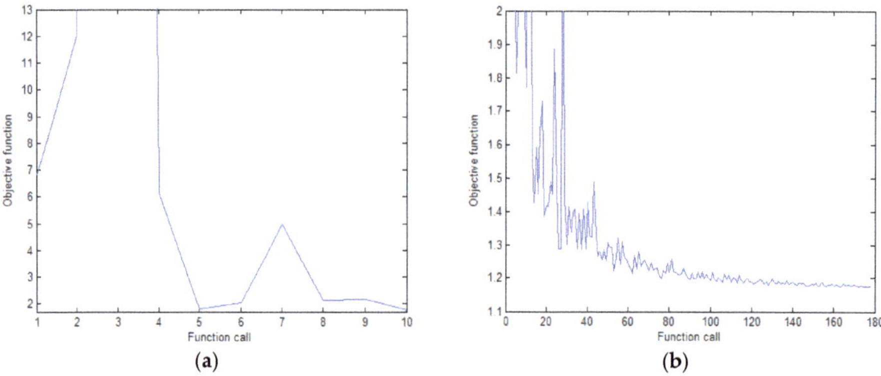

(a)   (b)

Figure 6. The objective function change during building of the simplex (**a**) and during the optimization (**b**). Different scales in plots (**a**,**b**) are chosen to better show the change in the objective function when building the initial simplex.

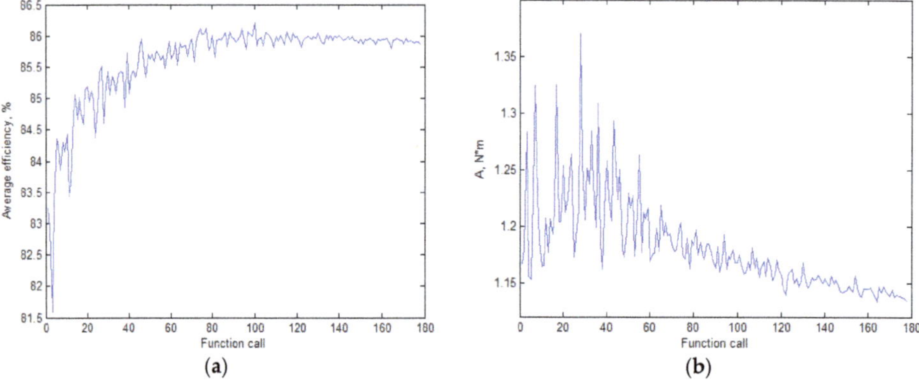

(a)   (b)

Figure 7. (**a**) Variation of the average efficiency during the optimization; (**b**) variation of the torque oscillations during the optimization.

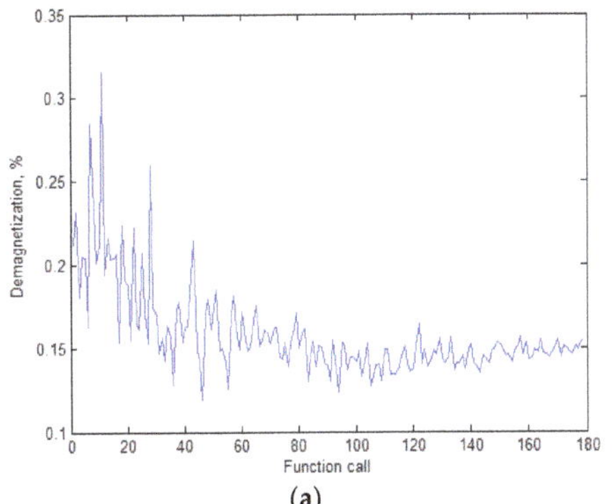

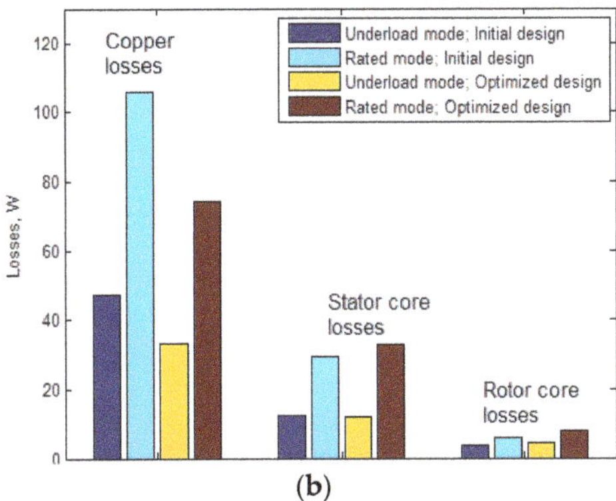

(a)   (b)

Figure 8. (**a**) Variation of the demagnetization percentage during the optimization; (**b**) losses of both FRMs in two modes.

Table 4. The FRM characteristics before and after optimization.

Parameter	Before Optimization		After Optimization	
Rotational speed, rpm	14,400	18,000	14,400	18,000
Electric frequency, kHz	0.96	1.2	0.96	1.2
Current, A (RMS)	7.3	11	4.8	7.2
Efficiency, %	83.9	82.6	86.6	85.2
Total losses, W	72.2	156	58.2	128.5
Mechanical power, W	386	754	386	754
Electric power, W	449	895	435	868
Duty cycle	0.213	0.449	0.3	0.69
Minimal instantaneous torque, N·m	0.048	0.07	0.063	0.068
$AMinDT$, N·m	0.21	0.33	0.193	0.33
$PPTR$, N·m	0.510	0.729	0.479	0.718
A, N·m		1.16		1.12
$PPTR$, % of the average value	199	182	187	180
Magnets deterioration, %	0.056	0.21	0.059	0.15
Stator core mass, g		340		260
Rotor core mass, g		96		105
Magnets volume, cm^3		5.5		5.9
Copper mass, g		239		295

It can be seen from Table 4 that a design that does not satisfy the optimization constraints can be chosen as an initial approximation; its average efficiency was only 83.2% while the constraint was <η> > 84%. The total losses of the FRM after optimization decreased by 1.2 times. The average efficiency after optimization was 85.9%, and the efficiency constraint was satisfied. In the initial design, the volume of demagnetized magnets was unacceptably high and equal to 0.21%. After optimization, it was equal to 0.15%, while the constraint was not more than 0.3%. Thus, the conditions for demagnetization were met with a margin. The value of the parameter A, which describes the oscillations of the torque, was reduced by 4%.

5. Conclusions

The cost of bonded magnets is lower compared with sintered magnets. Furthermore, in contrast with sintered magnets, bonded magnets are dielectric. Therefore, there is no need for segmentation of bonded magnets when using them in a motor, which leads to a lower motor price and a simpler manufacturing technology for a high-speed motor. However, bonded magnets have lower remanence and coercivity. In this study, the problem of developing a flux reversal motor (FRM) with bonded magnets was considered, in which the proposed FRM with bonded magnets has approximately the same efficiency as an FRM with sintered magnets [11]. In addition to this requirement, the peak-to-peak value of torque ripple (*PPTR*) is minimized and the minimum instantaneous torque value in two operating modes is maximized.

Due to the lower coercivity of bonded magnets, it is necessary to avoid their demagnetization in the optimized design. For this purpose, a procedure has been developed for the optimal design of a high-speed single-phase FRM with bonded magnets.

The main challenge in this work was the development of a single-phase flux reversal motor with an asymmetrical rotor which was introduced in [11], which has low torque ripple and no negative torque periods when applying bonded magnets, considering a new constraint on the volume of demagnetized magnets. The constraint must be taken into account to enable the use of bonded magnets for this motor. For this, in particular, the optimization criterion (2) with soft constraints was composed.

The optimization criterion was constructed in such a way so as to maximize the efficiency, reduce torque ripple, and reduce the volume of the demagnetized bonded magnets. The one-criterion Nelder–Mead method was applied in this work to optimize the FRM design. After the optimization, the total losses of the FRM decreased by 1.2 times. The demagnetization constraint was met with a margin. The torque oscillations A were reduced by 4%.

In future works, we plan to compare various optimization methods in the design of a single-phase flux reversal motor with an asymmetrical rotor and to manufacture and test an experimental prototype of the FRM.

Author Contributions: Conceptual approach, V.P. and V.D.; data curation, V.D. and V.K.; software, V.D. and V.P.; calculations and modeling, V.P., V.D. and V.K.; writing—original draft, V.P., V.D. and V.K.; visualization, V.D. and V.K.; review and editing, V.P., V.D. and V.K. All authors have read and agreed to the published version of the manuscript.

Funding: This research received no external funding.

Data Availability Statement: All data are contained within the article.

Acknowledgments: The authors thank the editors and reviewers for their careful reading and constructive comments.

Conflicts of Interest: The authors declare no conflict of interest.

References

1. Bentouati, S. Permanent magnet brushless dc motors for consumer products. In Proceedings of the 9th International Conference on Electrical Machines and Drives, Canterbury, UK, 1–3 September 1999; pp. 118–122. [CrossRef]
2. Lee, J.; Lee, E.; Kim, J. Design of the single phase SRM for the blower considering self-starting. In Proceedings of the 2005 International Conference on Electrical Machines and Systems, Nanjing, China, 27–29 September 2005; pp. 667–670. [CrossRef]
3. Dmitrievskii, V.; Prakht, V.; Pozdeev, A.; Klimarev, V.; Mikhalitsyn, A. Single-phase Flux reversal motor for Angular grinder. In Proceedings of the 8th IET International Conference on Power Electronics, Machines and Drives (PEMD 2016), Glasgow, UK, 19–21 April 2016; pp. 1–6. [CrossRef]
4. Ostovic, V. Performance comparison of U-Core and round-stator single-phase permanent-magnet motors for pump applications. *IEEE Trans. Ind. Appl.* **2002**, *38*, 476–482. [CrossRef]
5. Lee, W.; Han, D.; Sarlioglu, B. Comparative Performance Analysis of Reference Voltage-Controlled Pulse Width Modulation for High-Speed Single-Phase Brushless DC Motor Drive. *IEEE Trans. Power Electron.* **2017**, *33*, 4560–4568. [CrossRef]
6. Hu, H.-J.; Cao, G.-Z.; Huang, S.-D.; Wu, C.; Peng, Y.-P. Drive circuit-based torque-ripple suppression method for single-phase BLDC fan motors to reduce acoustic noise. *IET Electr. Power Appl.* **2019**, *13*, 881–888. [CrossRef]
7. Gieras, J. Design of high-speed permanent magnet machines. *Przeglad Elektrotechniczny* **2015**, *95*, 1–8. [CrossRef]
8. Zhang, Z.; Tang, X.; Wang, D.; Yang, Y.; Wang, X. Novel Rotor Design for Single-Phase Flux Switching Motor. *IEEE Trans. Energy Convers.* **2018**, *33*, 354–361. [CrossRef]
9. Chen, Y.; Chen, S.; Zhu, Z.Q.; Howe, D.; Ye, Y.Y. Starting Torque of Single-Phase Flux-Switching Permanent Magnet Motors. *IEEE Trans. Magn.* **2006**, *42*, 3416–3418. [CrossRef]
10. Dmitrievskii, V.; Prakht, V.; Kazakbaev, V.; Sarapulov, S. Optimal Design of a High-Speed Single-Phase Flux Reversal Motor for Vacuum Cleaners. *Energies* **2018**, *11*, 3334. [CrossRef]
11. Dmitrievskii, V.; Prakht, V.; Kazakbaev, V.; Golovanov, D. Optimum Design of High-Speed Single-Phase Flux Reversal Motor with Reduced Torque Ripple. *Appl. Sci.* **2020**, *10*, 6024. [CrossRef]

12. Jeong, K.; Ahn, J. Design and characteristics analysis of a novel single-phase hybrid SRM for blender application. *J. Electr. Eng. Technolo.* **2018**, *13*, 1996–2003. [CrossRef]

13. Lukman, G.; Hieu, P.; Jeong, K.; Ahn, J. Characteristics analysis and comparison of high-speed 4/2 and hybrid 4/4 poles switched reluctance motor. *Machines* **2018**, *6*, 4. [CrossRef]

14. Dmitrievskii, V.; Prakht, V.; Kazakbaev, V. Novel Rotor Design for High-Speed Flux Reversal Motor. *Energy Rep.* **2020**, *6*, 1544–1549. [CrossRef]

15. Zhou, Y.; Zhou, L.; Hu, B.; Li, R. Design and performance analysis of permanent magnet flux-switching motors using segmental permanent magnets. *IEICE Electron. Exp.* **2019**, *16*, 1–6. [CrossRef]

16. Yamazaki, K.; Fukushima, Y. Effect of Eddy-Current Loss Reduction by Magnet Segmentation in Synchronous Motors with Concentrated Windings. *IEEE Trans. Ind. Appl.* **2011**, *47*, 779–788. [CrossRef]

17. Wills, D.; Kamper, M. Reducing PM eddy current rotor losses by partial magnet and rotor yoke segmentation. In Proceedings of the XIX International Conference on Electrical Machines—ICEM 2010, Rome, Italy, 6–8 September 2010; pp. 1–6. [CrossRef]

18. Kobayashi, M.; Morimoto, S.; Sanada, M.; Inoue, Y. Basic Study of PMASynRM with Bonded Magnets for Traction Applications. In Proceedings of the 2018 International Power Electronics Conference (IPEC-Niigata 2018-ECCE Asia), Niigata, Japan, 20–24 May 2018; pp. 2802–2807. [CrossRef]

19. Constantinides, S. Magnet Selection, Sintered & Bonded NdFeB Magnets, Arnold Magnetic Technologies. 15–17 October 2003. Available online: https://www.arnoldmagnetics.com/wp-content/uploads/2017/10/Magnet-Selection-Constantinides-Gorham-2003-psn-hi-res.pdf (accessed on 12 December 2020).

20. Tsunata, R.; Takemoto, M.; Ogasawara, S.; Watanabe, A.; Ueno, T.; Yamada, K. Development and evaluation of an axial gap motor with neodymium bonded magnet. In Proceedings of the 2016 XXII International Conference on Electrical Machines (ICEM), Lausanne, Switzerland, 4–7 September 2016; pp. 272–278. [CrossRef]

21. Borghi, C.; Casadei, D.; Cristofolini, A.; Fabbri, M.; Serra, G. Minimizing torque ripple in permanent magnet synchronous motors with polymer-bonded magnets. *IEEE Trans. Magn.* **2002**, *38*, 1371–1377. [CrossRef]

22. Ferraris, L.; Pošković, E.; La Cascia, D. Design optimization for the adoption of bonded magnets in PM BLDC motors. In Proceedings of the IECON 2014—40th Annual Conference of the IEEE Industrial Electronics Society, Dallas, TX, USA, 29 October–1 November 2014; pp. 476–482. [CrossRef]

23. Abad, V.; Sagredo, J.; Gonzalez, J. FEA Analysis and Optimization of Rotor Models in Permanent-Magnet Synchronous Motors fitted with Bonded Rare-Earth Magnets. *Ren. Energy Power Qual. J.* **2019**, *17*, 37–42. [CrossRef]

24. Cupertino, F.; Pellegrino, G.; Gerada, C. Design of synchronous reluctance machines with multiobjective optimization algorithms. *IEEE Trans. Ind. Appl.* **2014**, *50*, 3617–3627. [CrossRef]

25. Zăvoianu, A.; Bramerdorfer, G.; Lughofer, E.; Silber, S.; Amrhein, W.; Klementac, E. Hybridization of multi-objective evolutionary algorithms and artificial neural networks for optimizing the performance of electrical drives. *Eng. Appl. Artific. Intell.* **2013**, *26*, 1781–1794. [CrossRef]

26. Krasopoulos, C.; Beniakar, M.; Kladas, A. Robust Optimization of High-Speed PM Motor Design. *IEEE Trans. Magn.* **2017**, *53*, 1–4. [CrossRef]

27. Dmitrievskii, V.; Prakht, V.; Kazakbaev, V. IE5 Energy-Efficiency Class Synchronous Reluctance Motor with Fractional Slot Winding. *IEEE Trans. Ind. Appl.* **2019**, *55*, 4676–4684. [CrossRef]

28. Prakht, V.; Dmitrievskii, V.; Kazakbaev, V. Optimal Design of Gearless Flux-Switching Generator with Ferrite Permanent Magnets. *Mathematics* **2020**, *8*, 206. [CrossRef]

29. Bonded Magnets, Abbreviated Product List, Arnold Magnetic Technologies, 2013. Available online: https://www.arnoldmagnetics.com/wp-content/uploads/2017/10/Plastiform-Bonded-Magnets-Abbreviated-Product-List-11-11-13.pdf (accessed on 12 December 2020).

30. He, L.; Gilbert, M.; Johnson, T.; Pritchar, T. Conceptual design of AM components using layout and geometry optimization. *Comput. Math. Appl.* **2019**, *78*, 2308–2324. [CrossRef]

31. Non-oriented Electrical steel, Posco, product Catalogue. 2019. Available online: http://www.steel-n.com/e-sales/pdf/en/e_electrical_pdf_NO_2020.pdf (accessed on 12 December 2020).

32. Nelder, J.; Mead, R. A Simplex Method for Function Minimization. *Comput. J.* **1965**, *7*, 308–313. [CrossRef]

33. Prakht, V.; Dmitrievskii, V.; Kazakbaev, V.; Sarapulov, S. Steady-state model of a single-phase flux reversal motor. In Proceedings of the IEEE 58th International Scientific Conference on Power and Electrical Engineering of Riga Technical University (RTUCON 2017), Riga, Latvia, 12–13 October 2017; pp. 1–5. [CrossRef]

 *mathematics*

Article

Comparative Evaluation for an Improved Direct Instantaneous Torque Control Strategy of Switched Reluctance Motor Drives for Electric Vehicles

Mahmoud Hamouda [1,2,*], Amir Abdel Menaem [3], Hegazy Rezk [4,5], Mohamed N. Ibrahim [6,7,8] and László Számel [2]

1 Electrical Engineering Department, Mansoura University, 35516 Mansoura, Egypt
2 Department of Electric Power Engineering, Budapest University of Technology and Economics, H-1521 Budapest, Hungary; szamel.laszlo@vet.bme.hu
3 Department of Automated Electrical Systems, Ural Power Engineering Institute, Ural Federal University, 620002 Yekaterinburg, Russia; abdel.menaem@urfu.ru
4 College of Engineering at Wadi Addawaser, Prince Sattam Bin Abdulaziz University, Wadi Aldawaser 11991, Saudi Arabia; hegazy.hussien@mu.edu.eg
5 Electrical Engineering Department, Faculty of Engineering, Minia University, 61111 Minia, Egypt
6 Electrical Engineering Department, Kafrelshiekh University, 33511 Kafr El-Sheikh, Egypt; m.nabil@ugent.be
7 Department of Electromechanical, Systems and Metal Engineering, Ghent University, 9000 Ghent, Belgium
8 FlandersMake@UGent–Corelab EEDT-MP, 3001 Leuven, Belgium
* Correspondence: m_hamouda26@mans.edu.eg

Citation: Hamouda, M.; Abdel Menaem, A.; Rezk, H.; Ibrahim, M.N.; Számel, L. Comparative Evaluation for an Improved Direct Instantaneous Torque Control Strategy of Switched Reluctance Motor Drives for Electric Vehicles. *Mathematics* **2021**, *9*, 302. https://doi.org/10.3390/math9040302

Academic Editor: Alessandro Niccolai

Received: 30 December 2020
Accepted: 1 February 2021
Published: 4 February 2021

Abstract: Due to the expected increase in the electric vehicles (EVs) sales and hence the increase of the price of rare-earth permanent magnets, the switched reluctance motors (SRMs) are gaining increasing research interest currently and in the future. The SRMs offer numerous advantages regarding their structure and converter topologies. However, they suffer from the high torque ripple and complex control algorithms. This paper presents an improved direct instantaneous torque control (DITC) strategy of SRMs for EVs. The improved DITC can fulfill the vehicle requirements. It involves a simple online torque estimator and a torque error compensator. The turn-on angle is defined analytically to achieve wide speed operation and maximum torque per ampere (MTPA) production. Moreover, the turn-off angles are optimized for minimum torque ripples and the highest efficiency. In addition, this paper provides a detailed comparison between the proposed DITC and the most applicable torque control techniques of SRMs for EVs, including indirect instantaneous torque control (IITC), using torque sharing function (TSF) strategy and average torque control (ATC). The results show the superior performance of the proposed DITC because it has the lowest torque ripples, the highest torque tor current ratio, and the best efficiency over the low and medium speed ranges. Moreover, the comparison shows the advantages of each control technique over the range of speed control. It provides a very clear overview to develop a universal control technique of SRM for EVs by merging two or more control techniques.

Keywords: switched reluctance motor; direct instantaneous torque control; numerical analysis; optimization

1. Introduction

Electric vehicles (EVs) are the way forward for green transportation and for establishing a low-carbon economy [1–3]. The simple and robust structure, low cost, less maintenance, high reliability, fault-tolerant, high efficiency, high-speed capability, and large constant power-speed ratio make switched reluctance motors (SRMs) a strong candidate with real chances on the market for vehicle propulsion [4–6].

SRMs do not suffer from the drawbacks noted in DC, induction, and permanent magnets (PMs) machine drives. They offer great robustness of construction. In addition, SRMs have none of the mechanical problems at high speeds that beset other drives. The lack of PMs or rotor winding also reduces cost and offers increased high-speed operation capability [1]. Furthermore, the SRM drives have a highly reliable converter topology. The stator windings are connected in series with the switches preventing the shoot-through faults to which the AC rotating field machine's converters are exposed [7,8]. The low rotor inertia allows high torque per inertia ratio and fast response. In addition, the robust rotor construction raises the maximum operating speed and the permissible rotor temperature. SRM has an inherent four-quadrant operation that meets the demands of EVs propulsion. However, the main obstacles that limit the usage of SRMs in high-performance variable-speed applications are the high torque ripple and the complicated control [9,10].

The machine design, in the early stages, is used to reduce torque ripple. The machine design is effective only over a limited speed range [11]. A wider operating range, but still limited, can be achieved by current or flux profiling [12,13]. However, the profiling techniques require time-consuming pre-calculations to find the optimal current or flux profiles. Therefore, the instantaneous torque control (ITC) is gaining interest in the areas of torque ripple reduction for SRM drives [14]. The indirect ITC (IITC) is an effective strategy for torque ripple reduction in SRMs. It uses a torque sharing function (TSF) to distribute torque between motor phases. In [14], the minimization of both the copper loss and the derivatives of current references were the main objective of the used offline TSF. However, the limitations of offline TSF are treated using an online TSF [15]. A proportional and integral compensator with torque error is added to the torque reference. The structure of TSF, in addition to the torque estimator, complicates the control algorithm. In [16], the torque-speed capability was improved using an adaptive TSF. Over the speed range, the turn-on angles were controlled without adjustment of the shape of TSF. However, the TSF had a lower torque to current rations. In [17], the phase current was compensated in the demagnetizing period to reduce torque ripple and extend speed range. In [18], the reference current shape was modified to reduce the current tracking error. Despite the effectiveness of TSF for torque ripple reduction in SRM drives, it is meant only for low-speed operation. In addition, it used a torque inverse model that is not a straightforward transformation. The torque is a function of position and current. Developing analytical expressions for real-time implementation is not an easy task [19]. Moreover, the fitting accuracy affects the control performance. Look-up tables can be employed for such problems but require additional memory for data storage [20].

On the other hand, the average torque control (ATC) is an advantageous solution for EVs. It has many advantages compared to ITC [21–23]. ATC has a much simple structure; it has a higher torque per ampere ratio, needs a discrete rotor position, and can be used for the entire speed range. However, ATC has a relatively large amount of torque ripple compared to ITC because it controls only the turn-on (θ_{on}) and turn-off (θ_{off}) angles. The optimization of θ_{on} and θ_{off} angles is achieved mainly for torque ripple reduction, efficiency improvement, or copper loss minimization. In [24], the maximization of average torque per ampere was aimed. The torque ripple that does not suit several applications, including EVs, was ignored. Therefore, secondary objectives, including torque-ripple reduction, copper losses minimization, and efficiency improvement, were included [23,25,26].

On the other hand, the direct ITC (DITC) is an effective solution for torque ripple reduction of SRM drives. It controls the torque directly because it employs a hysteresis torque controller. In [27], the DITC of SRM was introduced. The terminal quantities (flux and current) were used to estimate the instantaneous motor torque. However, the integration of phase voltage for flux calculation limited the operation of DITC for low speeds. In addition, the errors in signal processing, phase resistance, and analog to digital converters cause an integration offset. Moreover, the built DITC is effective only till base speed as long as the back EMF is less than or equal to DC voltage. Furthermore, fixed switching angles were used, which affects the generated torque and efficiency.

In [28], the low-speed limits were handled by the online estimation of motor torque as a function of current and rotor position in the form of lookup tables. However, the lookup tables require large memory to store data. In addition, the effect of commutation angles was not included. In [29], a simplified DITC of SRM was achieved. The inner control loops of current and flux were excluded, which eliminates the fault-tolerant advantage of SRM converter topologies. Moreover, the torque was estimated using flux and current data in the form of lookup tables. These tables require large memory to store data and have the problems of flux estimation errors in [27]. In [30], the torque ripple was reduced by adding a PI controller before the torque hysteresis regulator. Three conduction zones were defined, which complicated the control algorithm. Moreover, more fluctuations of DC voltage are expected due to energy return in zone 1. In [31,32], a five-level converter was used to reduce the torque ripple of SRM drives based on DITC. However, the dynamic balance of DC-link capacitor voltage has to be considered, and appropriate switching states have to be generated. In [33], a multi-level power converter was proposed based on a modular converter and three-level switch module. The proposed converter complicates the control algorithm and increases the cost and dissipated heat. In [34], an optimized DITC was achieved based on an adaptive dynamic commutation strategy. The turn-on angle was modified by a torque error regulator. The turn-off angle was defined according to the phase current endpoint detector. However, due to continuous changes of operating conditions in EVs, the variation of commutation angles does not suit their applications very well. Moreover, the maximum torque per ampere (MTPA) production is not guaranteed by forcing phase current to decay at the aligned position using the endpoint detector.

This paper presents an improved DITC strategy of SRMs for EVs. The proposed control uses a simple online torque estimator and calculates the instantaneous motor torque as a function of current and position to avoid flux integration errors and improve the low-speed operation capability. Furthermore, a torque error compensator is added to compensate for the torque ripple. This, in turn, reduces the torque ripple and extends the smooth torque-speed range. Moreover, the control parameters are optimized for the best performance including maximum torque per ampere (MTPA), minimum torque ripples, extended speed operation, and high efficiency. First, the turn-on (θ_{on}) angle is calculated analytically for the MTPA production. Second, an optimization-based method is set for the turn-off (θ_{off}) angle. The optimization aims to achieve the minimum torque ripples and the maximum efficiency at each operating point. The cost function is calculated within the steady-state machine simulation model. The required torque to current conversions is obtained from the finite element analysis (FEA) data of studied 8/6 SRM. Furthermore, this paper provides a detailed comparison between the proposed DITC and the most applicable torque control techniques of SRMs for EVs including the IITC and the ATC. Each control strategy (IITC and ATC) is optimized for the best performance. The optimization details are included in the next sections.

The paper organization is as follows: Section 2 shows the machine modeling and performance indices. The proposed control, the optimization problem, the solution method, and the optimization results are involved in Section 3. The simulation verification is presented in Section 4. Finally, Section 5 provides the conclusions drawn from this research.

2. Machine Modeling

Due to the double saliency of SRMs, the flux-linkage $\lambda(i,\theta)$, inductance $L(i,\theta)$, and torque $T(i,\theta)$ have nonlinear relations with current (i) and position (θ). Equation (1) shows the voltage equation. The electromagnetic torque of kth phase (T_k) and the total electromagnetic torque (T_e) with q-phases can be represented by Equation (2). The mechanical dynamics is shown by Equation (3) [25].

$$v_k = Ri_k + \frac{\partial \lambda_k(i_k,\theta)}{\partial t}; \quad \lambda_k(i_k,\theta) = \int (v_k - Ri_k)\ dt \tag{1}$$

$$T_k = \frac{1}{2}\frac{\partial L_k}{\partial \theta}i_k^2; \quad T_e = \sum_{k=1}^{q} T_k \tag{2}$$

$$T_e - T_L = B\omega + J\frac{d\omega}{dt} \tag{3}$$

where J is the inertia, B is the frictional coefficient, ω is the rotor speed, and T_L is the load torque.

The finite element method (FEM) is employed to generate the magnetic characteristics of the studied 8/6 SRM. The FEM data are used in form of look-up tables to build the machine model in MATLAB/Simulink [25]. The studied motor is 4 kW, 1500 r/min, 600 V, 8/6 poles, 4 phases SRM. The flux linkage and torque characteristics are shown in Figure 1a,b, respectively. These figures show only a few curves for simplification, while the complete flux and torque characteristics are calculated for current [0:1:50]A and position of [0:0.5:30]°. The unaligned and aligned positions are defined by angle $\theta = 0°$ and angle $\theta = 30°$, respectively. As seen, both the flux and torque have highly nonlinear characteristics.

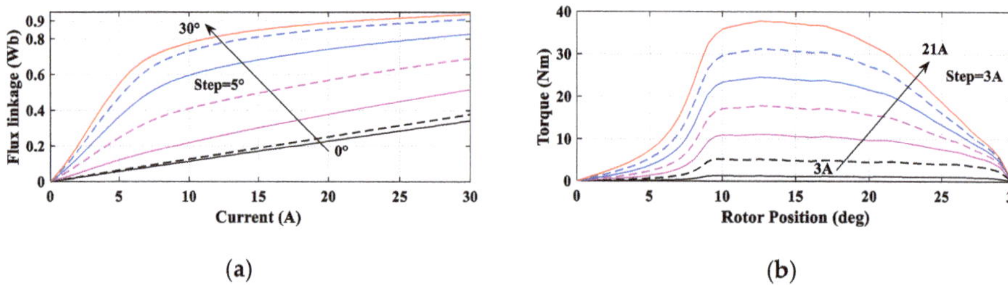

Figure 1. The finite element method (FEM)-calculated characteristics—(**a**) flux linkage and (**b**) torque.

Performance Indices

The performance indices include the calculation of average torque (T_{av}), torque ripple (T_r), average supply current (I_{av}), efficiency (η), mechanical output power (P_m), the total harmonic distortion (THD) of phase current, switching frequency of converter (f_{sw}), and root mean square (RMS) supply current (I_{RMS}) [4,10,29].

The torque ripple (T_r) of SRM is calculated from the maximum and minimum instantaneous torque values (T_{max} and T_{min}) as expressed by Equation (4). The average torque (T_{av}) is calculated over one electric cycle (τ).

$$T_r = \frac{T_{max} - T_{min}}{T_{av}}; \quad T_{av} = \frac{1}{\tau} \int_0^\tau T_e(t)dt \tag{4}$$

The efficiency (η) and average supply current (I_{av}) can be expressed as

$$\eta = \frac{\omega\, T_{av}}{V_{DC}\, I_{av}}; \quad I_{av} = \frac{1}{\tau} \int_0^\tau i_s(t)dt \tag{5}$$

The mechanical output power can be estimated as follows:

$$P_m = T_{av}(t)\cdot\omega(t) \tag{6}$$

Equation (7) is the most adopted for spectrum performance for THD of phase current [35].

$$THD = \sqrt{\frac{I_{rms}^2 - I_{1,rms}^2}{I_{1,rms}^2}} \tag{7}$$

where $I_{1,rms}$ represents the root mean square (RMS) value of the fundamental component of phase current. I_{rms} depicts the RMS value of phase current.

The RMS supply current (I_{RMS}) and switching frequency (f_{sw}) are seen by Equations (8) and (9), respectively.

$$I_{RMS} = \sqrt{\frac{1}{\tau} \int_0^\tau i_k^2(t)dt} \tag{8}$$

$$f_{sw} = \frac{1}{\tau} \int_0^\tau N_T dt \tag{9}$$

where N_T is the total number of switching of IGBTs over one electric period τ.

3. The Proposed Direct Instantaneous Torque Control (DITC)

Figure 2 shows the block diagram of the proposed DITC. It has an outer loop speed controller, middle loop torque controller, and inner loop current controller. The speed controller outputs a reference torque signal (T_{ref}). The torque error (ΔT) is the difference between T_{ref} and the estimated actual motor torque (T_{est}). ΔT is processed through a hysteresis torque controller that outputs the state signals. The reference current (i_{ref}) is calculated as a function of T_{ref} using a proposed torque to the current conversion scheme. In addition, a torque ripple compensator is added, it uses a PI controller (probably a P controller) to compensate for the torque errors. Moreover, the commutation angles (θ_{on} and θ_{off}) are estimated online for the best performance. A simple online torque estimator using third-order polynomial is used to avoid the required big memory for the look-up tables. The torque is estimated as a function of phase current and position, the details are included in [36].

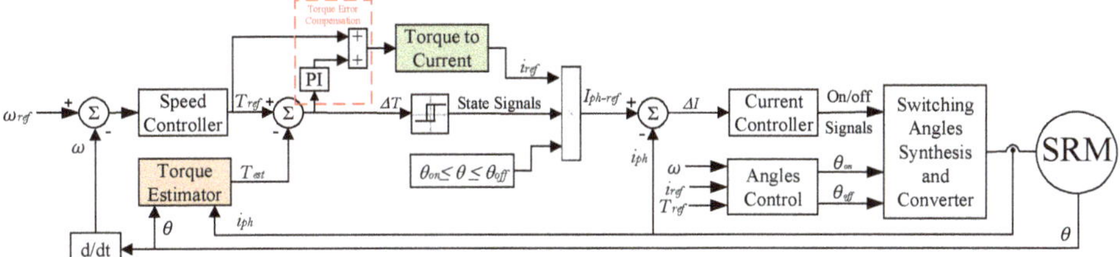

Figure 2. Block diagram of the proposed direct instantaneous torque control (DITC).

3.1. Torque to Current Conversion

Due to the high nonlinear torque characteristics of SRMs, the torque to current conversion is not a feedforward transformation. For a precise torque to current conversion, the control algorithm will be much complicated. However, in this case, the reference torque signal (T_{ref}) is required to be converted to a reference current signal (i_{ref}). This conversation can be implemented simply using polynomial fitting. This, in turn, helps to simplify the overall control algorithm.

For 8/6 SRM, the ideal conduction angle is 15°. Each phase will produce torque over 15°. Moreover, the conditions for maximum torque per ampere (MTPA) include the peak phase current to reach its reference value at the end of the minimum inductance zone (angle θ_m) [37]. Therefore, for the best torque production is achieved over a period [θ_m, $\theta_m + 15°$], as shown in Figure 3a. for each current magnitude, the average torque can be estimated from the FEM-calculated torque data. Then, polynomial fitting can be simply carried out, as seen in Figure 3b.

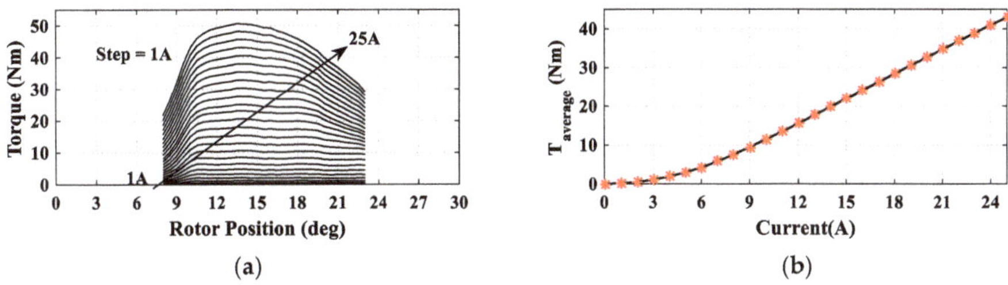

Figure 3. Torque to current conversion—(**a**) torque curves over most efficient 15° and (**b**) torque fitting.

3.2. Switching Angles Optimization

To achieve the MTPA conditions, the turn-on (θ_{on}) angle is calculated using Equation (10) [37]. This Equation determines the optimum θ_{on} to provide the most efficient operation. It considers accurately the effect of back-emf voltage at low and high speeds.

$$\theta_{on} = \theta_m + \frac{L_{eff}(i,\theta)}{R + k_{b-eff}\omega} ln\left(1 - i_{ref}\frac{R + k_{b-eff}\omega}{V_{DC}}\right) \tag{10}$$

where R is the phase resistance, and V_{DC} is the dc voltage.

On the other hand, an optimization problem is set for the turn-off (θ_{off}) angle to provide the minimum torque ripples, the lowest copper losses, and the highest efficiency. The objective function is provided by Equation (11) with a combination of torque ripple (T_r), copper losses (P_{cu}), and efficiency (η).

$$F_{obj}\left(\theta_{off}\right) = min\left(w_r\frac{T_r}{T_{rb}} + w_{cu}\frac{P_{cu}}{P_{cub}} + w_\eta\frac{\eta_b}{\eta}\right) \tag{11}$$

$$w_r + w_{cu} + w_\eta = 1 \tag{12}$$

where F_{obj} is the objective function. T_{rb} is the base value of torque ripples. P_{cub} is the base value of the copper loss. η_b is the base value of efficiency. w_r is the weight factor of torque ripples. w_{cu} is the weight factor of copper loss. η_r is the weight factor of efficiency.

Figure 4 shows the flowchart of the developed searching algorithm. At each operating point, defined by the reference torque and speed, the turn-off angle (θ_{off}) is changed in small steps. The simulation model is employed to calculate the torque ripple, copper loss, and efficiency at each step. At the end of the search, the minimum torque ripple, the minimum copper losses, and the maximum efficiency are defined as the base values (T_{rb}, P_{cub}, and η_b). The turn-off angle (θ_{off}) is varied from $\theta_{off\text{-}min} = 15°$ to $\theta_{off\text{-}max} = 28°$ in steps of $0.2°$. Then, the optimum angle is defined using Equation (11). This procedure is repeated several times according to the desired speeds and torque levels. In this paper, the torque is changed with a step of 2 Nm. The speed step is taken as 200 r/min. Figure 5 presents the optimum turn-off angles. As noted, for a given motor speed, the turn-off angle is almost constant. It decreases with increasing motor speed.

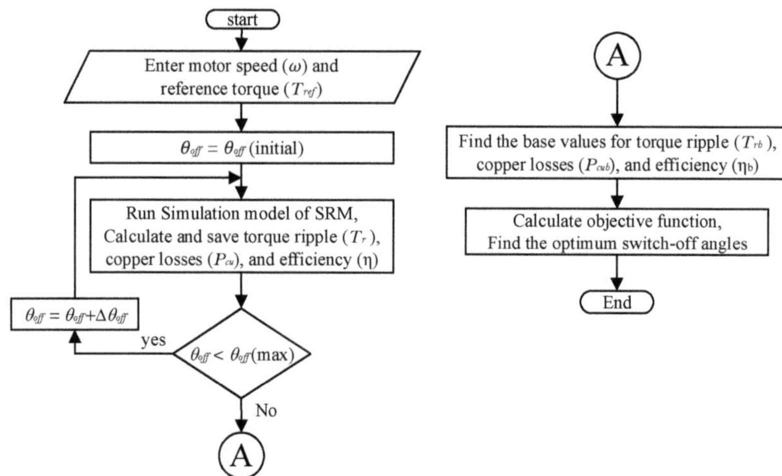

Figure 4. The flowchart of the searching algorithm.

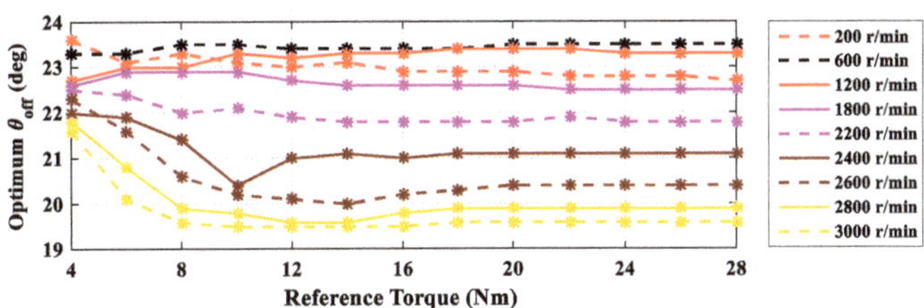

Figure 5. The optimized turn-off angles.

The weight factors (w_r, w_{cu}, and η_r) are chosen according to the desired level of optimization. The weighting factors are $w_r = 0.4$, $w_{cu} = 0.3$, and $\eta_r = 0.3$.

3.3. Simulation Results of the Proposed DITC

The simulation results of the proposed DITC are provided in Figure 6. A sudden change of the commanded reference speed is made at 0.3 and 0.6 sec (Figure 6a). The load torque has a constant value of 17 Nm. The motor can track the desired speed efficiently. The generated torque has a very good profile as illustrated in Figure 6b. Till speed of 2000 r/min, the amount of torque ripple is very minor (Figure 6c); as the motor speed increases, the torque ripples increase. The θ_{on} and θ_{off} angle have smooth variation along with the speed and torque level, as shown in Figure 6d,e, respectively. The mechanical output power and the total motor efficiency are provided in Figure 6f,g, respectively. The system has very good efficiency. As the motor speed increases, the efficiency also increases.

4. The Other Torque Control Techniques of SRM

This section involves the most applicable torque control techniques of SRM drives for EVs. It provides the ATC, followed by the IITC.

4.1. Average Torque Control (ATC)

The block diagram of the adopted ATC is shown in Figure 7 [22,23]. The outer loop speed control provides the reference torque (T_{ref}). The torque error (ΔT) is processed by the torque controller (PI) that outputs i_{ref}. The switching angles (θ_{on} and θ_{off}) are estimated as functions of motor speed (w) and reference torque/current.

4.1.1. Switching Angles Optimization

The optimization aims to achieve the lowest torque ripple, the lowest copper losses, and the highest efficiency. Three groups multi-objective optimization function is used as follows:

$$F_{obj}\left(\theta_{on}, \theta_{off}\right) = min\left(w_r\frac{T_r}{T_{rb}} + w_{cu}\frac{P_{cu}}{P_{cub}} + w_\eta\frac{\eta_b}{\eta}\right) \tag{13}$$

$$w_r + w_{cu} + w_\eta = 1 \tag{14}$$

subject to

$$\theta_{on}^{min} \leq \theta_{on} \leq \theta_{on}^{max}; \quad \theta_{off}^{min} \leq \theta_{off} \leq \theta_{off}^{max} \tag{15}$$

where θ_{on}^{min} and θ_{on}^{max} are the minimum and the maximum limits of the θ_{on}, respectively. θ_{off}^{min} and θ_{off}^{max} are the minimum and the maximum limits of the θ_{off}, respectively.

The weight factors for (w_r, w_{cu}, and w_η) are determined according to the required optimization level. Due to the higher torque ripple of ATC, greater importance is directed to reduce torque ripples. The weight factors are set to $w_r = 0.6$, $w_{cu} = 0.2$, and $w_\eta = 0.2$.

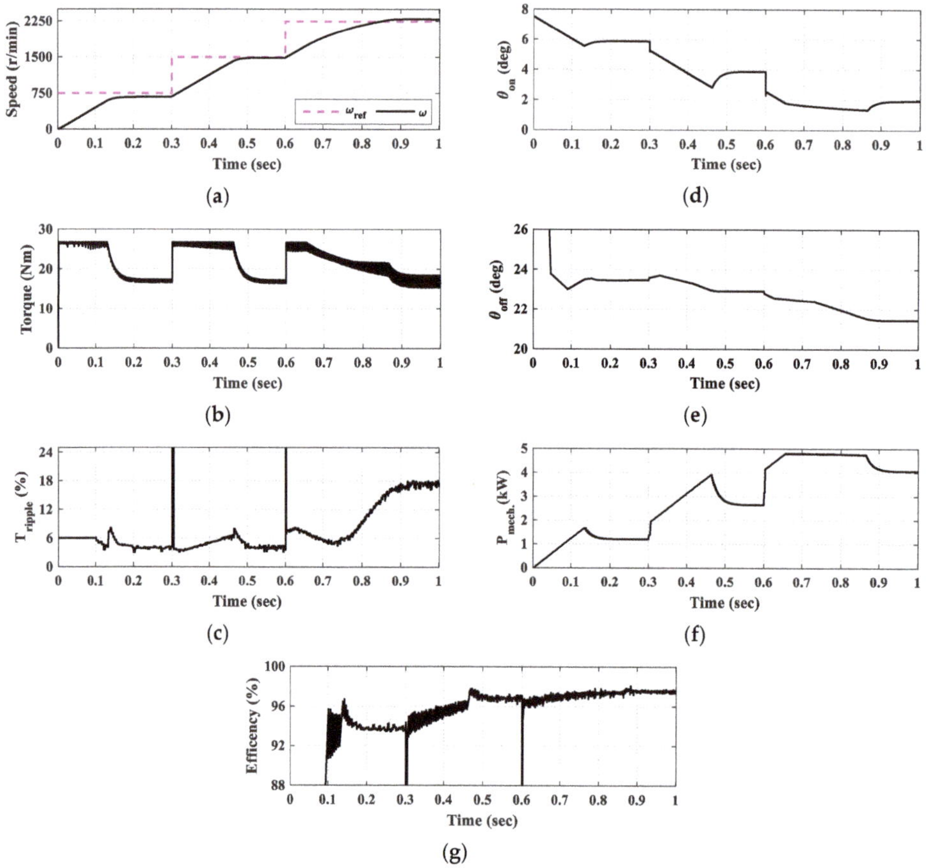

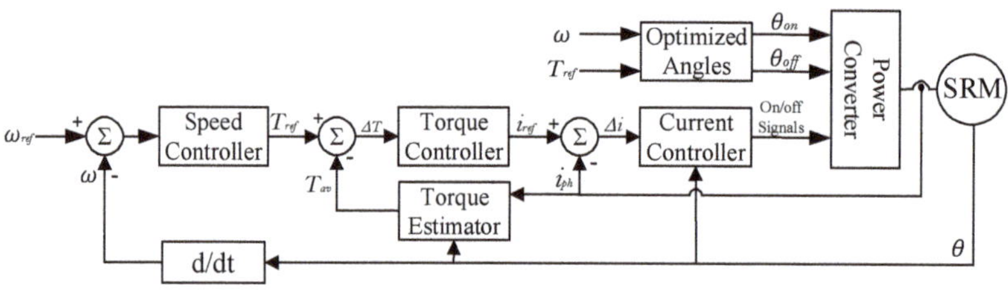

Figure 6. The simulation results for DITC—(**a**) motor speed, (**b**) total torque, (**c**) torque ripple, (**d**) turn-on angle, (**e**) turn-off angle, (**f**) output power, and (**g**) efficiency.

Figure 7. Block diagram of average torque control (ATC) technique.

4.1.2. Simulation Results of ATC

The simulation results for the proposed ATC are presented in Figure 8. A sudden change of the commanded reference speed is made at 0.4 sec and 0.9 sec (Figure 8a). The load torque has a constant value of 17 Nm. The motor can track the desired speed efficiently. The generated torque has a good profile as illustrated in Figure 8b. In general, as the motor speed increases, the torque ripples increase, as seen in Figure 8c. As seen, the torque ripple is high at very low speed. The θ_{on} and θ_{off} angles have adaptive and smooth variations along with motor speed and torque level, as shown in Figure 8d,e, respectively. The mechanical output power and the total motor efficiency are presented in Figure 8f,g, respectively. The system has very good efficiency, especially at higher speeds.

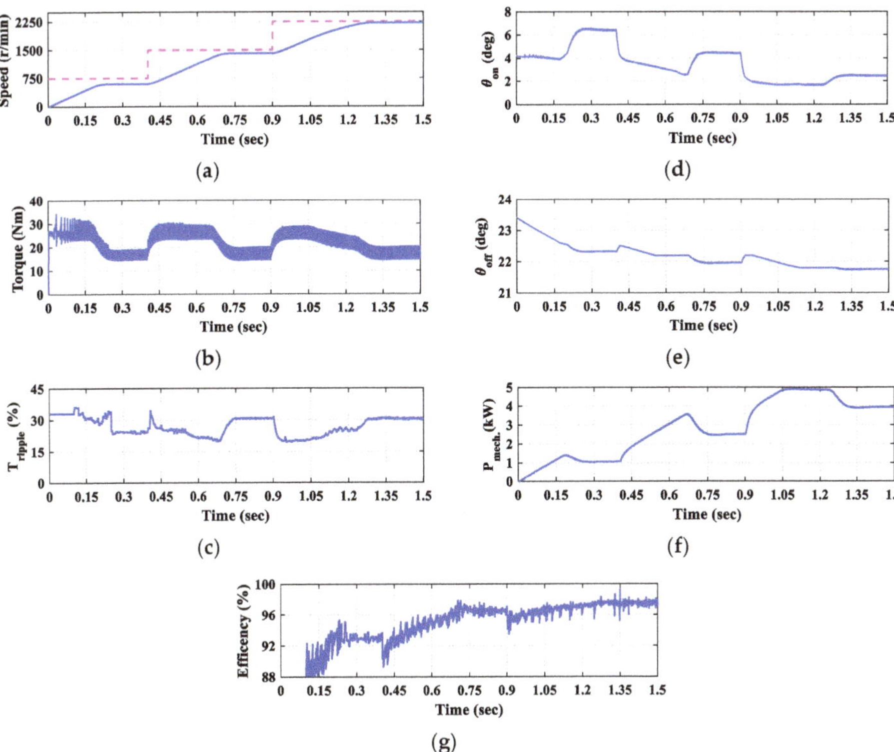

Figure 8. The simulation results for ATC—(**a**) motor speed, (**b**) total torque, (**c**) torque ripple, (**d**) turn-on angle, (**e**) turn-off angle, (**f**) output power, and (**g**) efficiency.

4.2. Indirect Instantaneous Torque Control (IITC)

The block diagram of the IITC is presented in Figure 9. It has an outer loop speed controller and an inner loop current controller. The middle loop torque controller contains a TSF and torque inverse model $i(T, \theta)$. Moreover, the MTPA is achieved through the proper estimation of turn-on angle (θ_{on}) using Equation (10). The torque to current conversion is also used here. Furthermore, a torque compensation is carried out to compensate for the torque ripple. The torque error (ΔT) is processed within the TSF to extend the operating speed range. The modified TSF is provided by Equation (16). The torque error is compensated with the incoming phase as it has the lower absolute changing rate of flux linkage with rotor position.

$$TSF(\theta) = \begin{cases} 0, & if(0 \leq \theta \leq \theta_{on}) \\ \frac{T_e}{2} - \frac{T_e}{2}\cos\frac{\pi}{\theta_{ov}}(\theta - \theta_{on}) + \Delta T, & if(\theta_{on} \leq \theta \leq \theta_{on} + \theta_{ov}) \\ T_e + \Delta T, & if\left(\theta_{on} + \theta_{ov} \leq \theta \leq \theta_{off}\right) \\ T_e - \left(\frac{T_e}{2} - \frac{T_e}{2}\cos\frac{\pi}{\theta_{ov}}(\theta - \theta_{on})\right), & if\left(\theta_{off} \leq \theta \leq \theta_{off} + \theta_{ov}\right) \\ 0, & if\left(\theta_{off} + \theta_{ov} \leq \theta \leq \theta_p\right) \end{cases} \tag{16}$$

where θ_p is the rotor period. θ_{ov} is the over-lap angle ($\theta_{ov} \leq 0.5\theta_p - \theta_{off}$).

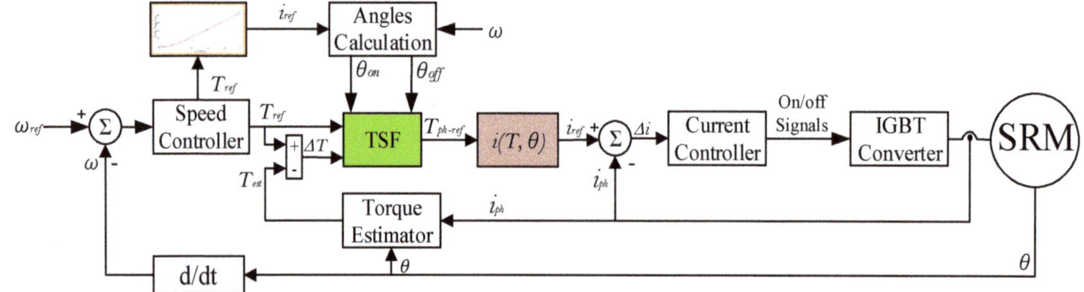

Figure 9. Block diagram of indirect instantaneous torque control (IITC) scheme.

Simulation Results of IITC

The simulation results for the proposed IITC are presented in Figure 10. A sudden change of the commanded reference speed is made at 0.3 sec and 0.6 sec (Figure 10a). The load torque has a constant value of 17 Nm. The motor can track the desired speed efficiently. The generated torque has a very good profile, as illustrated in Figure 10b. At low speed (below the base speed of 1500 r/min), the amount of torque ripple is very minor. As the motor speed increases, the torque ripples increase, as seen in Figure 10c. The θ_{on} angle has a smooth variation along with the speed and torque level, as shown in Figure 10d. The smooth change means lower disturbance and less noise. The mechanical output power and the total motor efficiency are presented in Figure 10e,f, respectively. The system has very good efficiency, especially at higher speeds.

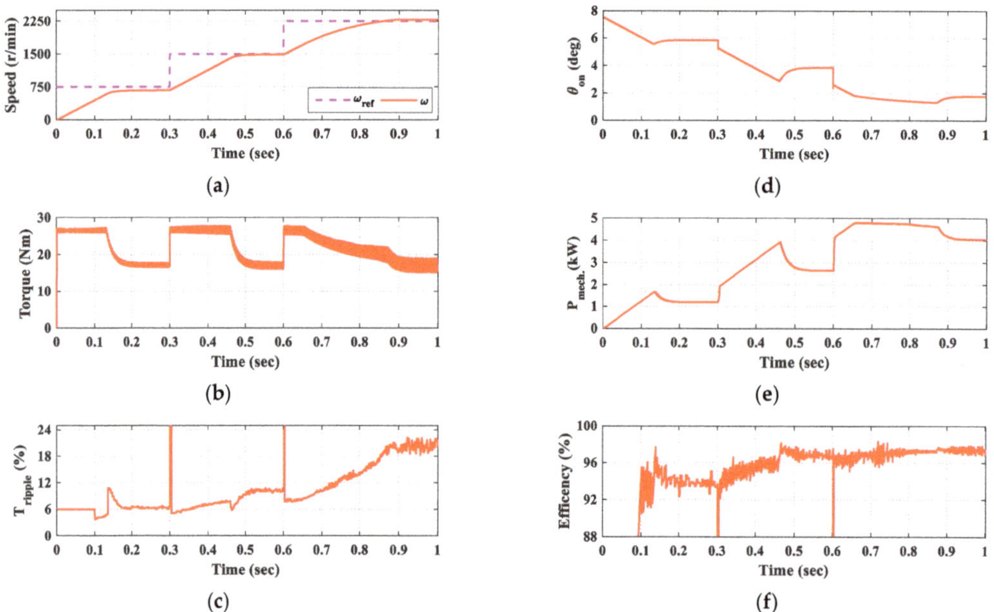

Figure 10. The simulation results for IITC—(**a**) motor speed, (**b**) total torque, (**c**) torque ripple, (**d**) turn-on angle, (**e**) output power, and (**f**) efficiency.

5. Comparative Analysis and Discussion

To develop the best control technique of SRMs for EV applications, a comparative study with a detailed analysis of control performance is essential to gain the benefits of each technique. This comparative study includes the DITC, the TSF with MTPA and ripple compensation, and the ATC.

The study was conducted under variable loading conditions that represent the actual EV load. The parameters for the simulated EV are included in [1]. The study was also achieved under the full load conditions because EVs have a continuous change in operating point. Therefore, the motor will be under the full load conditions in the acceleration times.

5.1. Under EV Loading

Figure 11 shows the steady-state characteristics under EV loading conditions. The load torque proportionally increases with motor speed, as shown in Figure 11a. As noted, the ATC has the capability to provide higher torque production under high speeds (beyond 2200 r/min). The DITC has the lowest torque ripples till the speed of 2500 r/min, as illustrated in Figure 11b. After that speed, the ATC provides the lowest torque ripples. The DITC has the highest Tav/I_{RMS} ratio till the speed of 2200 r/min, after that, the ATC provides the best Tav/I_{RMS} ratio (Figure 11c). The ATC provides the lowest switching converter frequency, as presented in Figure 11d. The maximum achievable switching frequency is less than 10 kHz that fits most of the industrial applications. The IITC provides the lowest THD of phase current, followed by DITC and then ATC, as illustrated by Figure 11e. After 2500 r/min, the ATC yields the lowest THD of phase current. As seen in Figure 11f, the DITC and ATC have the lowest $d\lambda/dt$. The efficiency curve is shown in Figure 11g. The DITC and ATC have higher efficiencies (almost the same) under low speeds (see Figure 11h), but the ATC provides higher efficiency.

5.2. Under Full Load Conditions

Figure 12 shows the steady-state characteristics under full load conditions. The motor is loaded with a constant load torque of 26 Nm until it reaches the base speed (1500 r/min) and then the torque decreases inversely with speed, as shown in Figure 12a. In general, the ATC has the capability to provide higher torque production at high speeds. The DITC has the lowest torque ripple at low speeds, while the ATC has a lower torque ripple at high speeds (Figure 12b). The best Tav/I_{RMS} ratio is obtained by the DITC (Figure 12c). The IITC shows a higher switching frequency for low speeds (Figure 12d) compared to Figure 11d. The IITC provides the lowest THD for phase current, followed by ATC and then DITC, as seen in Figure 12e. Figure 12f shows the efficiency curves. As noted, the DITC has the best efficiency at low speeds, while the ATC provides the highest efficiency at high speeds.

5.3. The Steady-State Torque Curves

The steady-state torque curves under different operating speeds are illustrated in Figure 13. As observed, the DITC has the smoothest torque profile and hence the lower torque ripple. Despite the higher torque ripples of ATC, its torque profile seems very smooth especially at low speeds, as shown in Figure 13a,b. As the speed increases, the torque ripple also increases. After 2000 r/min, the torque ripple appears even with DITC and TSF, as seen in Figure 13c,d.

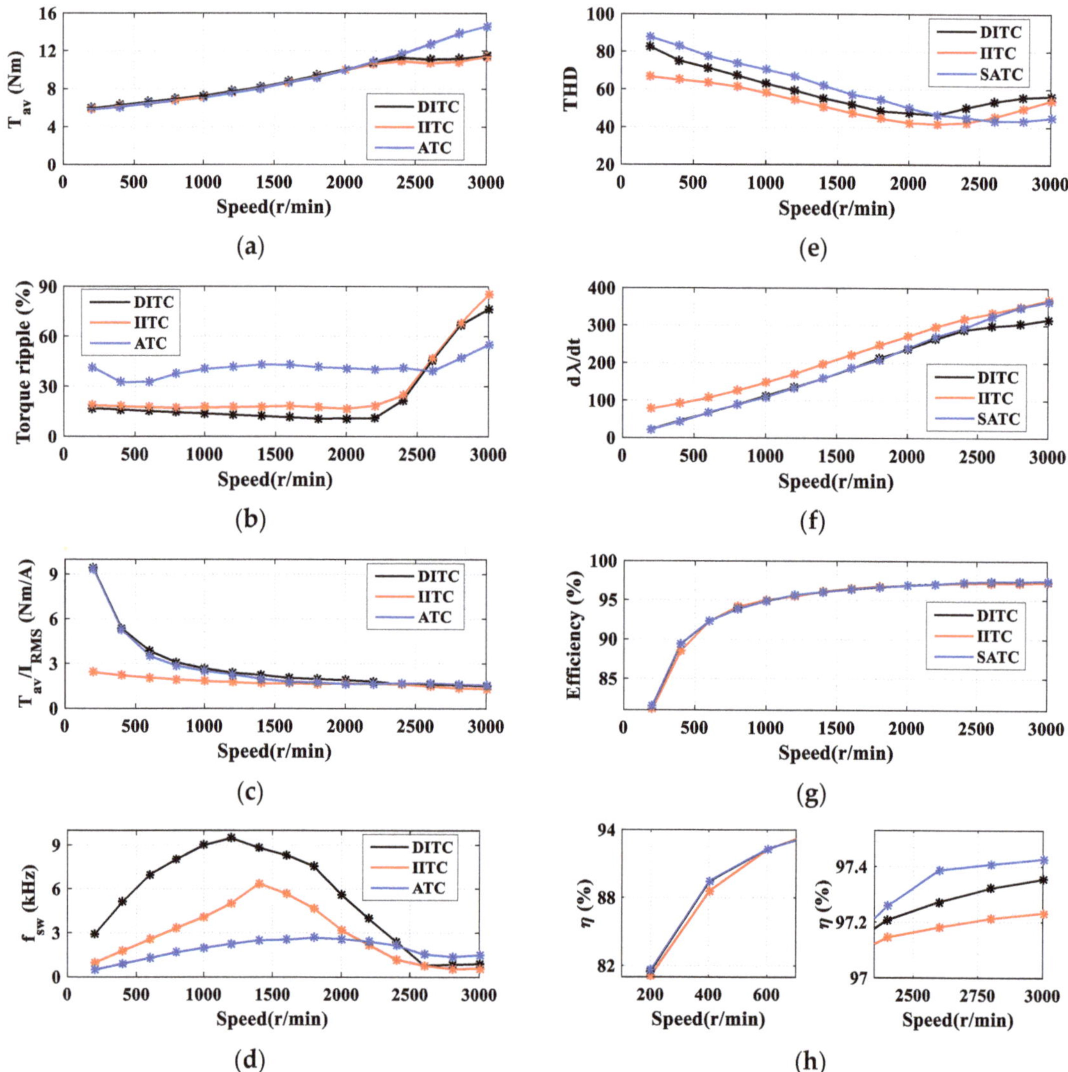

Figure 11. The steady-state characteristics under electric vehicle (EV) loading—(**a**) average torque, (**b**) torque ripple, (**c**) torque per current ratio, (**d**)switching frequency, (**e**) THD, (**f**) flux derivatives, (**g**) efficiency, and (**h**) zoom on efficiency.

5.4. Dynamic Torque Response

The dynamic torque performance of the three control techniques is illustrated in Figure 14. The control techniques are tested with a sudden change in reference torque signal from 5 Nm to 20 Nm at 0.05 sec. The DITC and IITC techniques have a fast dynamic response. The ATC shows a slower torque response, but still acceptable because it employs a PI controller that outputs reference current.

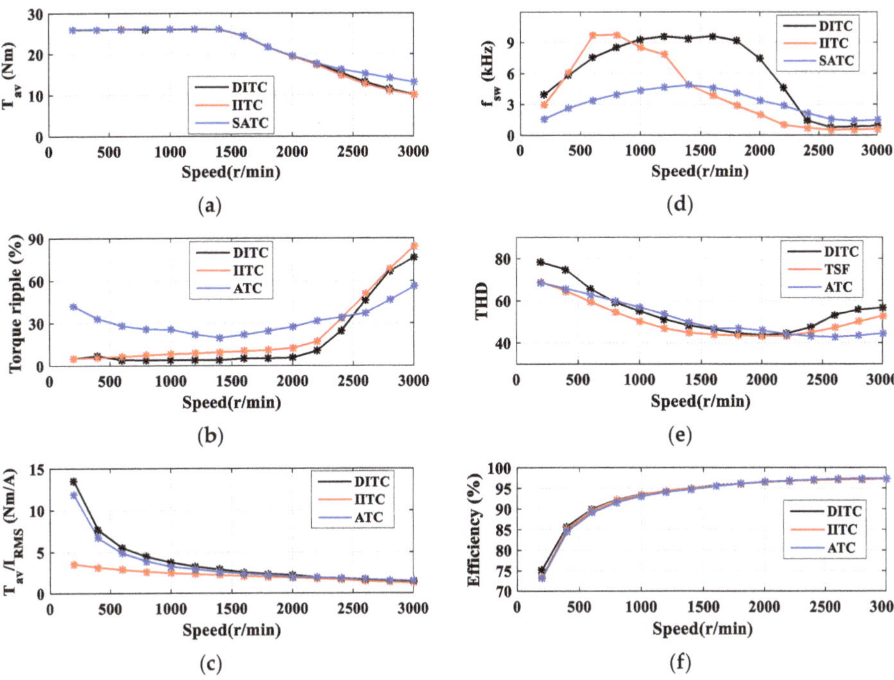

Figure 12. The steady-state characteristics under EV loading—(**a**) average torque, (**b**) torque ripple, (**c**) torque per current ratio, (**d**) switching frequency, (**e**) total harmonic distortion (THD), and (**f**) efficiency.

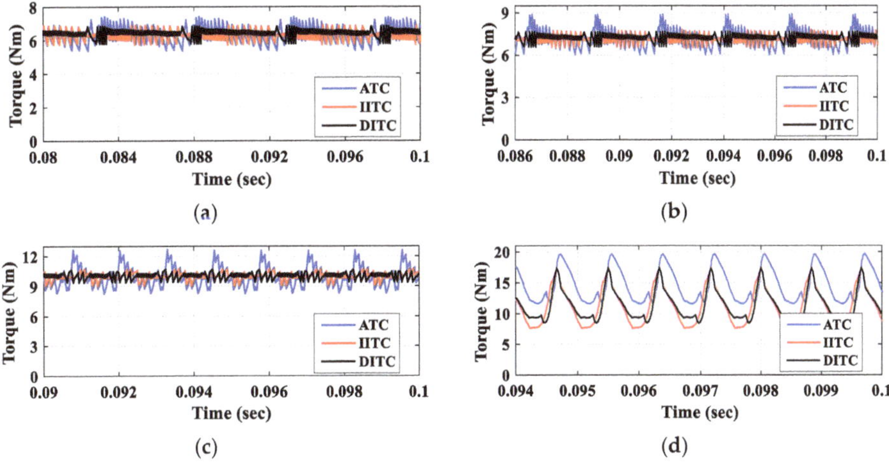

Figure 13. The torque curves at speed of (**a**) 500 r/min, (**b**) 1000 r/min, (**c**) 2000 r/min, and (**d**) 3000 r/min.

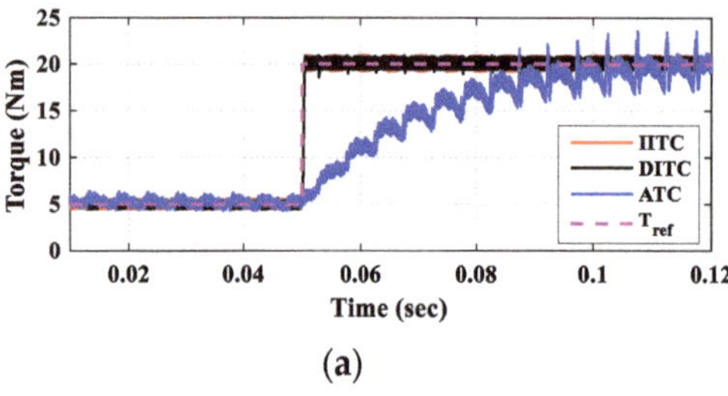 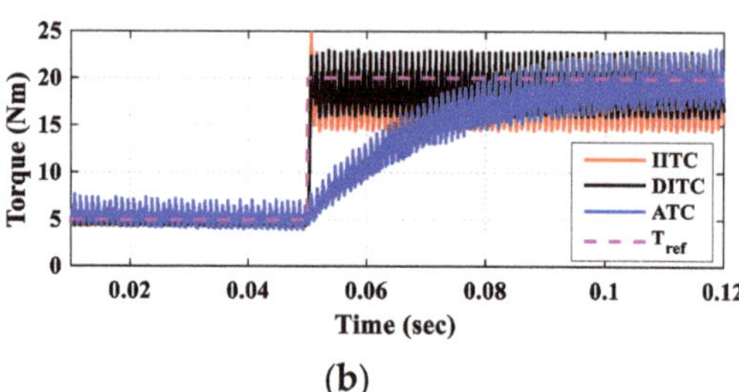

(a)   **(b)**

Figure 14. The dynamic torque curves at speed of (**a**) 500 r/min and (**b**) 2500 r/min.

6. Summary

Tables 1 and 2 show the detailed conclusion for the comparative study. As concluded from Table 1, the DITC shows the best overall performance because it has a less complex control algorithm, does not have a torque inverse model, provides a fast dynamic torque response, and has a dynamic torque response, which is the requirement of torque inverse model.

Table 1. Summary of comparison results.

Indices	Low Speed		
	DITC	**IITC**	**ATC**
Algorithm complexity	Moderate	Complex	Complex
Requirement of torque inverse model	No	Yes	No
Dynamic torque response	Fast	Fast	Slow
Requirement of online torque estimation	Yes	Yes	Yes
Required control period	Short	Short	Long

Table 2. Summary of comparison results over speed ranges.

Indices	Low Speed			High Speed		
	DITC	**IITC**	**ATC**	**DITC**	**IITC**	**ATC**
Average torque	High	High	High	Low	Low	High
Torque ripple	Low	Medium	High	Medium	Medium	Low
Switching frequency	High	Medium	Low	Medium	Low	High
Torque/current ratio	High	Low	Medium	High	Low	High
THD	Medium	Low	High	High	Medium	Low
Flux derivatives	Low	High	Low	Low	High	High
Efficiency	High	Medium	High	Medium	Medium	High

Table 2 provides a comparison regarding the peed ranges. In conclusion, for low speeds, the DITC provides the lowest torque, the highest torque/current ratio, the lowest flux derivatives, and the highest efficiency. However, it has a moderate value for the THD of phase current; it also has a higher switching frequency (<10 kHz) but still within the applicable range.

On the other hand, for high speeds, the ATC can provide high average torque production. It also shows the minimum torque ripple. Moreover, it has the highest torque/current ratio, the lower THD of phase current, and the highest efficiency. However, it has a higher switching frequency and higher flux derivatives.

7. Conclusions

This paper presents an improved DITC of SRM drives for EVs. The main concern is to satisfy the vehicle requirements including MTPA, minimum torque ripple, wide speed range, and high efficiency. First, the turn-on angle was estimated analytically to provide the MTPA conditions. Second, an optimization-based method was used to estimate the optimum turn-off angles that provide the lowest torque ripple, the lowest copper losses, and the highest efficiency. In addition, the proposed DITC compensates for the torque error to provide a lower torque ripple with extended speed operation. A torque-to-current conversion was conducted using the FEM-calculated torque data. Moreover, the IITC and the ATC techniques were implemented and compared to the proposed DITC. The results show the superior performance of the proposed DITC. As noted, the DITC can provide the lowest torque ripple, the highest torque to current ratio, and the best efficiency over the low and medium speed ranges. Moreover, the comparison presents a very good perspective to develop a universal control technique of SRM drives for EVs. This paper recommends a universal control that uses the proposed DITC over the low speeds and utilizes the ATC for the high speeds.

Author Contributions: Conceptualization, M.H.; Formal analysis, M.H.; Investigation, M.H. and M.N.I.; Methodology, M.H.; Project administration, L.S.; Software, M.H., A.A.M., H.R., and M.N.I.; Validation, M.H.; Visualization, H.R. and M.N.I.; Writing—original draft, M.H.; Writing—review and editing, A.A.M., H.R., M.N.I., and L.S. All authors have read and agreed to the published version of the manuscript.

Funding: This research received no external funding.

Conflicts of Interest: The authors declare no conflict of interest.

Abbreviations

ATC	Average torque control
DC	Direct current
DITC	Direct instantaneous torque control
DTC	Direct torque control
EVs	Electric vehicles
FEA	finite element analysis (FEA)
IITC	Indirect instantaneous torque control
ITC	Instantaneous torque control
MTPA	Maximum torque per ampere
PMs	Permanent magnets
RMS	Root Mean Square
SRM	Switched reluctance motor
THD	Total harmonic distortion
TSF	Torque sharing function

References

1. Chen, H.; Yan, W.; Gu, J.; Sun, M. Multiobjective Optimization Design of a Switched Reluctance Motor for Low-Speed Electric Vehicles with a Taguchi–CSO Algorithm. *IEEE/ASME Trans. Mechatron.* **2018**, *23*, 1762–1774. [CrossRef]
2. Yueying, Z.; Chuantian, Y.; Yuan, Y.; Weiyan, W.; Chengwen, Z.; Zhu, Y.; Yang, C.; Yue, Y.; Wei, W.; Zhao, C. Design and optimisation of an In-wheel switched reluctance motor for electric vehicles. *IET Intell. Transp. Syst.* **2019**, *13*, 175–182. [CrossRef]
3. Jiang, J.W.; Bilgin, B.; Emadi, A. Three-Phase 24/16 Switched Reluctance Machine for a Hybrid Electric Powertrain. *IEEE Trans. Transp. Electrif.* **2017**, *3*, 76–85. [CrossRef]
4. Zhu, J.; Cheng, K.W.E.; Xue, X. Design and Analysis of a New Enhanced Torque Hybrid Switched Reluctance Motor. *IEEE Trans. Energy Convers.* **2018**, *33*, 1965–1977. [CrossRef]
5. Bartolo, J.B.; Degano, M.; Espina, J.; Gerada, C. Design and Initial Testing of a High-Speed 45-kW Switched Reluctance Drive for Aerospace Application. *IEEE Trans. Ind. Electron.* **2017**, *64*, 988–997. [CrossRef]
6. Mousavi-Aghdam, S.R.; Feyzi, M.R.; Bianchi, N.; Morandin, M. Design and Analysis of a Novel High-Torque Stator-Segmented SRM. *IEEE Trans. Ind. Electron.* **2016**, *63*, 1458–1466. [CrossRef]
7. Nam, K.H. *AC Motor Control and Electrical Vehicle Applications*; CRC Press Taylor & Francis Group: Boca Raton, FL, USA, 2019.
8. Ehsani, M.; Gao, Y.; Emadi, A. *Modern Electric, Hybrid Electric, and Fuel Cell Vehicles: Fundamentals, Theory, and Design*; CRC Press: Boca Raton, FL, USA, 2010.
9. Husain, T.; Elrayyah, A.; Sozer, Y.; Husain, I. Unified Control for Switched Reluctance Motors for Wide Speed Operation. *IEEE Trans. Ind. Electron.* **2019**, *66*, 3401–3411. [CrossRef]

10. Husain, T.; Elrayyah, A.; Sozer, Y.; Husain, I. An efficient universal controller for switched-reluctance machines. In Proceedings of the 2013 28th Annual IEEE Applied Power Electronics Conference and Exposition (APEC), Long Beach, CA, USA, 17–21 March 2013; pp. 1530–1536.

11. Gan, C.; Wu, J.; Sun, Q.; Kong, W.; Li, H.; Hu, Y. A Review on Machine Topologies and Control Techniques for Low-Noise Switched Reluctance Motors in Electric Vehicle Applications. *IEEE Access* **2018**, *6*, 31430–31443. [CrossRef]

12. Mikail, R.; Husain, I.; Sozer, Y.; Islam, M.S.; Sebastian, T. Torque-Ripple Minimization of Switched Reluctance Machines Through Current Profiling. *IEEE Trans. Ind. Appl.* **2013**, *49*, 1258–1267. [CrossRef]

13. Shaked, N.; Rabinovici, R. New procedures for minimizing the torque ripple in switched reluctance motors by optimizing the phase-current profile. *IEEE Trans. Magn.* **2005**, *41*, 1184–1192. [CrossRef]

14. Ye, J.; Bilgin, B.; Emadi, A. An Offline Torque Sharing Function for Torque Ripple Reduction in Switched Reluctance Motor Drives. *IEEE Trans. Energy Convers.* **2015**, *30*, 726–735. [CrossRef]

15. Ye, J.; Bilgin, B.; Emadi, A. An Extended-Speed Low-Ripple Torque Control of Switched Reluctance Motor Drives. *IEEE Trans. Power Electron.* **2015**, *30*, 1457–1470. [CrossRef]

16. Suebsuang, P.J.S. An adaptive low-ripple torque control of switched reluctance motor for small electric vehicle. In Proceedings of the International Conference on Electrical Machines and Systems, Wuhan, China, 17–20 October 2008; pp. 3327–3332.

17. Chithrabhanu, A.; Vasudevan, K. Online compensation for torque ripple reduction in SRM drives. In Proceedings of the IEEE Transportation Electrification Conference (ITEC-India), Chicago, IL, USA, 22–24 June 2017; pp. 1–6.

18. Shirahase, M.; Morimoto, S.; Sanada, M. Torque ripple reduction of SRM by optimization of current reference. In Proceedings of the International Power Electronics Conference—ECCE ASIA, Sapporo, Japan, 21–24 June 2010; pp. 2501–2507.

19. Nasirian, V.; Kaboli, S.; Davoudi, A.; Moayedi, S. High-Fidelity Magnetic Characterization and Analytical Model Development for Switched Reluctance Machines. *IEEE Trans. Magn.* **2012**, *49*, 1505–1515. [CrossRef]

20. Hamouda, M.; Szamel, L. Torque Control of Switched Reluctance Motor Drives for Electric Vehicles. In Proceedings of the Automation and Applied Computer Science Workshop, Budapest, Hungary, 16 June 2017; pp. 9–20.

21. Jamil, M.U.; Kongprawechnon, W.; Chayopitak, N. Average Torque Control of a Switched Reluctance Motor Drive for Light Electric Vehicle Applications. *IFAC PapersOnLine* **2017**, *50*, 11535–11540. [CrossRef]

22. Cheng, H.; Chen, H.; Yang, Z. Average torque control of switched reluctance machine drives for electric vehicles. *IET Electr. Power Appl.* **2015**, *9*, 459–468. [CrossRef]

23. Petrus, V.; Pop, V.; Gyselinck, A.-C.; Martis, J.; Iancu, C. Average Torque Control of an 8/6 Switched Reluctance Machine for Electric Vehicle Traction. *J. Comput. Sci. Control Syst.* **2012**, *5*, 59.

24. Bose, B.K.; Miller, T.J.E.; Szczesny, P.M.; Bicknell, W.H. Microcomputer Control of Switched Reluctance Motor. *IEEE Trans. Ind. Appl.* **1986**, *22*, 708–715. [CrossRef]

25. Xue, X.D.; Cheng, K.W.E.; Lin, J.K.; Zhang, Z.; Luk, K.F.; Ng, T.W.; Cheung, N.C. Optimal Control Method of Motoring Operation for SRM Drives in Electric Vehicles. *IEEE Trans. Veh. Technol.* **2010**, *59*, 1191–1204. [CrossRef]

26. Nguyen, D.-M.; Bahri, I.; Krebs, G.; Berthelot, E.; Marchand, C.; Ralev, I.V.; Burkhart, B.; De Doncker, R.W. Efficiency Improvement by the Intermittent Control for Switched Reluctance Machine in Automotive Application. *IEEE Trans. Ind. Appl.* **2019**, *55*, 4167–4182. [CrossRef]

27. Inderka, R.; De Doncker, R. DITC-direct instantaneous torque control of switched reluctance drives. *IEEE Trans. Ind. Appl.* **2003**, *39*, 1046–1051. [CrossRef]

28. Fuengwarodsakul, N.; Menne, M.; Inderka, R.; De Doncker, R. High-Dynamic Four-Quadrant Switched Reluctance Drive Based on DITC. *IEEE Trans. Ind. Appl.* **2005**, *41*, 1232–1242. [CrossRef]

29. Chancharoensook, P. Direct instantaneous torque control of a four-phase switched reluctance motor. In Proceedings of the International Conference on Power Electronics and Drive Systems (PEDS), Taipei, Taiwan, 2–5 November 2009; pp. 770–777.

30. Castro, J.; Andrada, P.; Blanqué, B. Minimization of torque ripple in switched reluctance motor drives using an enhanced direct instantaneous torque control. In Proceedings of the XXth International Conference on Electrical Machines, Marseille, France, 2–5 September 2012; pp. 1021–1026.

31. Kianinezhad, M.M.B.M.S.R. Direct Instantaneous Torque Control of Switched Reluctance Motors Using Five Level Converter. In Proceedings of the 46th International Universities Power Engineering Conference, Soest, Germany, 5–8 September 2011.

32. Yang, Q.; Ma, M.; Chang, Z.; Zhang, X.; Lin, Z. Direct instantaneous torque control of switched reluctance motor with three-bridge five-level converter. *PEMD* **2016**, *2016*, 574. [CrossRef]

33. Song, S.; Peng, C.; Guo, Z.; Ma, R.; Liu, W. Direct Instantaneous Torque Control of Switched Reluctance Machine Based on Modular Multi-Level Power Converter. In Proceedings of the 22nd International Conference on Electrical Machines and Systems (ICEMS), Harbin, China, 11–14 August 2019.

34. Sun, Q.; Wu, J.; Gan, C. Optimized Direct Instantaneous Torque Control for SRMs with Efficiency Improvement. *IEEE Trans. Ind. Electron.* **2020**, *68*, 1. [CrossRef]

35. Sandre-Hernandez, O.; Rangel-Magdaleno, J.D.J.; Morales-Caporal, R. A Comparison on Finite-Set Model Predictive Torque Control Schemes for PMSMs. *IEEE Trans. Power Electron.* **2018**, *33*, 8838–8847. [CrossRef]

36. Hamouda, M.; Számel, L. Accurate Magnetic Characterization Based Model Development for Switched Reluctance Machine. *Period. Polytech. Electr. Eng. Comput. Sci.* **2019**, *63*, 202–212. [CrossRef]
37. Hamouda, M.; Számel, L. A new technique for optimum excitation of switched reluctance motor drives over a wide speed range. *Turk. J. Electr. Eng. Comput. Sci.* **2018**, *26*, 2753–2767. [CrossRef]

 *mathematics*

Article

Optimal Rotor Design of Synchronous Reluctance Machines Considering the Effect of Current Angle

Hegazy Rezk [1,2,*], Kotb B. Tawfiq [3,4,5], Peter Sergeant [3,4] and Mohamed N. Ibrahim [3,4,6]

1   College of Engineering at Wadi Addawaser, Prince Sattam Bin Abdulaziz University, Wadi Aldawaser 11991, Saudi Arabia
2   Electrical Engineering Department, Faculty of Engineering, Minia University, Minia 61111, Egypt
3   Department of Electromechanical, Systems and Metal Engineering, Ghent University, 9000 Ghent, Belgium; kotb.basem@ugent.be (K.B.T.); Peter.Sergeant@UGent.be (P.S.); m.nabil@ugent.be (M.N.I.)
4   FlandersMake@UGent—corelab EEDT-MP, 3001 Leuven, Belgium
5   Department of Electrical Engineering, Faculty of Engineering, Menoufia University, Menoufia 32511, Egypt
6   Electrical Engineering Department, Kafrelshiekh University, Kafrelshiekh 33511, Egypt
*   Correspondence: hr.hussien@psau.edu.sa

Abstract: The torque density and efficiency of synchronous reluctance machines (SynRMs) are greatly affected by the geometry of the rotor. Hence, an optimal design of the SynRM rotor geometry is highly recommended to achieve optimal performance (i.e., torque density, efficiency, and power factor). This paper studies the impact of considering the current angle as a variable during the optimization process on the resulting optimal geometry of the SynRM rotor. Various cases are analyzed and compared for different ranges of current angles during the optimization process. The analysis is carried out using finite element magnetic simulation. The obtained optimal geometry is prototyped for validation purposes. It is observed that when considering the effect of the current angle during the optimization process, the output power of the optimal geometry is about 3.32% higher than that of a fixed current angle case. In addition, during the optimization process, the case which considers the current angle as a variable has reached the optimal rotor geometry faster than that of a fixed current angle case. Moreover, it is observed that for a fixed current angle case, the torque ripple is affected by the selected value of the current angle. The torque ripple is greatly decreased by about 34.20% with a current angle of 45° compared to a current angle of 56.50°, which was introduced in previous literature.

Keywords: current angle; design of electric motors; flux-barriers; optimization; synchronous reluctance motor; torque ripple

Citation: Rezk, H.; Tawfiq, K.B.; Sergeant, P.; Ibrahim, M.N. Optimal Rotor Design of Synchronous Reluctance Machines Considering the Effect of Current Angle. *Mathematics* 2021, 9, 344. https://doi.org/10.3390/math9040344

Academic Editor: Jinfeng Liu

Received: 17 January 2021
Accepted: 5 February 2021
Published: 9 February 2021

Publisher's Note: MDPI stays neutral with regard to jurisdictional claims in published maps and institutional affiliations.

1. Introduction

Recently, interest in synchronous reluctance machines (SynRMs) has increased remarkably thanks to their advantages compared to other types of electrical machines [1–5]. They offer a good torque density, a high efficiency, and a wide range of operating speeds [6]. In addition to their simple and robust structure, they have no windings, cages, and permanent magnets in their rotor, resulting in very low rotor losses and hence good thermal management [2]. These advantages make SynRMs a good competitor compared to the other electric machines in several electric drive systems in different industrial applications such as hospitals and aerospace [7]. It is evident through the literature that the performance of SynRMs (torque ripple, average torque, efficiency, and power factor) greatly depends on the saliency ratio (the ratio between the direct and quadrature axis inductances) [8]. This ratio is a function of several parameters of the machine design such as the winding, magnetic material, and rotor flux-barriers [9–11]. Starting from the standard stator design of the induction machine, the rotor flux barrier parameters are key elements in the performance of the SynRM. There are several parameters in the rotor as sketched in Figure 1. Therefore, it is evident that an optimization process is necessary to optimally select the parameters of the SynRM rotor.

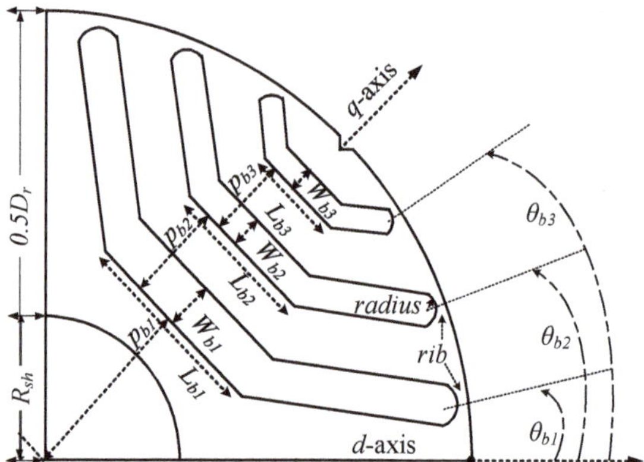

Figure 1. Rotor geometry of one pole of synchronous reluctance machines (SynRM).

Through literature, several research papers about the optimization of SynRMs can be found [12–29]. Various optimization schemes have been reported and applied to SynRM design. For example, in [12], the rotor of the SynRM was optimized using a multi-objective differential evolution algorithm for high-speed applications focusing on selecting the barrier angles and the magnetic insulation ratio. The other geometrical parameters of the rotor were derived from these two parameters (the barrier angles and the magnetic insulation ratio). In [13], the optimal number of flux barriers and rotor poles of the SynRM were optimally selected by applying a weighted-factor using a multi-objective optimization technique. A circular rotor shape was used in [14] to maximize the torque and to minimize the torque ripple of the SynRM using a multi-objective genetic algorithm. In this design, a circular rotor was adopted in order to minimize the number of parameters and to get a time-efficient optimization process. Three different geometries for rotor flux barriers (one rotor has a circular flux barrier and the other two rotors use a rectangular flux barrier) were studied and compared in [15]. The design process uses the same optimization algorithm proposed in [14]. It was shown that the rectangular flux barrier has fewer structural challenges and lower inertia at high speed. In addition, it is preferred especially for machines with inserted permanent magnet (PM inside the flux-barriers. In [16], an alternative technique for the development of asymmetric flux barriers with rotor skewing is proposed in combination with design optimization to enhance average torque and reduce torque ripple. Although the torque ripple in this method is below 3%, the number of optimization variables in this method is relatively high between 29 and 37. This results in a slower and complicated optimization process. In [17], the rotor of the SynRM was optimized using three different popular optimization algorithms (simulated annealing, differential evolution, and genetic algorithm) to minimize the torque ripple and maximize the torque per Joule loss ratio. The differential evolution algorithm has shown the best result in terms of repeatability of the results and convergence time. It was demonstrated in [18–21] that the rotors with skewing techniques have reduced torque ripple significantly. In [22], the rotor of the SynRM was optimized to maximize the saliency ratio and minimize the thickness of the iron ribs. The rotor was made ribless in [23] to obtain an improved power factor, torque, and efficiency.

In [24], a generalized formula was proposed to select the widths and angles of the flux-barriers considering additional factors such as stator and rotor slot opening and number of slots. However, the torque ripple is still high. Furthermore, a preliminary design for the flux-barrier widths was introduced in [25] without considering the influence of the different number of stator slots. The effect of the number of stator slots was considered in [26] and the torque ripple was reduced from 23.38% to 12.3%. All the previous studies about the rotor design of the SynRM did not consider the current angle as a design variable during the design and optimization procedures. The current angle was fixed based on a rule of thumb or primary simulation i.e., in the range of 45° to 60° as in [26]. In [27], a simultaneous structural and magnetic topology optimization technique was developed for the rotor of the SynRM using solid isotropic with material penalization. The total structural compliance, torque ripple, and the average torque were simultaneously considered in this method. In [28], a new technique was proposed to design the rotor of the SynRM. A symmetrical rotor geometry with fluid shaped barriers was used in this optimization method. The optimal design in this method was chosen using the communication between MATLAB and Flux 2D. In [29], a line start SynRM was optimized using an optimization topology that uses the normalized Gaussian network. The computational time was reduced in this method as it separates out unpromising geometries. The effect of the current angle on the final optimal geometry of the SynRM has not been investigated before as far as we know.

This paper studies the effect of considering the current angle during the optimization process on the final optimal geometry of the rotor of the SynRM. Different cases are analyzed and compared for different ranges of current angles during the optimization process. This way, in some cases, the saturation level in the machine is enforced during the optimization process by varying the current angle range. Finite element magnetic simulation is carried out and compared for the optimal geometries. Finally, experimental results are conducted to validate the simulation results.

2. Design Optimization of SynRMs

2.1. Hybrid PSOGWO Technique

In this paper, the hybrid particle swarm optimizer and grey wolf optimizer (PSOGWO) algorithm was used to determine the best parameters in order to obtain the optimal rotor design of the SynRM. The following paragraphs briefly describe the core idea and the updating process of the PSO, GWO, and hybrid PSOGWO.

PSO was originally proposed by Kennedy and Eberhart to simulate the social behavior of a flock of birds [30,31]. In order to determine the best solution, every particle, representing a candidate solution, updates continuously its position and velocity. The following relation can be used to estimate the new step size of each particle.

$$v_i^{t+1} = \overbrace{w.v_i^t}^{first\_section} + \overbrace{C_1.r_1.(Pbest_i - x_i^t)}^{second\_section} + \ldots\ldots \atop \underbrace{C_2.r_2.(Gbest - x_i^t)}_{third\_section} \tag{1}$$

$$x_i^{t+1} = x_i^t + v_i^{t+1} \tag{2}$$

where w is the inertia factor; C_1 and C_2 denote the cognitive and the social coefficients; r_1 and r_2 denote random; t is the iteration number; i is the particle number; $Pbest$ is the local best; $Gbest$ is the global best.

The first section of (1) provides the exploration capability of the PSO. Whereas, the second section moves the particle towards the best position ever achieved by itself. The last section of (1) moves the particle according to the best position achieved by all the particles in the population. The core idea of GWO is extracted from the behavior of grey wolves. GWO simulates the hunting process and the leadership hierarchy of grey wolves [32]. Grey wolves exist at the highest level of the food chain and are regarded as predators.

The hunting mechanism contains two chief sections: tracking and catching the prey, then encircling and attacking the prey until movement stops. During the hunting process, prey is encircled by the grey wolves. To simulate the encircling behavior, the next relations can be considered [32]:

$$D = |C * X_p(t) - X(t)| \tag{3}$$

$$X(t+1) = X_p(t) - A * D \tag{4}$$

where t is the current iteration; X_p and X denote the position of the prey and the location of grey wolves, respectively.

A and C denote the coefficients vectors that are estimated using the following relations:

$$A = a * (2 * r_1 - 1) \tag{5}$$

$$C = 2 * r_2 \tag{6}$$

r_1 and r_2 are random values; a is constant that reduces linearly from 2 to 0 over the optimization process.

The process update of grey wolves is carried out based on the following relation;

$$\begin{cases} D_\alpha = |C_1 * X_\alpha(t) - X(t)| \\ D_\beta = |C_2 * X_\beta(t) - X(t)| \\ D_\delta = |C_3 * X_\delta(t) - X(t)| \end{cases} \tag{7}$$

For every iteration, the best three wolves are represented by X_α, X_β, and X_δ;

$$\begin{cases} X_1 = |X_\alpha - a_1 * D_\alpha| \\ X_2 = |X_\beta - a_2 * D_\beta| \\ X_3 = |X_\delta - a_3 * D_\delta| \end{cases} \tag{8}$$

Finally, the updated position of the prey is provided by the average of three values of positions assessed as the best solutions:

$$X_p(t+1) = \frac{X_1 + X_2 + X_3}{3} \tag{9}$$

The fundamental idea of the hybrid PSOGWO is to integrate the capability of social thinking of the PSO with the local search ability of the GWO. A PSO suffers from shortcomings like catching the local minimum. Therefore, to avoid this disadvantage, the GWO was used to reduce the chance of trapping on the local minimum. Moreover, the GWO has the advantage of preserving a balance between exploitation and exploration during the optimizing procedure. More details about the mathematical modeling and physical meaning of the hybrid PSOGWO can be found in [33].

2.2. Optimization Process

In the optimization process, a stator of a standard induction machine of 5.5 kW with the parameters listed in Table 1 was employed. The stator geometry was kept fixed during the optimization process. Based on the number of stator slots and poles, the number of rotor flux-barriers could be identified which was selected to be three per pole [34,35]. Twelve rotor parameters, θ_{b1}, θ_{b2}, θ_{b3}, W_{b1}, W_{b2}, W_{b3}, L_{b1}, L_{b2}, L_{b3}, p_{b1}, p_{b2}, and p_{b3}, sketched in Figure 1 were considered during the optimization process. To avoid the conflicts in the obtained geometry, some constraints were made as shown in Figure 1 and Table 2. As mentioned before, the main core of this paper is to study the influence of considering the current angle during the optimization process on the final optimal geometry of the SynRM. Therefore, in this research, different ranges of the current angle were considered (five cases) as in Table 3. The ranges of the current angles were selected based on the fact that the current angle of the maximum torque of the SynRMs equaled 45° (i.e., d-axis current = q-axis current) when neglecting the saturation effect. Nevertheless, when considering the saturation effect, the current angle deviated from 45°. Therefore, in this paper, we tried to enforce different ranges of the current angle to around 45°

to determine the impact on the final optimal geometry; this will be shown in the next paragraphs. Although the range of case 5 locates within the case 4 range, there were different optimal geometries obtained based on the two cases. This was why we were trying to narrow the search region of the current angle as in case 5 and to increase this range as in case 4 and even to keep the current angle fixed as in case 3.

Table 1. Parameters of the SynRM.

Parameter	Value	Parameter	Value
Stator inner diameter	110 mm	Air gap length	0.3 mm
Stator outer diameter	180 mm	Slots	36
Rotor outer diameter	109.4 mm	poles	4
Shaft diameter	35 mm	Rated frequency	100 Hz
Axial length	140 mm	Rated power	5.5 kW
Rotor flux barriers per pole	3	Number of phases	3
Stator/Rotor steel	M270-50A/M330-50A	Rms rated current	12.3 A

Table 2. Rotor variables upper and lower limits.

Variable	Lower Limit	Upper Limit
θ_{b1}	5°	9.3°
θ_{b2}	15°	20°
θ_{b3}	25°	30°
W_{b1}	6 mm	8.3 mm
W_{b2}	5 mm	6.5 mm
W_{b3}	3 mm	4 mm
L_{b1}	20 mm	30 mm
L_{b2}	20 mm	25 mm
L_{b3}	10 mm	16 mm
p_{b1}	20 mm	23 mm
p_{b2}	9 mm	13.6 mm
p_{b3}	8 mm	11.8 mm
$radius_{1,2,3}$	25% of $W_{b1,2,3}$	

Table 3. Range of current angle for different cases.

Case Number	Range of Current Angle
Case 1	30°: 40°
Case 2	40°: 45°
Case 3	45°
Case 4	45°: 65°
Case 5	50°: 55°

The hybrid PSOGWO algorithm presented before was implemented to obtain the optimal rotor geometrical parameters and the current angle of each case in order to maximize the output torque and minimize the torque ripple of the machine. The cost function of the optimization is given as follows:

$$cost\ function = T_r^2 + \frac{1}{T_{av}} \tag{10}$$

where, T_r and T_{av} are the torque ripple in percent and the average torque of the SynRM.

The flow chart of the optimization loop is described in Figure 2. The finite element model (FEM) of the machine, in which the equations that represent the machine were solved numerically, was coupled with the PSOGWO technique to obtain the optimal geometry [1]. The losses were determined as in [8]. Later on, FEM is used to evaluate the performance of the obtained optimal machine.

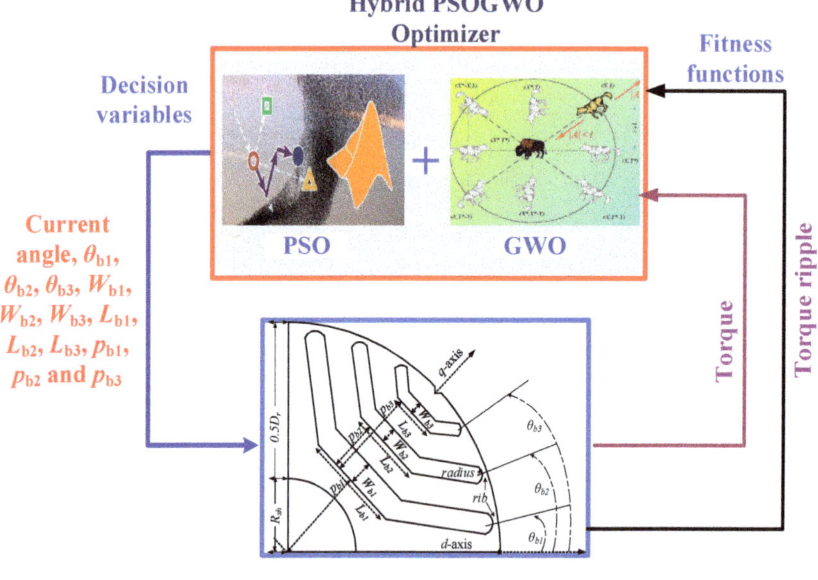

Figure 2. Flow chart of the optimization process.

As mentioned before, five different ranges of the current angle were considered (see Table 3). For each case, the range of the current angle was set and the optimization process was completed. The number of designs for each case was 210. Figure 3 shows the variation of current angle versus iteration number during the optimization process for different cases. The cost function versus the iteration number is reported in Figure 4. Figure 5 shows some performance indicators of the SynRM for different cases. Notice that the results in this section were obtained at different current angles. Therefore, it was not possible to compare the performance indicators of the five cases. However, this will be done in the next section. Moreover, it was proved from Figure 5 that the average torque and the power factor were greatly affected by the value of the current angle in a specific case (between different designs), while the impact on the torque ripple was lower. Figure 6 and Table 4 reveal that the iron volume of the obtained rotor geometry depended on the considered current angle during the optimization process. Figure 6 shows that the third case, which considered a fixed value for the current angle (45°), gave the largest rotor iron volume, while the fourth case, which considered sufficient range of current angle variation in which the current angle of maximum torque and minimum torque ripple of the SynRM existed, gave the lowest iron volume of the rotor. The rotor iron volume of the third case was about

11% higher than that of the fourth case. This meant that the inertia of the SynRM based on the fourth case was lower resulting in a fast-dynamic machine.

Table 4. Final optimal geometry for the SynRM for different case studies.

	Case 1	Case 2	Case 3	Case 4	Case 5
Current angle range [Deg.]	30°: 40°	40°: 45°	45°	45°: 65°	50°: 55°
Optimal angles [Deg.] θ_{b1}, θ_{b2}, θ_{b3}	9.3°, 16.27° and 27.65°	7.36°, 19.42° and 25.98°	7.14°, 19.89° and 26.39°	8.82°, 16.3° and 28.39°	7.84°, 18.66° and 25.69°
Optimal widths [mm] W_{b1}, W_{b2}, W_{b3}	8.3, 5.8 and 3.425	7.7, 6.49 and 3.44	6.76, 5.05 and 3.49	8.3, 5.8 and 4	6.71, 6.5 and 3
Optimal lengths [mm] L_{b1}, L_{b2}, L_{b3}	25.2, 24.52 and 12.38	30, 21.5 and 10.26	25.6, 22.7 and 11.5	30, 22.32 and 15.41	26.82, 21.5 and 16
Optimal positions [mm] p_{b1}, p_{b2}, p_{b3}	22.06, 5.3 and 4.75	22.29, 3.13 and 3	22.71, 3.9 and 3.17	22.93, 3.74 and 3	20.95, 4.3 and 3.7
Rotor iron volume [m^3]	1.780×10^{-4}	1.763×10^{-4}	1.932×10^{-4}	1.736×10^{-4}	1.842×10^{-4}

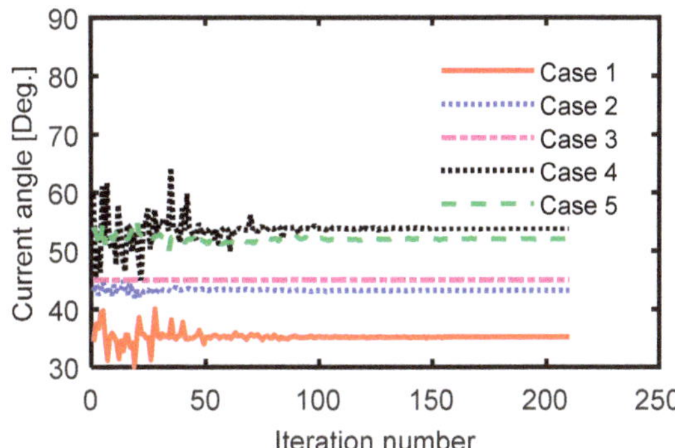

Figure 3. Variation of current angle versus iteration number for different cases.

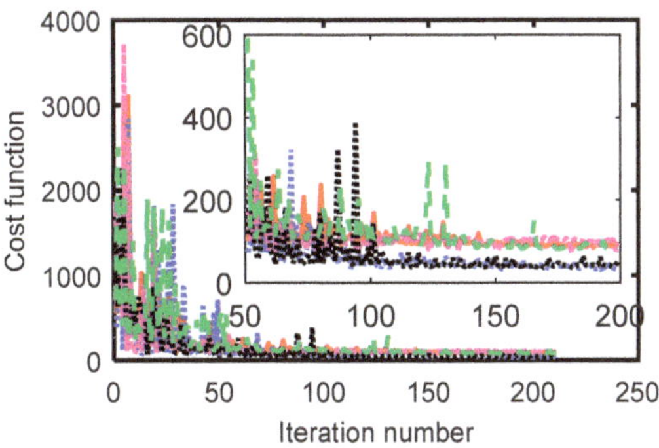

Figure 4. Cost function at different iteration number of the optimization technique.

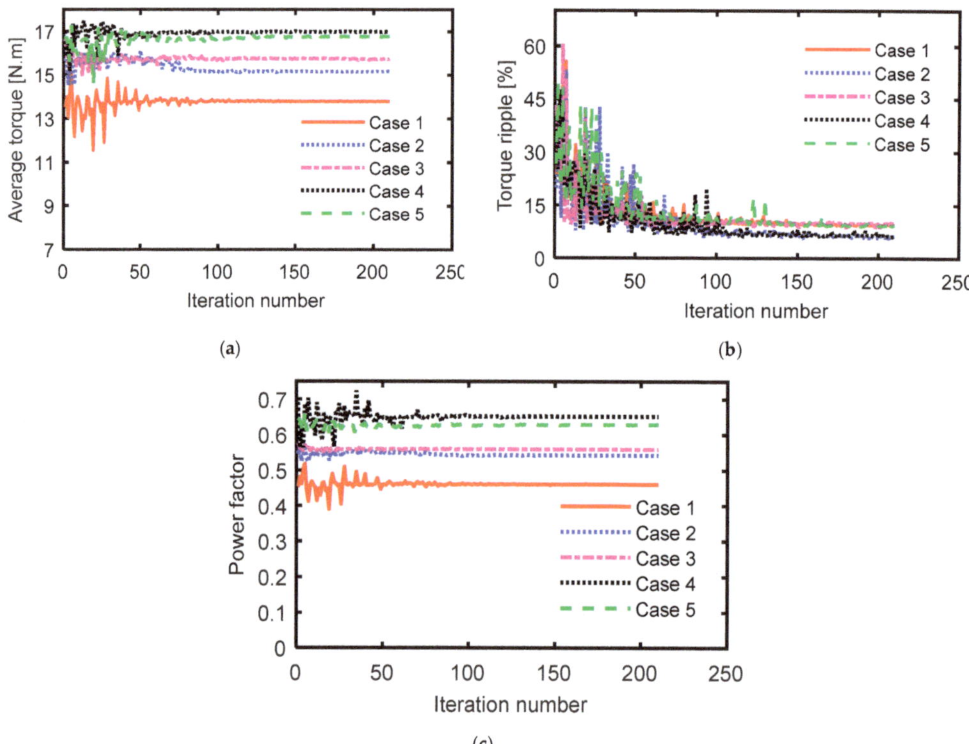

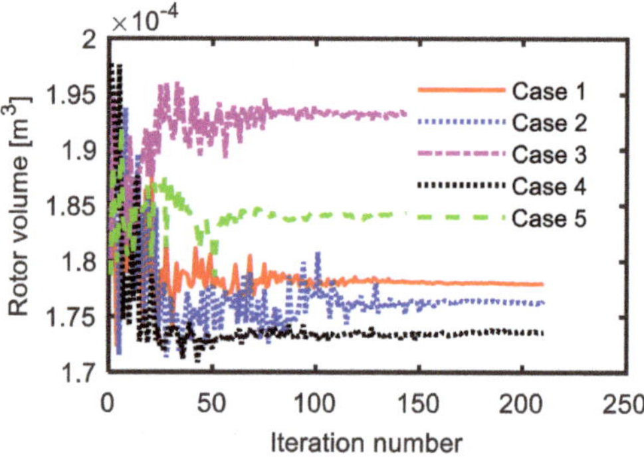

Figure 5. (**a**) Average torque, (**b**) torque ripple, and (**c**) power factor versus iteration number for different cases.

Figure 6. Rotor iron volume versus iteration number for different cases.

Figures 3–6 show that the steady-state response of the optimization process was not delayed considering the current angle, while in some cases, the response became even faster. The optimization process of cases 1 and 4 reached its steady-state after about 80 and 60 iterations respectively compared to about 70 iterations for the third case. In contrast, the fifth case had a slower performance as shown in the zoomed view of Figure 4.

Figures 7–10 and Table 4 show the final optimal geometry of the rotor flux-barrier angles, lengths, widths, and positions for different cases. In Figure 7, it is found that the flux-barrier angles were greatly varied when considering different ranges of current angle during the optimization process. For example, in the first case (with current angle range from 30°: 40°), the flux-barriers angles changed by about 2° to 6° compared to the optimal geometry proposed in [26] and by about 2° to 3.5° compared to the third case. However, when the current angle was kept fixed to 45° in the third case during the optimization process, the obtained optimal geometry for this case was different compared to the optimal geometry presented in [26] which used a current angle equal to 56.50°. This proved that the optimized geometry was sensitive to the chosen value of the current angle for fixed current angle cases.

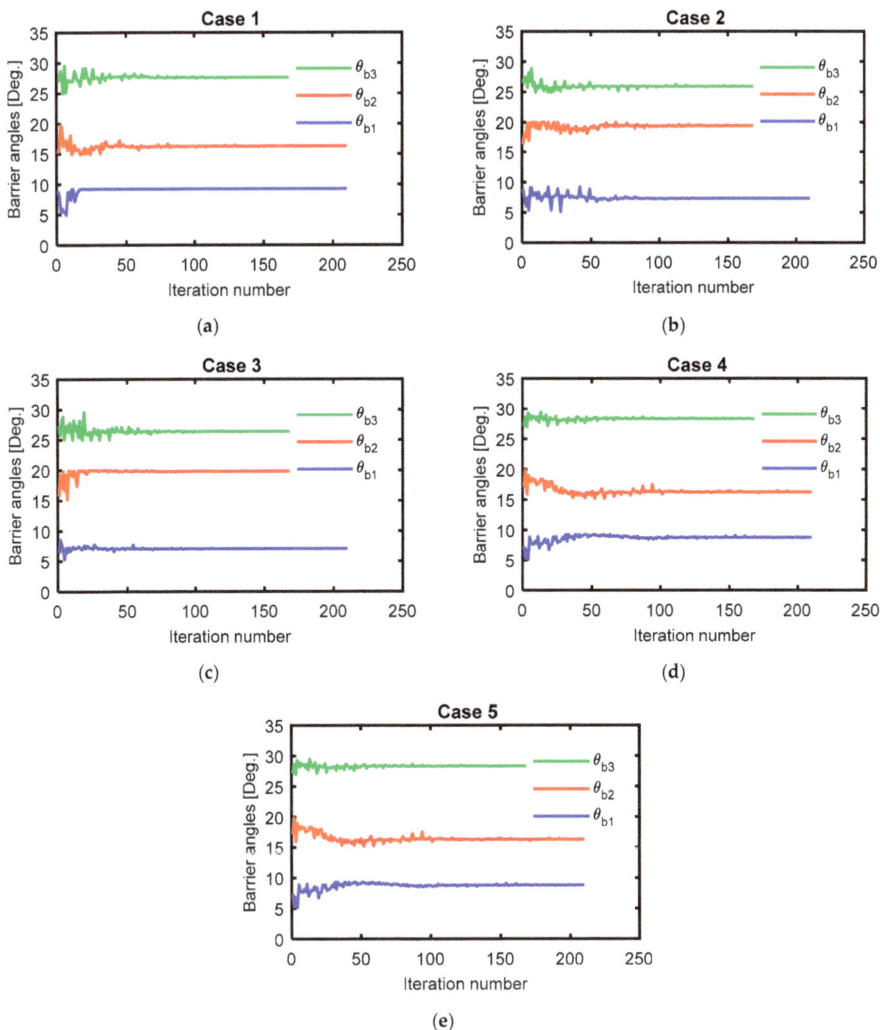

Figure 7. Optimal geometry of flux barrier angles for different cases, (**a**) case 1, (**b**) case 2, (**c**) case 3, (**d**) case 4 and (**e**) case 5.

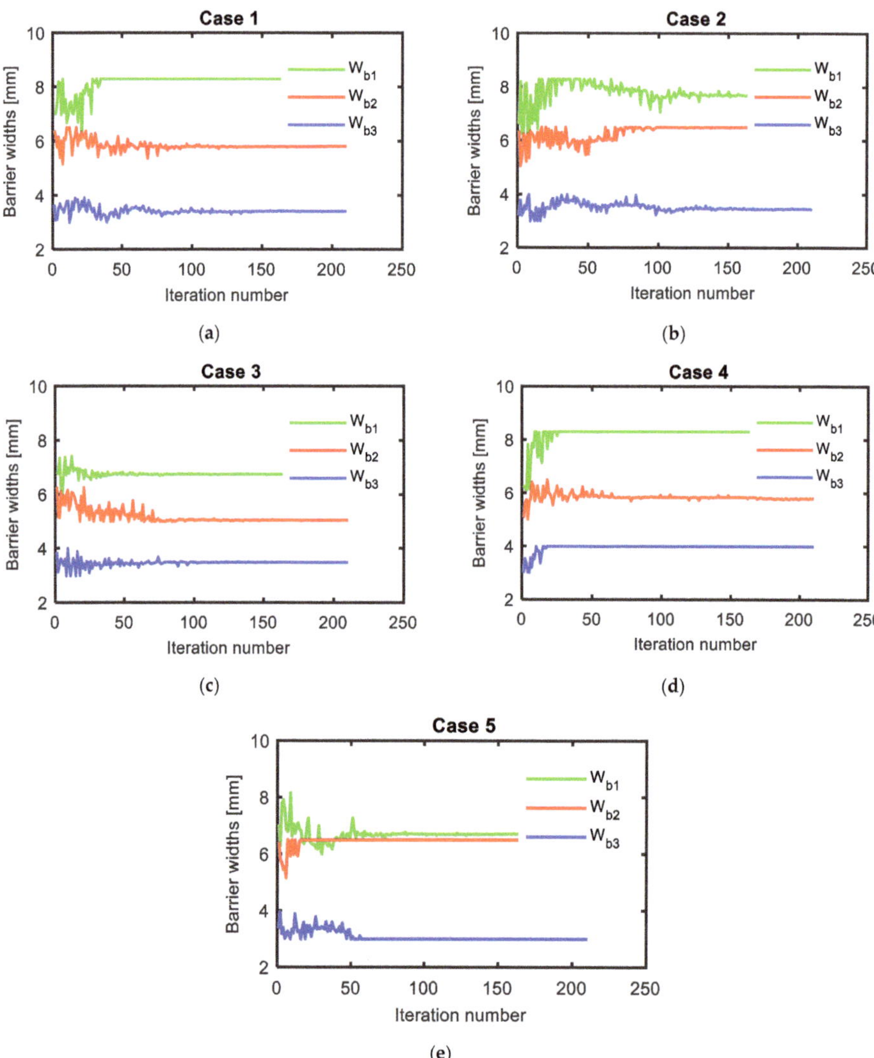

Figure 8. Optimal geometry of flux barrier widths for different cases, (**a**) case 1, (**b**) case 2, (**c**) case 3, (**d**) case 4 and (**e**) case 5.

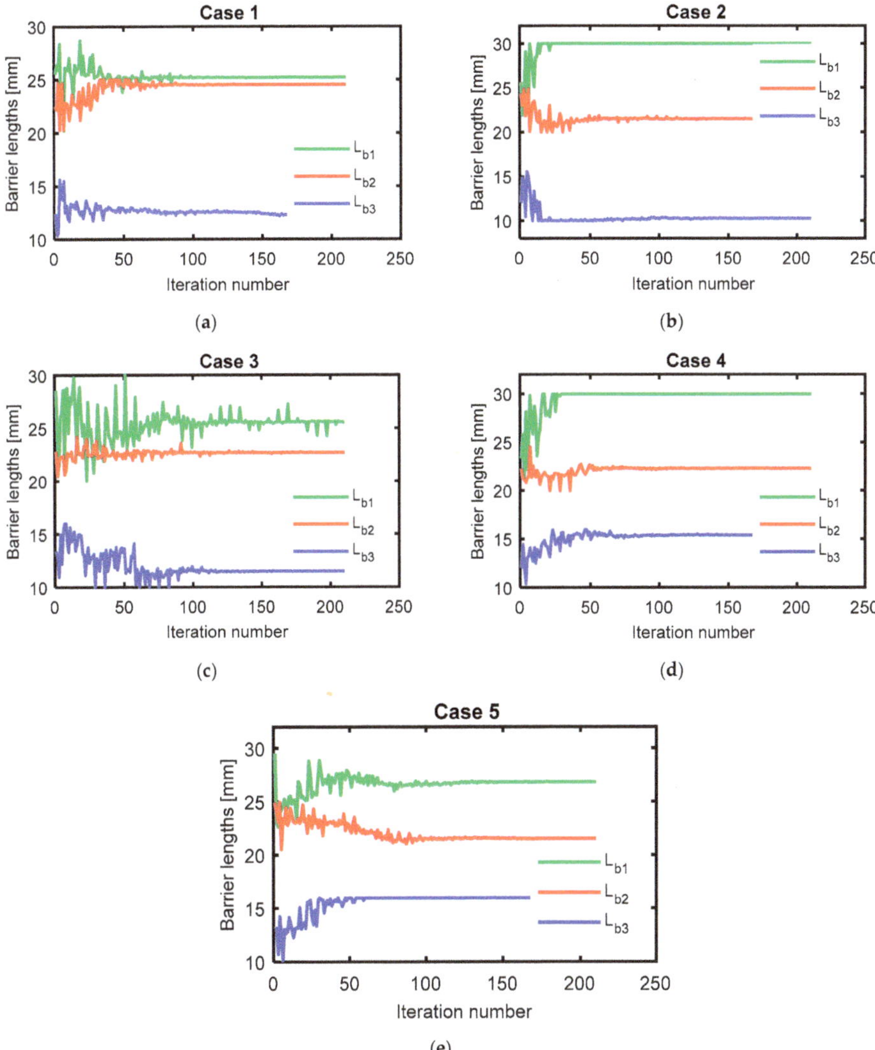

Figure 9. Optimal geometry of flux barrier lengths for different cases, (**a**) case 1, (**b**) case 2, (**c**) case 3, (**d**) case 4 and (**e**) case 5.

In addition, the optimal dimensions of the flux-barriers widths were also varied with current angles as shown in Figure 8. The flux-barriers optimal widths, W_{b1}, W_{b2}, and W_{b3} were changed by about 1.54, 0.75, and 0.7 mm respectively for the first case compared to their values in the third case. In addition, the flux-barriers optimal lengths, L_{b1}, L_{b2}, and L_{b3} were changed by about 0.40, 1.82, and 0.88 mm respectively for the first case compared to their values in the third case as shown in Figure 9. Moreover, Figure 10 shows that the flux-barriers optimal positions, p_{b1}, p_{b2}, and p_{b3} were changed by about 0.65, 1.40, and 1.58 mm respectively for the first case compared to their values in the third case.

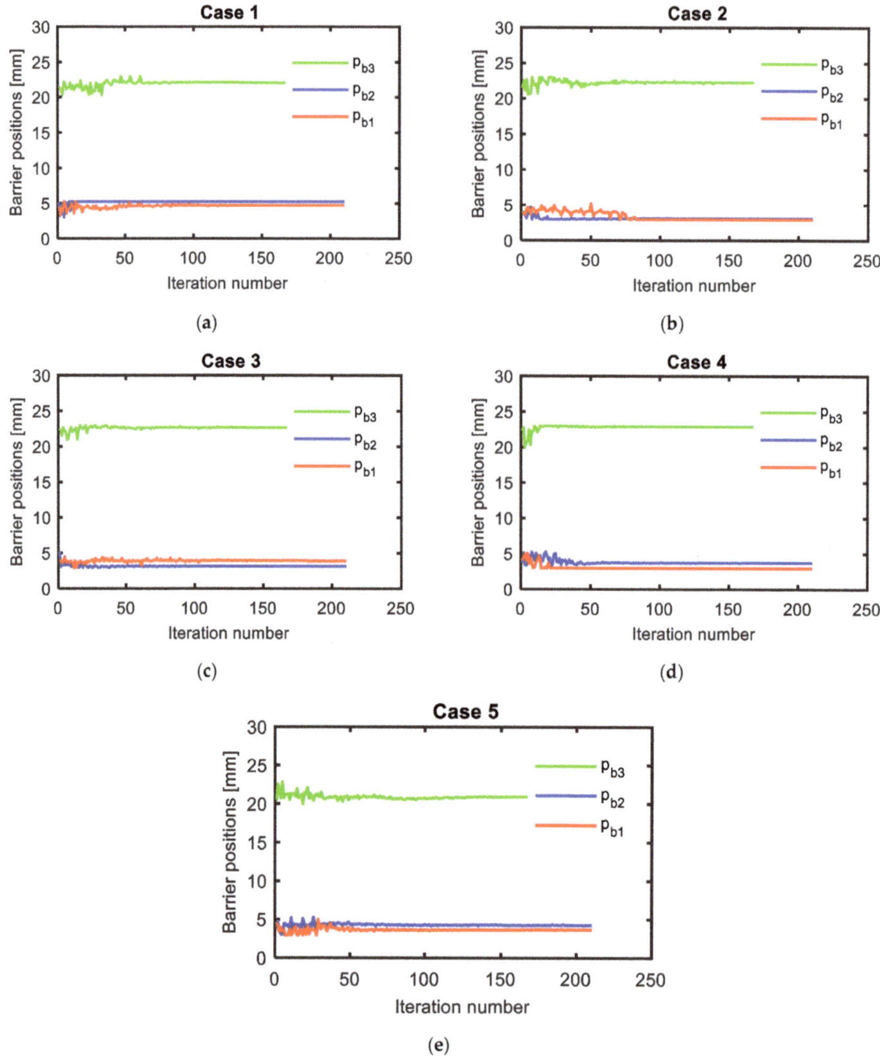

Figure 10. Optimal geometry of flux barrier positions for different cases, (**a**) case 1, (**b**) case 2, (**c**) case 3, (**d**) case 4 and (**e**) case 5.

3. Performance Analysis of SynRM

The performance of the optimal geometry of the rotor of the three-phase SynRM for different case studies was studied and compared using finite element magnetic simulations. The optimal geometry for each case is shown in Table 5. Figure 11a shows the output power of the three-phase SynRM for different cases at rated conditions (speed = 3000 rpm and RMS current = 12.23 A) and at different current angles. It was been found from Table 5 and Figure 11a that the first case gave the highest output power and the second case gave the lowest output power at the rated condition and at the optimal current angle. The optimal current angle was the angle that maximized the output power. The optimal current angle was 52.11° for both the first, the second, and the third case as shown in Figure 11a and Table 5, while it was 56.8° for the other cases. The output power in the first case was 5.65% higher than the second case. Moreover, the first case had about 3.32% higher output power compared to the third case. Note that the current angle in the third case was fixed during the optimization process while the effect of the current angle was considered in the first case as discussed in the previous section.

Table 5. Performance of the optimal geometry for different case studies of the three-phase SynRM using finite element model (FEM) simulation.

	Case 1	Case 2	Case 3	Case 4	Case 5
Optimal current angle	52.11°	52.11°	52.11°	56.8°	56.8°
Output power at optimal current angle [W]	5385	5097	5212	5273	5221
Torque ripple at optimal current angle [%]	7.4	6.5	7.9	5.85	10.58
Power factor at optimal current angle	0.6297	0.6185	0.6182	0.6628	0.6555
Saliency ratio at optimal current angle [%]	5.36	4.84	4.8	5.25	5.06

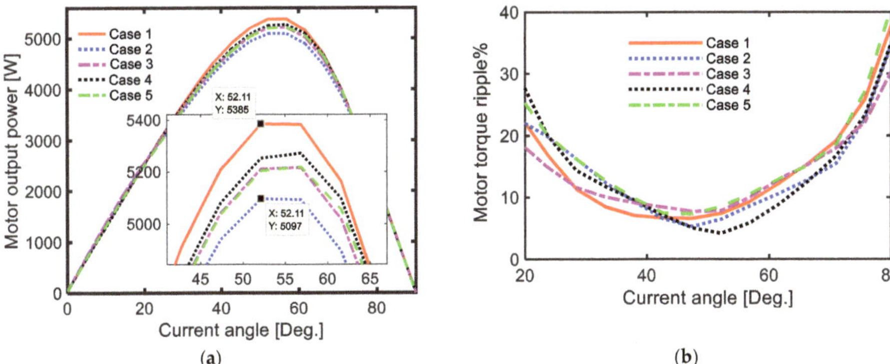

Figure 11. (**a**) Motor output power and (**b**) torque ripple at different current angles and at rated conditions for the optimal geometry of different cases.

Figure 11b shows the torque ripple of the three-phase SynRM for different cases at the rated conditions and at different current angles. The torque ripple decreased with the increase in current angle till it reached its minimum value then it increased again. It was found that the torque ripple had the lowest value in the fourth case, about 5.85%, while the fifth case gave the maximum value of the torque ripple: about 10.58%. The torque ripple for the first case was 7.4%. However, the torque ripple in the third case which used a fixed current angle during the optimization process was about 7.9%. The chosen value of the current angle for the fixed current angle cases significantly affected the obtained torque ripple at the optimal current angle. This was highly obvious in [26], which used a current angle equal to 56.5° and the obtained torque ripple with the optimal angle in [26] was about 12%.

To summarize, the much higher output power and lower torque ripple of our work compared to [26] justified research of the current angle in the geometrical optimization process, as was the goal of this paper.

Figure 12 shows the power factor and the saliency ratio for the optimal geometry for different cases studied at rated conditions and at different current angles. It is noted that the fourth case had the highest value of the power factor. It was 0.6628. This was due to its higher optimal current angle compared to the first case. The power factor of the first case was 0.6297. Figure 12b shows that the first case gave the highest saliency ratio and the second case gave the lowest saliency ratio. In addition, the saliency ratio of the first case was about 11.7% higher than its value in the third case. The distribution of flux density at the same instant of the optimal geometry is shown in Figure 13 for the various study cases.

It was been found that the rotor geometry calculated from the first case had less saturated area compared to other studied cases.

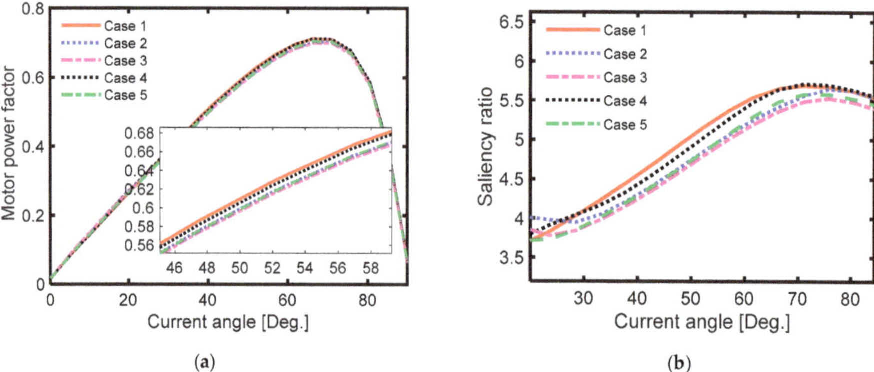

Figure 12. (**a**) Motor power factor and (**b**) saliency ratio at different current angles and at rated conditions for the optimal geometry of different cases.

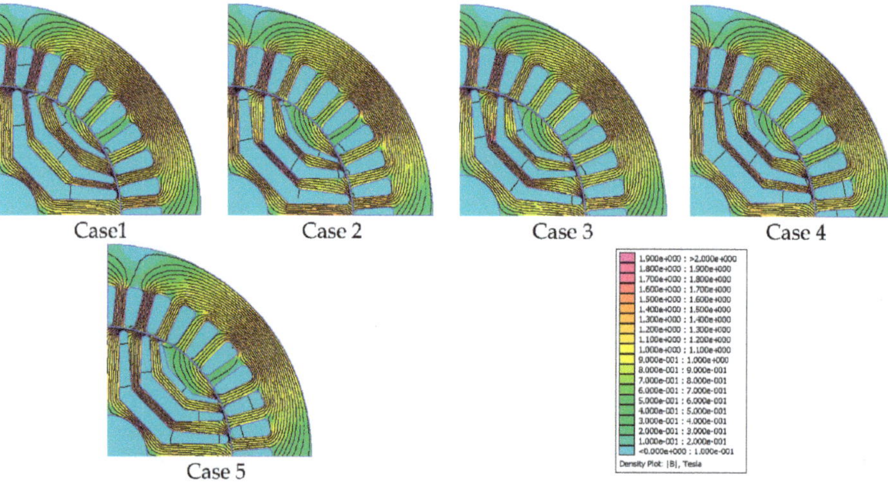

Figure 13. Flux density distribution of the optimal geometry for different cases.

From the previous analysis and discussion, it was evident that considering the current angle during the optimization process of SynRMs was beneficial. Besides, it also observed that the range of the current angle played a role in the maximum output torque and torque ripple value.

4. Experimental Results

Figure 14a shows a complete experimental test bench to validate the simulation results presented before. It consisted of a 5.5 kW three-phase SynRM with a 36-slot stator and four-pole rotor as in case 4. The stator and the rotor are shown in Figure 14b. The three-phase SynRM was coupled with a 10 kW three-phase induction motor. The SynRM implemented in this paper required a control system, as it had no rotor cage. Hence, the desired speed of the three-phase SynRM was achieved using an induction motor, as the three-phase SynRM worked in the mode of torque control. A three-phase inverter based on space vector modulation with a 6.6 kHz switching frequency and 600 V DC bus voltage was used to control the three-phase SynRM. A digital signal processing (DSP1103) was used to obtain the required switching pulses. Incremental encoder and torque sensor were used to measure the rotor speed and the average torque respectively. The input electrical power for the three-phase SynRM was computed using a power analyzer.

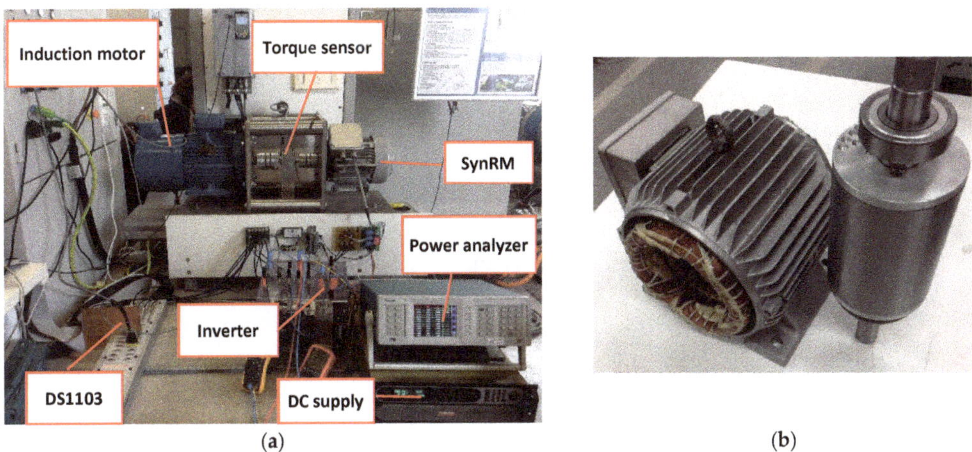

Figure 14. (**a**) The complete experimental setup and (**b**) three-phase SynRM stator and rotor.

To validate the implemented simulation model, various measurements were obtained on the prototype. Figure 15 shows the simulated and measured value of the SynRM output torque at half the rated speed and current (speed = 1500 rpm and RMS current= 6.1 A) at different current angles. There was good agreement between the simulated and average values. The average torque and power factor measured and simulated values at different line currents including overloading to double the rated current, at an optimal current angle, and at one-third the rated speed as shown in Figures 16a and 16b respectively. The torque was linearly varied with the current as shown in Figure 16a. Figure 16b shows that there was a step-change in the power factor. This was due to the change of optimal current angle with line current.

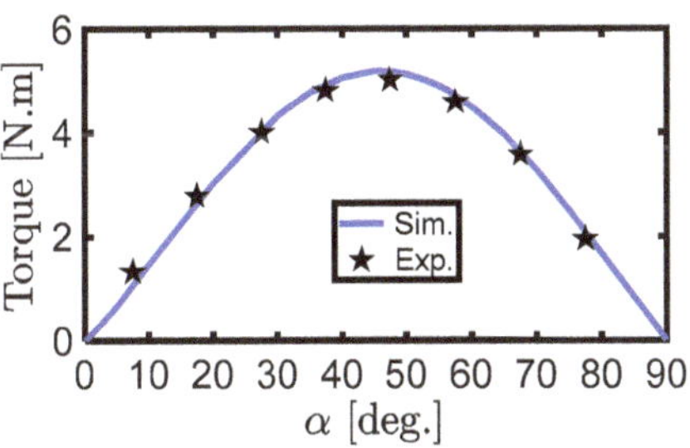

Figure 15. Simulated and measured output torque at different current angles at half the rated current and half the rated speed.

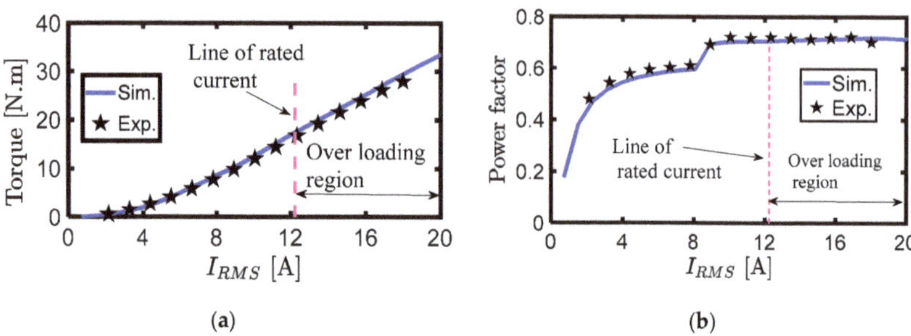

Figure 16. (**a**) Average output torque and (**b**) power factor at different line currents, at an optimal current angle, and at $N_r = 1000$ rpm.

Figure 17 shows the measured and simulated values for the efficiency and total losses of the SynRM at different line currents, at rated speed, and at an optimal current angle. There was a slight difference between the simulated and measured values of efficiency and losses. This was due to neglecting the effect of mechanical and switching losses and the inaccurate iron loss simulation model parameters. Figure 18 shows the efficiency map of the complete drive system at different rotor speeds and at optimal current angles including the flux weakening region.

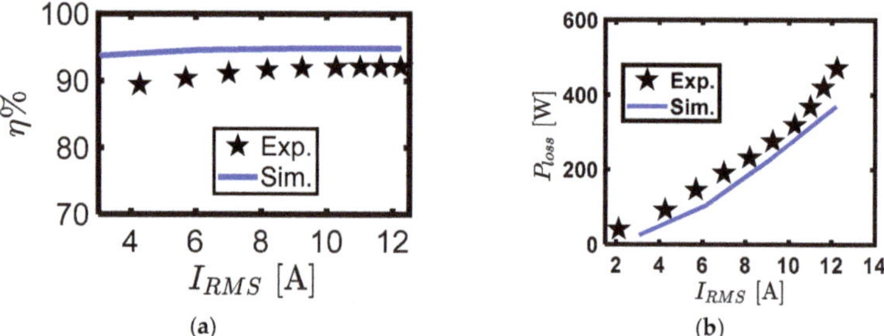

Figure 17. (**a**) Efficiency, (**b**) SynRM total losses at different line currents, optimal current angle, and at rated speed.

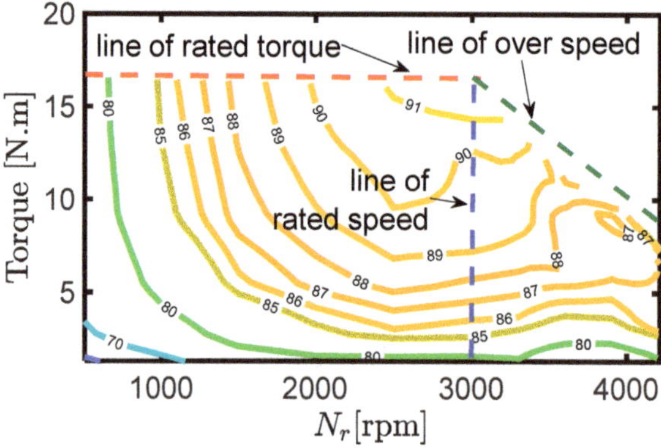

Figure 18. Efficiency map of the complete drive system at optimal current angles.

5. Conclusions

This paper has investigated the influence of considering the current angle on the final optimal geometry when designing the rotor of a synchronous reluctance motor. Five cases for different ranges of current angles have been studied and compared using finite element simulation. It is proved that considering the current angle on the SynRM rotor design is effective to obtain improved performance of the SynRM. It is observed that the output power is increased by about 3.32% when the current angle is considered as a variable during the rotor design optimization process compared to the fixed current angle case. In addition, the selection of the appropriate current angle value for fixed current angle cases has a significant influence on the torque ripple of the optimized rotor geometry. Further, it is also noticed that the range of the current angle plays a role in the maximum output torque and torque ripple value. Moreover, it is found that for a fixed current angle case (angle = 45°), the torque ripple of the optimal geometry is greatly decreased by 34.20% compared to a case of current angle equal to 56.50°, which is the reference case in literature [26]. In the end, a test bench for a 5.5 kW three-phase SynRM has been carried out to validate the simulated results.

Author Contributions: Conceptualization, K.B.T., M.N.I., H.R. and P.S., validation, K.B.T., M.N.I., H.R., writing—original draft preparation, K.B.T., M.N.I., H.R., writing—review and editing, M.N.I., H.R., and P.S. All authors have read and agreed to the published version of the manuscript.

Funding: This research received no external funding.

Conflicts of Interest: The authors declare no conflict of interest.

References

1. Bacco, G.; Bianchi, N.; Mahmoud, H. A Nonlinear Analytical Model for the Rapid Prediction of the Torque of Synchronous Reluctance Machines. *IEEE Trans. Energy Convers.* **2018**, *33*, 1539–1546. [CrossRef]
2. Tawfiq, K.B.; Ibrahim, M.N.; El-Kholy, E.E.; Sergeant, P. Refurbishing three-phase synchronous reluctance machines to multiphase machines. *Electr. Eng.* **2020**. [CrossRef]
3. Ibrahim, M.N.; Sergeant, P.; Rashad, E.E.M. Combined Star-Delta Windings to Improve Synchronous Reluctance Motor Performance. *IEEE Trans. Energy Convers.* **2016**, *31*, 1479–1487. [CrossRef]
4. Bianchi, N.; Bolognani, S.; Bon, D.; Pré, M.D. Torque Harmonic Compensation in a Synchronous Reluctance Motor. *IEEE Trans. Energy Convers.* **2008**, *23*, 466–473. [CrossRef]
5. Tawfiq, K.B.; Ibrahim, M.N.; El-Kholy, E.E.; Sergeant, P. Performance Improvement of Existing Three Phase Synchronous Reluctance Machine: Stator Upgrading to 5-Phase With Combined Star-Pentagon Winding. *IEEE Access* **2020**, *8*, 143569–143583. [CrossRef]
6. Ibrahim, M.N.F.; Sergeant, P.; Rashad, E.M. Relevance of Including Saturation and Position Dependence in the Inductances for Accurate Dynamic Modeling and Control of SynRMs. *IEEE Trans. Ind. Appl.* **2017**, *53*, 151–160. [CrossRef]
7. Vagati, A. *Synchronous Reluctance Drives: A New Alternative (Tutorial Course Notes)*; IEEE Press: Piscataway, NJ, USA, 1994; Chapter 6; pp. 1–27.
8. Ibrahim, M.N.; Sergeant, P.; Rashad, E.M. Synchronous Reluctance Motor Performance Based on Different Electrical Steel Grades. *IEEE Trans. Magn.* **2015**, *51*, 1–4. [CrossRef]
9. Zhao, W.; Zhao, F.; Lipo, T.A.; Kwon, B.-I. Optimal Design of a Novel V-Type Interior Permanent Magnet Motor with Assisted Barriers for the Improvement of Torque Characteristics. *IEEE Trans. Magn.* **2014**, *50*, 1–4. [CrossRef]
10. Davoli, M.; Bianchini, C.; Torreggiani, A.; Immovilli, F. A design method to reduce pulsating torque in PM assisted synchronous reluctance machines with asymmetry of rotor barriers. In Proceedings of the IECON 2016—42nd Annual Conference of the IEEE Industrial Electronics Society, Florence, Italy, 24–26 October 2016; pp. 1566–1571.
11. Baek, J.; Bonthu, S.; Choi, S. Design of five-phase permanent magnet assisted synchronous reluctance machine for low output torque ripple applications. *IET Electr. Power Appl.* **2016**, *10*, 339–346. [CrossRef]
12. Babetto, C.; Bacco, G.; Bianchi, N. Synchronous Reluctance Machine Optimization for High-Speed Applications. *IEEE Trans. Energy Convers.* **2018**, *33*, 1266–1273. [CrossRef]
13. Howard, E.; Kamper, M.J. Weighted Factor Multiobjective Design Optimization of a Reluctance Synchronous Machine. *IEEE Trans. Ind. Appl.* **2016**, *52*, 2269–2279. [CrossRef]
14. Cupertino, F.; Pellegrino, G.; Armando, E.; Gerada, C. A SyR and IPM machine design methodology assisted by optimization algorithms. In Proceedings of the 2012 IEEE Energy Conversion Congress and Exposition (ECCE), Raleigh, NC, USA, 15–20 September 2012; pp. 3686–3691. [CrossRef]

15. Pellegrino, G.; Cupertino, F.; Gerada, C. Automatic Design of Synchronous Reluctance Motors Focusing on Barrier Shape Optimization. *IEEE Trans. Ind. Appl.* **2015**, *51*, 1465–1474. [CrossRef]
16. Howard, E.; Kamper, M.J.; Gerber, S. Asymmetric Flux Barrier and Skew Design Optimization of Reluctance Synchronous Machines. *IEEE Trans. Ind. Appl.* **2015**, *51*, 3751–3760. [CrossRef]
17. Cupertino, F.; Pellegrino, G.; Gerada, C. Design of Synchronous Reluctance Motors With Multiobjective Optimization Algorithms. *IEEE Trans. Ind. Appl.* **2014**, *50*, 3617–3627. [CrossRef]
18. Vagati, A.; Pastorelli, M.A.; Francheschini, G.; Petrache, S.C. Design of low-torque-ripple synchronous reluctance motors. *IEEE Trans. Ind. Appl.* **1998**, *34*, 758–765. [CrossRef]
19. Muteba, M.; Twala, B.; Nicolae, D.V. Torque ripple minimization in synchronous reluctance motor using a sinusoidal rotor lamination shape. In Proceedings of the 2016 XXII International Conference on Electrical Machines (ICEM), Lausanne, Switzerland, 4–7 September 2016; pp. 606–611.
20. Arafat, A.; Choi, S. Active current harmonic suppression for torque ripple minimization at open phase faults in a five-phase PMa-SynRM. *IEEE Trans. Ind. Electron.* **2019**, *66*, 922–931. [CrossRef]
21. Bianchi, N.; Degano, M.; Fornasiero, E. Sensitivity analysis of torque ripple reduction of synchronous reluctance and interior PM motors. *IEEE Trans. Ind. Appl.* **2015**, *51*, 187–195. [CrossRef]
22. Donaghy-Spargo, C.M. Electromagnetic–mechanical design of synchronous reluctance rotors with fine features. *IEEE Trans. Magn.* **2017**, *53*, 1–8. [CrossRef]
23. Bao, Y.; Degano, M.; Wang, S.; Chuan, L.; Zhang, H.; Xu, Z.; Gerada, C. A Novel Concept of Ribless Synchronous Reluctance Motor for Enhanced Torque Capability. *IEEE Trans. Ind. Electron.* **2020**, *67*, 2553–2563. [CrossRef]
24. Taghavi, S.; Pillay, P. A novel grain-oriented lamination rotor core assembly for a synchronous reluctance traction motor with a reduced torque ripple algorithm. *IEEE Trans. Ind. Appl.* **2016**, *52*, 3729–3738. [CrossRef]
25. Moghaddam, R.R. Synchronous Reluctance Machine (SynRM) in Variable Speed Drives (VSD) Applications. Ph.D. Thesis, KTH Royal Institute of Technology, Stockholm, Sweden, 2011.
26. Ibrahim, M.N.; Sergeant, P.; Rashad, E. Simple Design Approach for Low Torque Ripple and High Output Torque Synchronous Reluctance Motors. *Energies* **2016**, *9*, 942. [CrossRef]
27. Guo, F.; Brown, I.P. Simultaneous Magnetic and Structural Topology Optimization of Synchronous Reluctance Machine Rotors. *IEEE Trans. Magn.* **2020**, *56*, 1–12. [CrossRef]
28. Uberti, F.; Frosini, L.; Szabo, L. An Optimization Procedure for a Synchronous Reluctance Machine with Fluid Shaped Flux Barriers. In Proceedings of the 2020 International Conference on Electrical Machines (ICEM), Gothenburg, Sweden, 23–26 August 2020; pp. 389–395. [CrossRef]
29. Lolova, I.; Barta, J.; Bramerdorfer, G.; Silber, S. Topology optimization of line-start synchronous reluctance machine. In Proceedings of the 2020 19th International Conference on Mechatronics—Mechatronika (ME), Prague, Czech Republic, 2–4 December 2020; pp. 1–7. [CrossRef]
30. Eberhart, R.; Kennedy, J. A new optimizer using particle swarm theory. In Proceedings of the Sixth International Symposium on Micro Machine and Human Science, MHS'95, Nagoya, Japan, 4–6 October 1995; pp. 39–43.
31. Abdalla, O.; Rezk, H.; Ahmed, E.M. Wind driven optimization algorithm based global MPPT for PV system under non-uniform solar irradiance. *Sol. Energy* **2019**, *180*, 429–444. [CrossRef]
32. Mohamed, A.; Diab, A.A.Z.; Rezk, H. Partial shading mitigation of PV systems via different meta-heuristic techniques. *Renew. Energy* **2019**, *130*, 1159–1175. [CrossRef]
33. Şenel, F.A.; Gökçe, F.; Yüksel, A.S.; Yiğit, T. A novel hybrid PSO–GWO algorithm for optimization problems. *Eng. Comput.* **2019**, *35*, 1359–1373. [CrossRef]
34. Wang, K.; Zhu, Z.Q. Torque ripple reduction of synchronous reluctance machines optimal slot/pole and flux-barrier layer number combinations. *Int. J. Comput. Math. Electr. Electron. Eng.* **2015**, *34*, 3–17. [CrossRef]
35. Palmieri, M.; Perta, M.; Cupertino, F.; Pellegrino, G. Effect of the numbers of slots and barriers on the optimal design of synchronous reluctance machines. In Proceedings of the 2014 International Conference on Optimization of Electrical and Electronic Equipment (OPTIM), Brasov, Romania, 22–24 May 2014.

Article

Robust Sensorless Model-Predictive Torque Flux Control for High-Performance Induction Motor Drives

Ahmed G. Mahmoud A. Aziz [1,2], Hegazy Rez [3] and Ahmed A. Zaki Diab [1,*]

[1] Electrical Engineering Department, Faculty of Engineering, Minia University, Minia 61111, Egypt; A.G.mahmoud@mhiet.edu.eg
[2] El Minia High Institute of Engineering and Technology, Minia 61111, Egypt
[3] College of Engineering at Wadi Addawaser, Prince Sattam Bin Abdulaziz University, Wadi Aldawaser 11991, Saudi Arabia; hr.hussien@psau.edu.sa
* Correspondence: a.diab@mu.edu.eg; Tel.: +20-102-177-7925

Abstract: This paper introduces a novel sensorless model-predictive torque-flux control (MPTFC) for two-level inverter-fed induction motor (IM) drives to overcome the high torque ripples issue, which is evidently presented in model-predictive torque control (MPTC). The suggested control approach will be based on a novel modification for the adaptive full-order-observer (AFOO). Moreover, the motor is modeled considering core losses and a compensation term of core loss applied to the suggested observer. In order to mitigate the machine losses, particularly at low speed and light load operations, the loss minimization criterion (LMC) is suggested. A comprehensive comparative analysis between the performance of IM drive under conventional MPTC, and those of the proposed MPTFC approaches (without and with consideration of the LMC) has been carried out to confirm the efficiency of the proposed MPTFC drive. Based on MATLAB® and Simulink® from MathWorks® (2018a, Natick, MA 01760-2098 USA) simulation results, the suggested sensorless system can operate at very low speeds and has the better dynamic and steady-state performance. Moreover, a comparison in detail of MPTC and the proposed MPTFC techniques regarding torque, current, and fluxes ripples is performed. The stability of the modified adaptive closed-loop observer for speed, flux and parameters estimation methodology is proven for a wide range of speeds via Lyapunov's theorem.

Keywords: induction motor; model predictive; sensorless; high performance

Citation: Aziz, A.G.M.A.; Rez, H.; Diab, A.A.Z. Robust Sensorless Model-Predictive Torque Flux Control for High-Performance Induction Motor Drives. *Mathematics* **2021**, *9*, 403. https://doi.org/10.3390/math9040403

Academic Editor: Vladimir Prakht

Received: 19 January 2021
Accepted: 15 February 2021
Published: 18 February 2021

Publisher's Note: MDPI stays neutral with regard to jurisdictional claims in published maps and institutional affiliations.

1. Introduction

Several schemes for speed-sensorless vector-controlled induction motor (IM) drives have been suggested for nearly a decade. Via flux and speed sensors positioned within the machine will deteriorate the machine's robustness and raise the associated maintenance expenses. Speed sensor cost is within the same range as the price of motor itself, at least for machines with ratings below 10 kW. In many applications, sensor mounting to the motor is an obstacle. The majority of estimators proposed in sensorless IM drives for combined flux and also speed estimation could be divided into three categories: (a) Model-reference-adaptive system (MRAS) [1,2]. Although MRAS-based estimators are favored due to simplicity, ease of implementation, and documented stability [3], they have certain drawbacks in the low-speed zone, where open-loop integration can result in instability due to stator resistance misestimating [4]. (b) Full-order observers (FOO) [5] providing both stator current and stator or rotor flux estimates. In FOO, alteration is made throughout the error between both measured stator current and its estimated value that is often used to adjust estimated speed in an adaptation law. (c) Reduced-order observers (ROO) [6], that are only rotor or stator flux estimators. It is the potential to make the ROO essentially sensorless [7], i.e., speed will not appear in the observer equations. The major drawback for sensorless velocity drives is their poor performance when IM operates at lowish velocity range or near zero-velocity point. However, low-velocity and zero-velocity operating case

49

of IMs is the popular for industrial applications, like hoists (sometimes operating at 0.5 Hz to 1 Hz), marble cutting machines (operating at 0.3 Hz sometimes) and mine carrier cars (sometimes operating at 0 Hz). Therefore, sensorless drives performance at low-velocity ranges necessities to be improved.

To achieve an active decoupled control for flux and electromagnetic torque, the direct torque control (DTC) was offered to replace the field-oriented control (FOC) in the drive domain. DTC provides a simpler scheme, faster response, and lower machine parameter dependency than FOC. Besides that, it does not include coordinates transformation or current regulations [8,9]. The key problem of this technique is the excessive amount of flux and also electromagnetic torque ripples created through the variable switching frequency due to discrete nature of the hysteresis comparators and voltage vector selection look-up-table (LUT) [10].

In order to address some weaknesses in classical DTC techniques, an effective control approach called model predictive control (MPC), has recently been proposed. MPC's key advantages are its ability to recognize various control targets, the straightforward management of nonlinearities and control limits of the model, and easy implementation [11–14]. Recent works have implemented MPC method to avoid limitations of classical LUT-based DTC of IM drives [15–18]. The proposed MPDTC techniques implemented in these studies contributed to improving the drive's dynamic efficiency, but as an essential feature of the regulated variables, the ripple content was still present. There are different explanations for why the ripple phenomena in the MPDTC are presented. The foremost one would be originated in view of the technique principle of MPDTC itself. After the minimization of a cost function, it selects a voltage vector and inevitably applies this vector for the entire subsequent sampling duration, even though this is deleterious in certain situations, as it may occur that the regulated quantity (i.e., the torque) is therefore pushed out of reasonable limits before the end of the loop.

Different methods have been established to eliminate unwanted ripples, such as splitting the sampling interval into two parts and adding dual different voltage vectors in each interval [19–21], for example. This has essentially helped to limit the ripples. Otherwise, the complexities and computational burden of those methods are amplified and this would affect the dynamic digital control response, which is intended to be very rapid in many numbers of industrial applications.

Model-predictive current control (MPCC) was presented in [22,23] as a solution to limit the use of the cost function weighting value. Two prediction horizons were suggested in [23] to predict stator currents, which resulted in the reduction of the accompanying noise. However, measurement time was increased, otherwise amplifying the commutations of the inverter.

Due to its intuitive definition, high versatility, and simple incorporation of constraints, model-predictive torque control (MPTC) has recently gained significant attention within the academic and industrial communities [24,25]. Despite the MPTC's intuitive definition and rapid response, as stator flux and also torque control variables have different amplitudes and units, a proper stator flux weighting factor should be built to achieve adequate performance [13,26,27].

Model-predictive flux control (MPFC) is suggested in [28] and compared with the conventional MPTC in [29] to prevent nontrivial weighting factor adjustment effort for MPTC. While in [13], the prediction errors are translated into ranking values for different control variables. As the finest one, the voltage vector leading to the minimum average ranking value is chosen.

Model-Predictive torque and flux control (MPTFC) [30] for variable speed drives is a significant member of the finite control set model-predictive control (FCS-MPC) family. A significant level of robustness to model parameter deviations could be achieved by directly controlling machine magnetization and electromagnetic torque [31]. Throughout the cost function, the expected errors of torque and stator flux magnitude are enhanced. A corresponding objective function, subject to inverter and machine model, is minimized online to this end. This results in an optimum switch position (i.e., control input) applied to the power converter [32].

Another feature of IM drives is that they exhibit greater efficiency when working at rated load [33,34]. However, owing to the persistence of iron losses and a part of copper losses, efficiency deteriorates under light load operation. This factor would harm the machine service life; accordingly, a criterion should be adopted to mitigate losses, particularly at light loads. Techniques for efficiency optimization are typically categorized into two types: search methods and model-based methods centered on how to decide the steady-state-optimal control variable [35]. Search-based methods are techniques of perturbation and observation that push the control command in the direction of minimal power losses. If closed-form solutions are not available, machine losses could either be calculated via loss model or measured using a power analyzer. Search-based strategies still struggle from a slow convergence rate and undesirable torque dynamics and fluctuations [36], despite attempts to raise the convergence rate using mathematical algorithms.

As a consequence, approaches focused on search are limited to steady-state processes. Multiple flux modes exist for industrial applications. The control minimizing loss is only switched on at steady-state, and during transients, the rated flux is usually linked to fast torque dynamics. There is no clear identification of transitions between flux modes during torque transients [37]. When the stator flux linkage shifts rapidly during torque transients, major losses are caused. Consequently, for highly dynamic load profiles, the steady-state optimal solution causes even more losses than constant rated flux operation.

An adaptive observer for combined flux and speed guesstimate in rotor speed reference frame is suggested in this paper. Rather than the traditional rotor flux plus stator current IM model, the rotor and stator flux models are utilized in the rotor speed reference frame. This makes it probable to apply the suggested observer in both stator and rotor-flux vector-controlled IM drives and in DTC-IM drives. It should also be mentioned that both stator and rotor resistance values and estimated velocity are the uncertainties considered in this article.

The current study introduces an efficient MPTFC technique for the IM drive in which some of the MPTC shortcomings can be mitigated, in specific by reducing the ripples of torque, stator flux and current. Furthermore, a loss minimization criterion (LMC) is suggested to improve the drive efficiency, particularly under light loads and low-speed operations, extending the IM's lifetime. Drive efficiency is validated by comprehensive simulation testing, in which the feasibility of the MPTFC strategy is demonstrated. The combination of prediction and estimation, in this work, is the additional challenge of sensorless MPTFC approach versus traditional sensorless approaches. **The paper's contributions could be summarized as follows**; the work intends to eliminate and replace the mechanical sensor with novel soft sensorless algorithms to minimize the drive cost and boost its reliability. Moreover, the paper offers a novel predictive torque-flux control approach for IM drive. Representing the IM model with consideration of core-loss and illustration of the used 2−level voltage source inverter has been implemented. Developing a novel AFOO, which contributes to reducing the fluctuations and thereby enhancing the method torque-flux and prediction, with the compensation of core-loss, for; rotor speed estimation, stator flux estimation, rotor flux estimation, stator and rotor resistance estimation, stator current estimation, electromagnetic torque estimation has been introduced. The analysis and design of the proposed MPTFC strategy are described in a straightforward manner that clarifies the suggested drive basic operation. To minimalize the copper and iron losses, particularly at low-velocity and light loading, which enhances the IM efficiency, an LMC is provided in steady-state operation. In order to assess IM dynamics under the proposed MPTFC and MPTC approaches, comprehensive simulation tests are conducted. Test results approve the suggested MPTFC reliability to be used as a stronger alternative to the MPTC. All gain selections of controllers and controllers that have been designed to ensure overall drive stability shall be found in Appendix A.

The paper starts with the implementation of the IM mathematical model and 2−level inverter, with consideration of core-loss. Then, the proposed AFOO is designed and explained. The suggested MPTFC method is subsequently presented and explained in-depth, and the proposed LCM is then presented and systematically evaluated. Eventually, the complete system design and performance test results are introduced and discussed.

2. Mathematical Model of IM and Inverter

In the stationary ($\alpha - \beta$) reference frame, indicated by the apex s for the stator and rotor amounts, Figure 1 demonstrates the IM model. No transformation of stator amounts (indicated by index s) is necessary for this frame, while the rotor amount (index rs) simply refers to the stator. Even though all of them remain sinusoidal during steady-state activity, this model makes it easier to formulate the predictive torque and flux control model proposed. Stator and rotor voltage balance equations in Figure 1 are provided as in (1) and (2) in the continuous time domain [38,39]:

$$\mathrm{p}\,\vec{\Psi}_s = \vec{v}_s - R_s\,\vec{i}_s \tag{1}$$

$$\mathrm{p}\,\vec{\Psi}_{rs} = -R_r'\,\vec{i}_{rs} + j\,\omega_{me}\,\vec{\Psi}_{rs} \tag{2}$$

where p is a differential operator, $\omega_{me} = P\,\omega_m$, with P is pole pairs and ω_m is the shaft speed. Additionally, $L_{\sigma s}$ and R_s are stator leakage inductance and resistance, respectively; $L_{\sigma r}'$ and R_r' are rotor leakage inductance and resistance, respectively, both referred to the stator; and R_{fe}, L_m are the equivalent resistance representing the iron losses, and the magnetizing inductance, respectively. The relationships among currents and fluxes and are assumed linear (no saturation) and are listed as follows:

$$\vec{\Psi}_s = L_{\sigma s}\,\vec{i}_s + L_m\,\vec{i}_m = L_{\sigma s}\,\vec{i}_s + \vec{\Psi}_m \tag{3}$$

$$\vec{\Psi}_{rs} = L_{\sigma r}'\,\vec{i}_{rs} + L_m\,\vec{i}_m = L_{\sigma r}'\,\vec{i}_{rs} + \vec{\Psi}_m \tag{4}$$

where $\vec{\Psi}_m$ and $\vec{i}_m$ are magnetizing air-gap flux and magnetizing current, respectively. Furthermore,

$$R_{fe}\,\vec{i}_{fe} = L_m\,\mathrm{p}\,\vec{i}_m \tag{5}$$

$$\vec{i}_m + \vec{i}_{fe} = \vec{i}_s + \vec{i}_{rs} \tag{6}$$

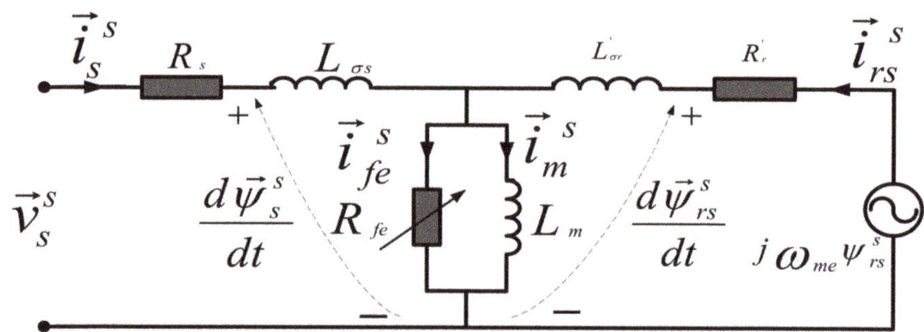

Figure 1. IM model in stationary ($\alpha - \beta$) reference frame taking iron losses into account.

With some management of the equations, Equation (2) can be replaced by

$$p\,\vec{i}_s = \frac{L_m}{L_m\,L'_{\sigma r} + L_{\sigma s}\left(L'_{\sigma r} + L_m\right)}\left\{\left(\frac{L'_{\sigma r} + L_m}{L_m}\right)\vec{v}_s - \left[R_s\left(\frac{L''_{\sigma r} + L_m}{L_m}\right) + R'_r\right]\vec{i}_s + R'_r\left(\vec{i}_m + \vec{i}_{fe}\right) - j\omega_{me}\,\vec{i}_r\right\} \tag{7}$$

Although dynamic IM performance is well represented in Equations (1) and (7), it is very multifaceted to appreciate. We suppose to rewrite the following Equations (8)–(11) to characterize the IM state-space model in terms of equivalent core-loss resistance, R_m, which is included in R_{fe} where R_m is presumed to be proportional to $\omega^{1.6}$ [40].

$$p\,i_{\alpha s} = -\left(\frac{R_s}{L_s\,\sigma} + \frac{R_r\,L_m^2}{L_s\,L_r^2\,\sigma}\right)i_{\alpha s} + \left[\frac{R_r\,L_m + R_m(sL_m - L_r)}{L_s\,L_r^2\,\sigma}\right]\Psi_{\alpha r} + \frac{L_m}{L_s\,L_r\,\sigma}\omega_r\,\Psi_{\beta r} + \frac{1}{L_s\,\sigma}v_{\alpha s} \tag{8}$$

$$p\,i_{\beta s} = -\left(\frac{R_s}{L_s\,\sigma} + \frac{R_r\,L_m^2}{L_s\,L_r^2\,\sigma}\right)i_{\beta s} + \left[\frac{R_r\,L_m + R_m(sL_m - L_r)}{L_s\,L_r^2\,\sigma}\right]\Psi_{\beta r} - \frac{L_m}{L_s\,L_r\,\sigma}\omega_r\,\Psi_{\alpha r} + \frac{1}{L_s\,\sigma}v_{\beta s} \tag{9}$$

$$p\,\Psi_{\alpha r} = \frac{R_r\,L_m}{L_r}\,i_{\alpha s} - \left(\frac{R_r - s}{L_r}\right)\Psi_{\alpha r} - \omega_r\,\Psi_{\beta r} \tag{10}$$

$$p\,\Psi_{\beta r} = \frac{R_r\,L_m}{L_r}\,i_{\beta s} - \left(\frac{R_r - s}{L_r}\right)\Psi_{\beta r} + \omega_r\,\Psi_{\alpha r} \tag{11}$$

where ω_r is the rotor speed, s is the slip and ω is the excitation frequency. Cross product of rotor and stator flux linkages vectors states the developed torque [28] as:

$$T_e = \frac{3}{2}\,P\,\frac{L_m}{L_r\,L_s - L_m^2}\left(\vec{\Psi}_s \times \vec{\Psi}_r\right) \tag{12}$$

For both MPTC and MPTFC methods, a 2−level voltage source inverter (VSI) is used in this work. The inverter topology and its feasible voltage vectors are shown in Figure 2. With every phase A, B and C the responding switching state S can be described as:

$$S = \frac{2}{3}\left(S_a + a\,S_b + a^2\,S_c\right) \tag{13}$$

where A, B and C are motor terminals, $a = e^{j2\pi/3}$, $S_i = 1$ means S_i on, S_i means off and DC-bus voltage is V_{dc}.

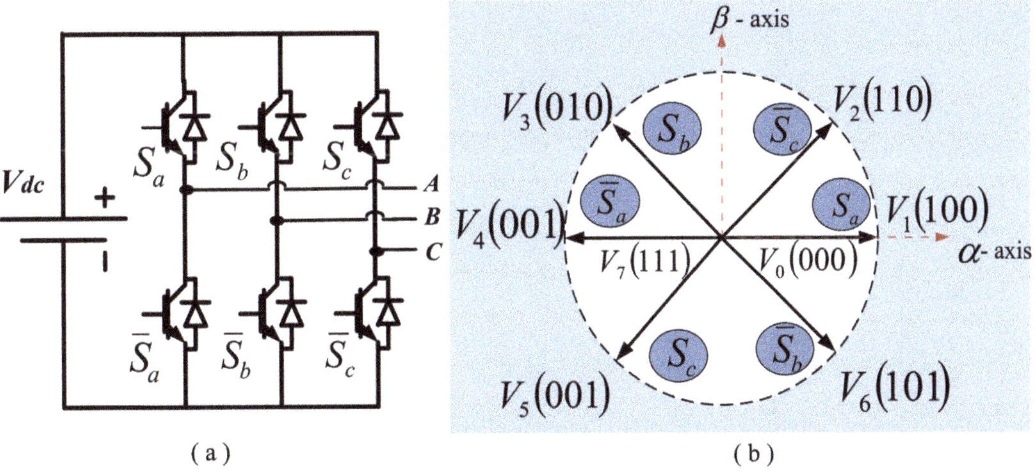

Figure 2. Inverter topology and its feasible voltage vectors (a) 2−level VSI circuit; (b) Voltage vectors.

The relation between inverter output voltage vector $\vec{v}_{s\alpha\beta,k}$ and the switching state S is stated as:

$$\vec{v}_{s\alpha\beta} = V_{dc}\,S \tag{14}$$

Voltage space vectors will regulate the output torque for the DTC technique, where two voltage vectors are selected in each sector to decrease or increase stator flux amplitude. There are eight switching states and seven separate voltage vectors V_0, V_1, V_2, ..., V_7 for a 2−level inverter-fed IM drive, as seen in Figure 2b. Cost function G is evaluated for each voltage vector value, and the vector generating minimum G is chosen as the best one.

3. Adaptive Full Order Observer Modification

In this work, accurate state estimation is a major step towards achieving good performance of MPTFC in dynamic implementation. Owing to its precision and insensitivity to parameter variance over a broad speed range [41], AFOO is approved for rotor speed-flux estimation. By compensating for the impact of core-loss, AFOO mathematical model can be revealed as [42]:

$$p\,\hat{x} = \hat{A}\,\hat{x} + B\,v_s + H\left(\hat{i}_s - i_s\right) + D\,\hat{\Psi}_r \tag{15}$$

where $\hat{x} = \left[\hat{i}_s\,\hat{\Psi}_r\right]^T$ is the estimated state for stator current and rotor flux, v_s is stator voltage vector, $I = \begin{bmatrix} 1 & 0 \\ 0 & 1 \end{bmatrix}$, $J = \begin{bmatrix} 0 & -1 \\ 1 & 0 \end{bmatrix}$, $B = \left[\frac{1}{L_s\sigma}I\,0\right]^T$, $[D]$ is the core-loss compensating term,

$$\hat{A} = \begin{bmatrix} \hat{A}_{11} & \hat{A}_{12} \\ \hat{A}_{21} & \hat{A}_{22} \end{bmatrix} = \begin{bmatrix} -\left(\frac{\hat{R}_s}{L_s\sigma} + \frac{\hat{R}_r L_m^2}{L_s L_r^2 \sigma}\right)I & \frac{L_m}{L_s L_r \sigma}\left(\frac{\hat{R}_r}{L_r}I - \hat{\omega}_r J\right) \\ \frac{\hat{R}_r L_m}{L_r} I & -\frac{\hat{R}_r}{L_r} I + \hat{\omega}_r J \end{bmatrix}, D = \begin{bmatrix} D_1\,I \\ D_2\,I \end{bmatrix} =$$

$R_m\left[\,-\frac{(L_r - s\,L_m)}{(\sigma L_s L_r^2)}I\quad -\frac{s}{L_r}I\,\right]^T$ and the observer gain matrix, see appendix, is given as:

$$H = \begin{bmatrix} h1 & h2 & h3 & h4 \\ -h2 & h1 & -h4 & h3 \end{bmatrix}^T \tag{16}$$

3.1. Stator Flux Estimation

DTC stator flux estimate is among the primary challenges that determine the accuracy and stability of the drive's operation, since it's not measured. The simplest solution is the pure integrator-based stator flux estimator voltage model, but it is susceptible to numerous problems [43]: (i) sensitivity in respecting to the dc-drift present in the pure integrator input that causes saturation of the integrator, and (ii) pure integrator initial conditions that cause an unwanted dc deviation in the estimation stator flux signal. We can rewrite AFOO in Equation (15) for IM model assumed in Equations (8)–(11) with the core-loss compensation concept as follows, in order to prevent the dc offset issue:

$$\begin{cases} p\,\vec{\hat{i}}_s = \hat{A}_{11}\,\vec{\hat{I}}_s + (\hat{A}_{12} + D_1)\,\vec{\hat{\Psi}}_r + B\,v_s + H_s\left(\hat{i}_s - i_s\right) \\ p\,\vec{\hat{\Psi}}_r = \hat{A}_{21}\,\vec{\hat{i}}_s + (\hat{A}_{22} + D_2)\,\vec{\hat{\Psi}}_r + B\,v_s + H_r\left(\hat{A}_s - i_s\right) \\ \vec{\hat{\Psi}}_s = L_1\hat{A}_s + L_2\,\vec{\hat{\Psi}}_r \end{cases} \tag{17}$$

where, $L_1 = \begin{bmatrix} \frac{L_s L_r - L_m^2}{L_r} & 0 \\ 0 & \frac{L_s L_r - L_m^2}{L_r} \end{bmatrix}$ and $L_2 = \begin{bmatrix} \frac{L_m}{L_r} & 0 \\ 0 & \frac{L_m}{L_r} \end{bmatrix}$.

The first benefit of the estimated stator flux in Equation (17) is that there is no need for motor speed statistics to estimate the flux. This reduces any additional errors, particularly at lower frequencies, associated with calculating or even measuring such signals. Another benefit is that estimated stator flux independent on machine resistances that improve drive reliability. Estimated stator flux $\vec{\hat{\Psi}}_s$ components in stationary $(\alpha - \beta)$ reference frame depending on estimated rotor flux $\vec{\hat{\Psi}}_r$ and stator current $\vec{\hat{i}}_s$ in Equation (17) can be stated as:

$$\begin{cases} \hat{\Psi}_{\alpha s} = \frac{L_m}{L_r}\hat{\Psi}_{\alpha r} + \frac{L_s L_r - L_m^2}{L_r}\hat{i}_{\alpha s} \\ \hat{\Psi}_{\beta s} = \frac{L_m}{L_r}\hat{\Psi}_{\beta r} + \frac{L_s L_r - L_m^2}{L_r}\hat{i}_{\beta s} \end{cases} \tag{18}$$

3.2. Rotor Speed Estimation

The cost and complexity of the system will be effectively minimized by a sensorless system where the velocity is determined instead of measured. One of the key reasons for the success of inverter fed IM drives is that it is possible to use any standard IM without modifications. It is suggested with current and voltage measuring devices that IM speed can be guessed without the need for speed or flux sensors to be mounted. Through traditional AFOO, rotor speed estimate is obtained via conventional adaptation law as in (15) [5,42]:

$$\hat{\omega}_r = K_{I\omega} \int_0^t \left[\hat{\Psi}_{\beta r} \left(i_{\alpha s} - \hat{i}_{\alpha s} \right) - \vec{\Psi}_{\alpha r} \left(i_{\beta s} - \hat{i}_{\beta s} \right) \right] dt \tag{19}$$

Like MRAs, the proposed scheme would be separated into two key components, which can be seen in Figure 3; each model is supplied with the same input signal $\vec{v}_s$. Additionally, the adaptive model is tuned by dual closed loops. The first takes into account the error among measured current $\vec{i}_s$ and estimated $\hat{\vec{i}}_s$ from (17), while the other considers the adaptively calculated speed (20). The first closed-loop is being responsible for compensating the offsets that have the major source in current sensors. The controller time constant is calibrated proportionally to a speed value for proper compensation over the entire range of IM rotor velocity. Moreover, for two main reasons, the controllers work slowly: to ensure a lack of effect on transient estimation and because it integrates errors through sinusoidal signals. The second closed-loop provides the rotor velocity $\hat{\omega}_r$, calculated on the basis of difference among estimated and measured currents $\Delta \vec{i}_s$ multiplied by estimated stator flux conjugate vector $\hat{\Psi}_s{}^*$ as per as (17).

$$\begin{cases} e_\omega = \mathrm{Im}\left(\hat{\vec{\Psi}}_s^* \, \Delta \vec{i}_s \right) = \hat{\Psi}_{\beta s}\left(i_{\alpha s} - \hat{i}_{\alpha s} \right) - \hat{\Psi}_{\alpha s}\left(i_{\beta s} - \hat{i}_{\beta s} \right) \\ \hat{\omega}_r = \left(K_{P\omega} + \frac{K_{I\omega}}{s} \right).e_\omega \end{cases} \tag{20}$$

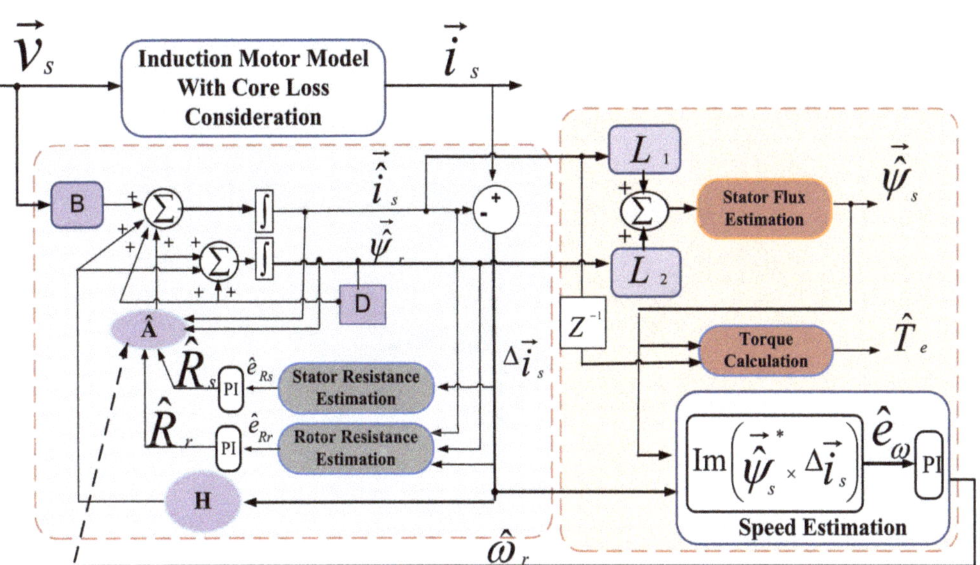

Figure 3. Proposed sensorless observer.

3.3. Other Parameter Estimation

IM stator resistance will vary thanks to temperature change during operation. To provide stator resistance estimation, adaptive control observer could be extended. According to the same Lyapunov's theory, stator resistance R_s can also be estimated like rotor speed via a PI controller [39].

$$\begin{cases} e_{R_s} = \hat{i}_{\alpha s}\left(\hat{i}_{\alpha s} - i_{\alpha s} \right) + \hat{i}_{\beta s}\left(\hat{i}_{\beta s} - i_{\beta s} \right) \\ \hat{R}_s = \left(K_{PR_s} + \frac{K_{IR_s}}{s} \right).e_{R_s} \end{cases} \tag{21}$$

In addition to the temperature variation effect on stator resistance, various parameters in the suggested observer will also adjust during operation. Owing to magnetic saturation, parameters L_s; L_r and L_m differ. Although magnetic saturation variance can be compensated for through the nonlinear magnetic model, rotor resistance variation would have a significant effect on the speed-accuracy of our adaptive observer.

It is recognized that the misestimating of R_r gives correct estimates of the rotor and stator fluxes during-steady state, but results in a speed misestimating [44]. Rotor resistance estimate can be incorporated into adaptive observer using the approach followed in [42] or IM thermal model. The influence of the core-loss on the estimation of both stator and rotor resistance and its compensation has been well defined in our previous work [45]. Stator current and rotor flux error estimation can give rotor resistance estimate utilizing Lyapunov theory via the suggested AFOO in (17) as:

$$\begin{cases} e_{R_r} = (\hat{\Psi}_{\alpha r} - L_m \hat{i}_{\alpha s})(i_{\alpha s} - \hat{i}_{\alpha s}) + (\hat{\Psi}_{\beta r} - L_m \hat{i}_{\beta s})(i_{\beta s} - \hat{i}_{\beta s}) \\ \hat{i}_r = \left(K_{PR_r} + \frac{K_{IR_r}}{s}\right) \cdot e_{R_r} \end{cases} \tag{22}$$

3.4. Torque Estimation

In this analysis, the proposed AFOO can also be used to determine the electromagnetic torque to reduce the error between the reference and desired motor output torque based on estimated stator current and flux. Delay between acquisition time and application time is important to be considered.

$$\hat{T}_e = \frac{3}{2} P \, \mathrm{Im} \left(\vec{\hat{\Psi}}_s^* \, \vec{\hat{i}}_{s,k-1} \right) \tag{23}$$

where k is the sampling time index.

As indicated in Figure 4, our proposed observer can easily estimate stator flux angle as:

$$\hat{\theta}_s = \tan^{-1}\left(\frac{\hat{\Psi}_{\beta s}}{\hat{\Psi}_{\alpha s}}\right) \tag{24}$$

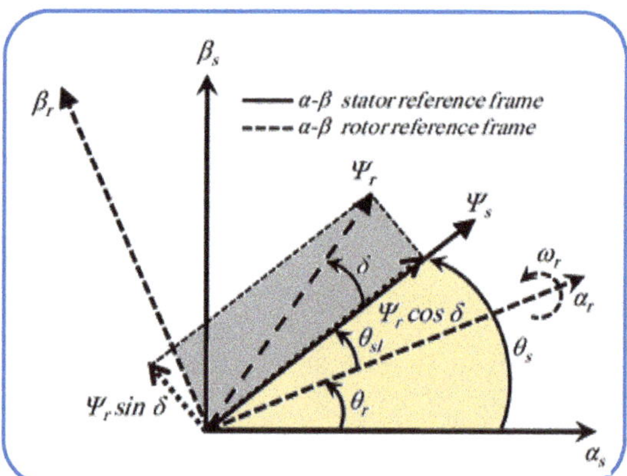

Figure 4. Stator and rotor flux vectors relationship in stationary reference frame.

The angular speed of stator flux linkage relative to rotor flux linkage may also be computed as:

$$\hat{\delta} = \tan^{-1}\left(\frac{\hat{\Psi}_{\beta r}}{\hat{\Psi}_{\alpha r}}\right) - \hat{\theta}_s \tag{25}$$

To incorporate the system (17) into the digital processor, it is crucial to acquire a discrete-time state-space representation of proposed system. It should be noted that matrix $\hat{A}$ depending on instantaneous calculated rotor speed value $\hat{\omega}_r$ making $\hat{A} = \hat{A}(\hat{\omega}_r(t))$ as a linear time-varying system. Time dependence implies that variation of ω_r, the offline numerical estimate of discrete-time equivalent system cannot be obtained. The solution would be to achieve a discrete-time varying method that can then be modified with the new estimated value of $\hat{\omega}_r$ at every sampling period.

A direct computation or Euler approximation (first-order sequence expansion) [46] are the more popular methods of obtaining the representation of sampled-data for the model. The approximation of Euler is an easy way of obtaining the discrete-time model with a similar response of dynamic behavior. The direct computation of state trajectory in (17) may not be simple as Euler approximation of the first order but gives a more precise representation in the discrete-time (for example, see [47] and references therein).

3.5. Conventional MPTC

For conventional MPTC, just one voltage vector is chosen and not applied till the next control time owing to the updated mechanism for modern microprocessors. The conventional MPTC diagram shown is in Figure 5. Using a PI controller, torque reference would be created via an outer speed control loop and stator flux reference is kept constant at asset value since field weakening and efficiency optimization process are not taken into account in this procedure [29]. In MPTC technique, torque and stator flux are being predicted for all possible voltage vectors supplied by the inverter based on the system model. Then, the best one is decided by minimizing an objective cost function consisting of a torque-flux tracking error.

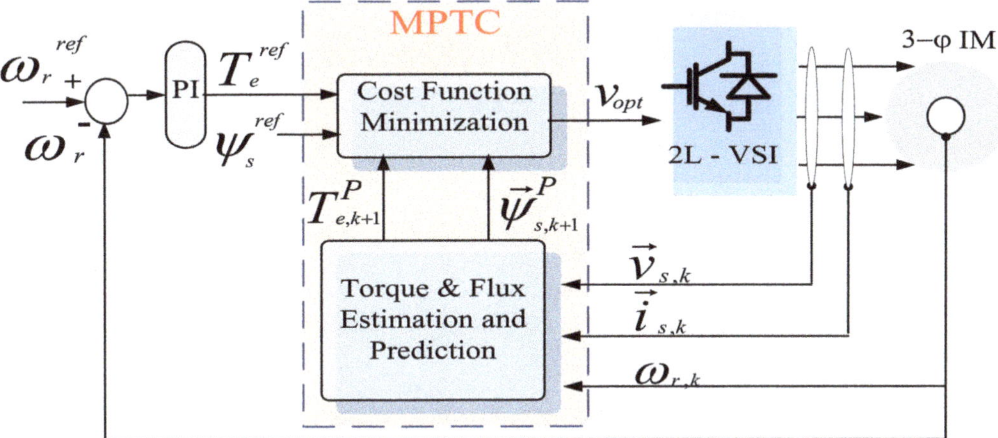

Figure 5. Conventional MPTC scheme.

3.6. Conventional MPFC

Torque and stator flux references are equivalently transformed into single reference for stator flux vector in this process. The MPFC diagram can be seen in Figure 6.

3.7. Proposed MPTFC

MPTC's cost function is generally defined as a linear combination with torque-flux errors to determining the finest voltage vector, that is defined as:

$$G_1 = \left| T_e^{ref} - T^{p}_{e,k+1} \right| + \lambda_\Psi \left| \Psi_s^{ref} - \vec{\Psi}^{p}_{s,k+1} \right| \tag{26}$$

where λ_Ψ is stator flux's weighting factor in MPTC. Stator flux weighting factor is calibrated in actual time to achieve minimum torque ripple, as suggested in [13]. The weighting factor λ_Ψ value has a decisive role in containing the unavoidable ripples if it properly chosen. Therefore, the λ_Ψ value must be correctly selected, which requires an online optimization process to choose the optimum value, which results in a further increase in the microcontroller's computational burden. To avoid the use of weighting factor in MPTC, a new stator flux reference based upon IM model is suggested in [28], which is equivalent to the original reference of torque and stator flux. To pick the finest voltage vector amongst the feasible ones, MPFC's cost function described in (27) is minimized.

$$G_2 = \left| \Psi_s^{ref} - \vec{\Psi}_{s,k+1}^p \right| \tag{27}$$

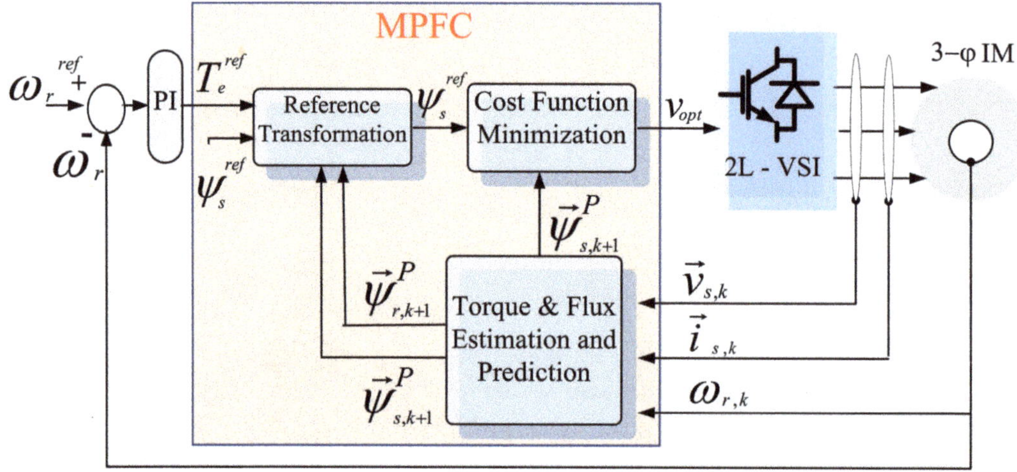

Figure 6. Conventional MPFC scheme.

It is clear that the weighting factor will be no longer needed, since only stator flux vector tracking error is involved. The key shortcoming of this technique is that it relies on reference flux angle estimate by using inverse-trigonometric function, which fails to estimate the angle during particular operating conditions [28].

Most often for MPTC and MPFC drives, stator flux reference Ψ_s^{ref} is usually set constant to its rated value as:

$$\Psi_s^{ref} = \left| \Psi_s^{ref} \right| \tag{28}$$

In order to accurately regulate torque and stator flux, we propose the MPTFC design in Figure 7 where the predictive values of stator flux, current and torque are based on suggested AFOO in (17).

MPTFC model predicts electromagnetic torque and stator flux vector's magnitude at next step in discrete-time $k + 1$ as a switch position function to determine v_{opt} at the present time step k. Prediction accuracy of MPCC and MPTFC, is main issue, as he process of prediction begins with the sampled of measured currents. Unless the measuring and sampling processes were not performed well, noise level will increase, which will then be reflected in the expected values and hence exacerbate voltage selection process, which eventually results in greater ripple content. To avoid that problem, we assume both prediction of torque-flux is dependent on stator currents estimate instead of the measured one.

Integrating estimated stator current in (17) with Euler method from $t = k\,T_s$ to $t = (k + 1)\,T_s$ and inserting (1) into it relates to the representation of discrete-time:

$$\vec{\hat{\Psi}}_{s,k+1}^p = \vec{\hat{\Psi}}_{s,k} + T_s \vec{v}_{s,k}(i) - \hat{R}_S\, T_s\, \vec{\hat{i}}_{s,k} \tag{29}$$

Hence, predicted stator flux appears as an estimated function.

$$\vec{\hat{i}}_{s,k+1}^p = \left(1 + \frac{T_s}{\tau_\sigma}\right)\vec{\hat{i}}_{s,k} + \frac{T_s}{\tau_\sigma + \tau_s}\left\{\frac{1}{R_\sigma}\left[\left(\frac{k_r}{\tau_r} - k_r j\,\hat{\omega}_r\right)\vec{\hat{\Psi}}_{r,k} + \vec{v}_{s,k}(i)\right]\right\} \tag{30}$$

where $\tau_r = \frac{L_r}{\hat{R}_r}$, $k_r = \frac{L_m}{L_r}$, $k_s = \frac{L_m}{L_s}$, $R_\sigma = \hat{R}_S + \hat{R}_r + k_r^2$, $\tau_\sigma = \frac{\sigma\,L_s}{R_\sigma}$, T_s is sampling-period and $\vec{v}_s(i)$ corresponding to 7 dissimilar voltage status, i is from 0 to 6.

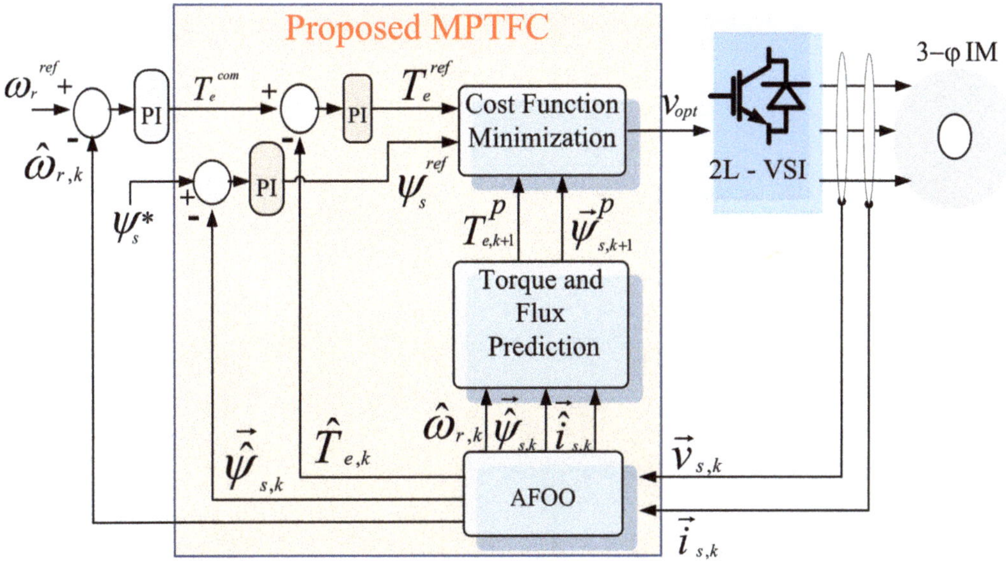

Figure 7. Proposed MPTFC diagram.

Predicted torque could be obtained after determining the predicted current and flux with:

$$T^P_{e,k+1} = \frac{3}{2}P\left[\vec{\hat{\Psi}}^p_{s,k+1} \times \vec{i}^{\,p}_{s,k+1}\right] \tag{31}$$

Therefore, a proper cost function would be developed for the proposed MPTF control centered on G_1 and G_2 as:

$$G = \lambda_T\left|T_e^{ref} - T^P_{e,k+1}\right| + \lambda_\Psi\left|\Psi_s^{ref} - \vec{\hat{\Psi}}^p_{s,k+1}\right| \tag{32}$$

where, torque-flux weighting coefficients are λ_T and λ_Ψ respectively. All feasible inverter topologies for the seven different flux and torque estimates are applied at every sampling stage. Then these samples are being used for determining the cost function. In one sampling period, a switching state that minimizes the G value is picked to be implemented. To drive 2−level-inverter, the switching states are then becoming the output. The flow chart of the whole process is seen within Figure 8.

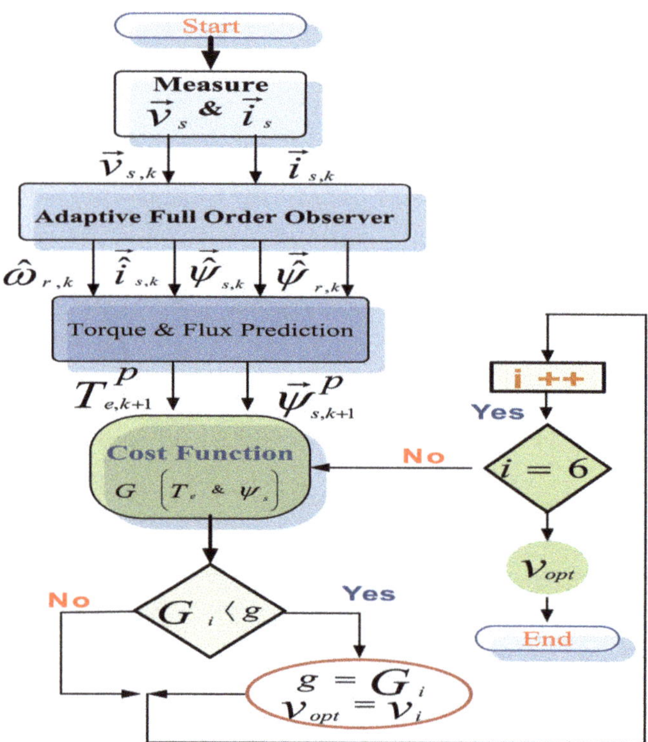

Figure 8. Proposed MPTFC flowchart with weighting factor optimization.

4. Optimal Steady-State Flux for Losses Minimization Criterion

To achieve an optimal stator flux reference for LMC, efficiency optimization is being discussed in this section. IM losses must be precisely defined to obtain a criterion from which whole losses could be minimized. Figure 1 presents the overall IM losses, which consisting of stator copper, rotor copper and core-losses; but not inverter, load, or mechanical losses.

$$P_{Loss} = 3\left[\, P_{cuS} + P_{cuR} + P_{fe}\,\right] \tag{33}$$

where P_{cuS}, P_{cuR} are stator and rotor copper losses and P_{fe} is core-loss.

$$P_{cuS} = R_s\, I_s{}^2 = R_s\left\{ I_m{}^2 + [I_r + \frac{R_m}{\omega L_m}I_m]^2 \right\} \tag{34}$$

$$P_{cuR} = R_r'\, I_r{}^2 = \left(\frac{L_m}{L_r}\right)^2 R_r\, I_r{}^2 \tag{35}$$

$$P_{fe} = R_{fe}\, I_{fe}{}^2 = \frac{\omega^2 L_m{}^3}{L_r R_m}\left[\frac{R_m}{\omega L_m}I_m\right]^2 \tag{36}$$

Whereas $R_m{}^2 \ll (\omega L_m)^2$, substituting (34) and (36) into (33) yields the total loss expression as:

$$P_{Loss} = 3\left\{ \left[R_s + \frac{R_m L_m}{L_r}\right]I_m{}^2 + \left[R_s + \left(\frac{L_m}{L_r}\right)^2 R_r\right]I_r{}^2 + \frac{2\,R_m R_s}{\omega L_m}I_m I_r \right\} \tag{37}$$

where $a = L_m/L_r$, $L_m' = a L_m$, $R_r' = a^2 R_r$, $R_{fe} = a \frac{(\omega L_m)^2}{R_m} = \frac{\omega^2 L_m^3}{L_r R_m}$, $I_{fe} = \frac{R_m}{\omega L_m} I_m$ and $I_s = \sqrt{I_m^2 + (I_r + I_{fe})^2}$. Direct FOC equations of IM that take core loss into accounts are:

$$\text{Rotorflux}: \quad \Psi_r' = L_m' I_m \tag{38}$$

$$\text{Electromagnetictorque}: \quad T_e = 3\,P\,\Psi_r'\,I_r \tag{39}$$

$$\text{Slipfrequency}: \quad \omega_{sl} = R_r \frac{\Psi_r'}{L_r} \tag{40}$$

The substitution of (38) and (39) into (37) helps us to rewrite the complete loss as:

$$P_{Loss} = 3\left\{ \left[R_s + \frac{R_m\,L_m}{L_r} \right] \frac{L_r^2}{L_m^4} \left| \Psi_r' \right|^2 + \left[R_s + \left(\frac{L_m}{L_r}\right)^2 R_r \right] \frac{T_e^2}{9\,P^2} \left| \Psi_r' \right|^{-2} + \left[\frac{2\,R_m\,R_s}{\omega L_m} \right] \frac{L_r\,T_e}{3\,P\,L_m^2} \right\} \tag{41}$$

This implies that P_{Loss} is a variable dependent on rotor flux Ψ_r'. Assuming that under constant load torque, none of machine parameters have any dependency upon rotor flux Ψ_r'. By setting a derivative of Equation (41) with respect to Ψ_r' to zero, we could obtain a rotor flux that provides the minimum losses.

$$\frac{\partial P_{Loss}}{\partial \Psi_r'} = 0 \tag{42}$$

$$\frac{\partial P_{Loss}}{\partial \left| \Psi_r' \right|} = 3\left\{ \begin{array}{l} 2\left[R_s + \frac{R_m\,L_m}{L_r} \right] \frac{L_r^2}{L_m^4} \left| \Psi_r' \right| - \\ 2\left[R_s + \left(\frac{L_m}{L_r}\right)^2 R_r \right] \frac{T_e^2}{9\,P^2} \left| \Psi_r' \right|^{-3} \end{array} \right\} \tag{43}$$

Solving Equation (43) gives us an appropriate rotor flux $\Psi_r'^{\,A}$ corresponding to the maximum efficiency point, where. $K = \left[\left(\frac{\hat{R}_s\,L_r^2 + \hat{R}_r\,L_m^2}{\hat{R}_s\,L_r^4 + L_m\,L_r^3\,R_m} \right)^{\frac{1}{4}} \frac{2\,R_m\,R_s}{\omega L_m} \right] \cdot \frac{L_m}{\sqrt{3}\,P}$.

$$\Psi_r'^{\,A} = K\,\sqrt{T_e} \tag{44}$$

Before applying $\Psi_r'^{\,A}$ to the direct FOC of IM, the coefficient with estimated parameters must be multiplied as follows:

$$\vec{\Psi}_{r,k}^{A} = \frac{L_r}{L_m} \Psi_r'^{\,A} = \frac{L_r}{L_m} \hat{K}\sqrt{T_e} \tag{45}$$

Corresponding suitable stator flux can also be determined based on the steady-state relationship among stator-rotor fluxes as [33]:

$$\vec{\Psi}_{s,k}^{A} = \sqrt{ \left(\frac{L_s}{L_m} \right)^2 \vec{\Psi}_{r,k}^{A} + \left(\frac{4\,\sigma\,L_s\,L_r}{3\,P\,L_m} \right)^2 \frac{T_e^2}{\left(\vec{\Psi}_{r,k}^{A} \right)^2} } \tag{46}$$

The optimum stator flux's reference amplitude required by suggested MPTFC in LMC is provided by Equation (46). Expression of (46) updates (28) as:

$$\Psi_s^{ref} = \left| \vec{\Psi}_{s,k}^{A} \right| \tag{47}$$

It thus completes the data necessary to apply (29) and so the cost function (32).

In order to calculate the efficiency of IM based on the total losses calculated in Equation (41), as in [47]:

$$\eta = \frac{P_{Out}}{P_{Out} + P_{Loss}} \tag{48}$$

where η is the efficiency and the output power is $P_{Out} = \omega_r\,T_e$.

5. System Layout

Complete approach layout is shown in Figure 9. Stator voltages and currents are the measured quantities, that used by suggested AFOO to obtain the estimated values of; rotor velocity $\hat{\omega}_{r,k}$, rotor flux $\hat{\vec{\Psi}}_{r,k}$, stator current $\hat{\vec{i}}_{s,k}$, stator flux $\hat{\vec{\Psi}}_{s,k}$ and torque $\hat{T}_{e,k}$ to predict stator flux $\hat{\vec{\Psi}}_{s,k+1}^{p}$ and corresponding predicted torque $T_{e,k+1}^{P}$. Estimated states are fed back toward the stator flux outer loop and inner loop of the torque to produce stator flux and torque references for cost function in Equation (32). The LMC algorithm provides optimum stator flux's amplitude; alternatively, stator flux reference could be followed through fed back flux loop. Optimum voltage vector is selected through applying cost function G on the base of the references and predicted values of both torque and stator flux to complete the suggested MPTFC of IM drive.

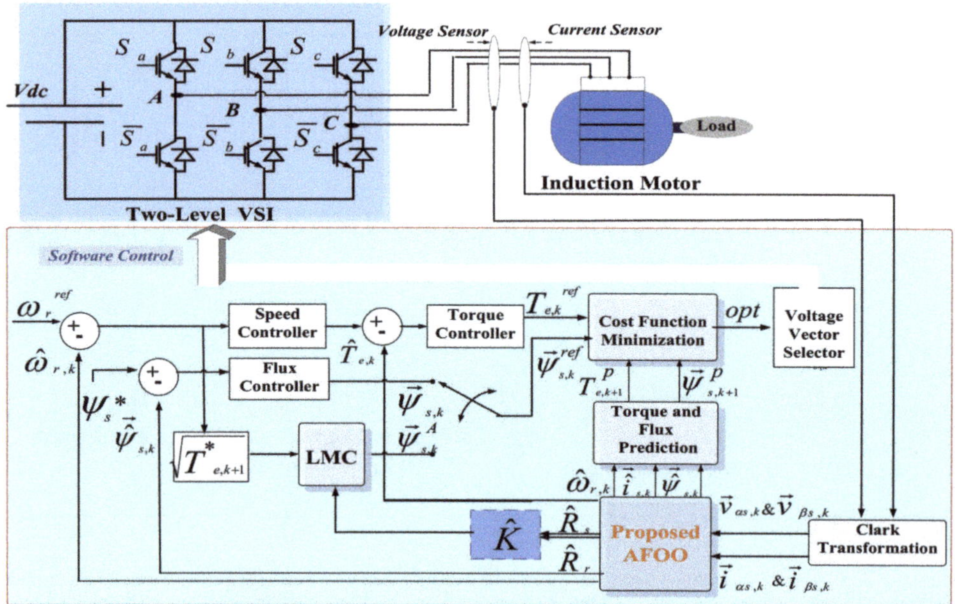

Figure 9. Proposed MPTFC layout with LMC for IM drive.

6. Results and Discussion

The suggested MPTFC approach is simulated in MATLAB/Simulink environment to verify its effectiveness. For simplicity, the proposed MPTFC performance is compared to conventional MPTC that utilize cost function G1 in Equation (26) and constant stator flux reference $\left|\Psi_s^{ref}\right| = 0.71$ wb. MPTC technique is referred to as method I, and the proposed one of MPTFC with the modified AFOO is referred to as method II in the following phrases, respectively. Motor parameters are mentioned in Table 1.

Table 1. IM Parameters.

Symbol	Parameters	Values
v_s	Rated voltage	380 V
n_p	No. pole pairs	1
f	Rated frequency	50 Hz
R_s	Stator resistance	1.2 Ω
R_r	Rotor resistance	1 Ω
L_s	Stator self-inductance	175×10^{-3} H
L_r	Rotor self-inductance	175×10^{-3} H
L_m	Magnetizing inductance	170×10^{-3} H
J	Moment of inertia	0.062 Kgm^2
Ψ_{sn}	Nominal stator flux	0.71 wb
T_n	Nominal torque	20 Nm
R_m	Core-resistance	2.186 KΩ
T_s	Sampling-time	4×10^{-5} s

The following hints about the simulating system should be noted; the IM has been considered nonlinear and takes the core-loss influence into account. Moreover, the proposed observer-based on the mentioned equations in the paper has been implemented using MATLAB® and Simulink® environment. Furthermore, the inverter has been built from the Simulink libraries of Simscape Electrical Specialized Power Systems library. However, no extra band-limited white noise was introduced into the signals.

Figure 10 shows starting response from standstill to (100 rps) for the both methods. Stator flux is first developed using preexcitation, and torque is restricted to 100 percent rated value (20 Nm) during the accelerating stage. An external load with 50% of the nominal value (10 Nm) is suddenly applied at ($t = 0.7$ s) to the tested IM. In MPTC, actual speed is compared with speed reference unlike in the proposed system which utizes estimate speed.

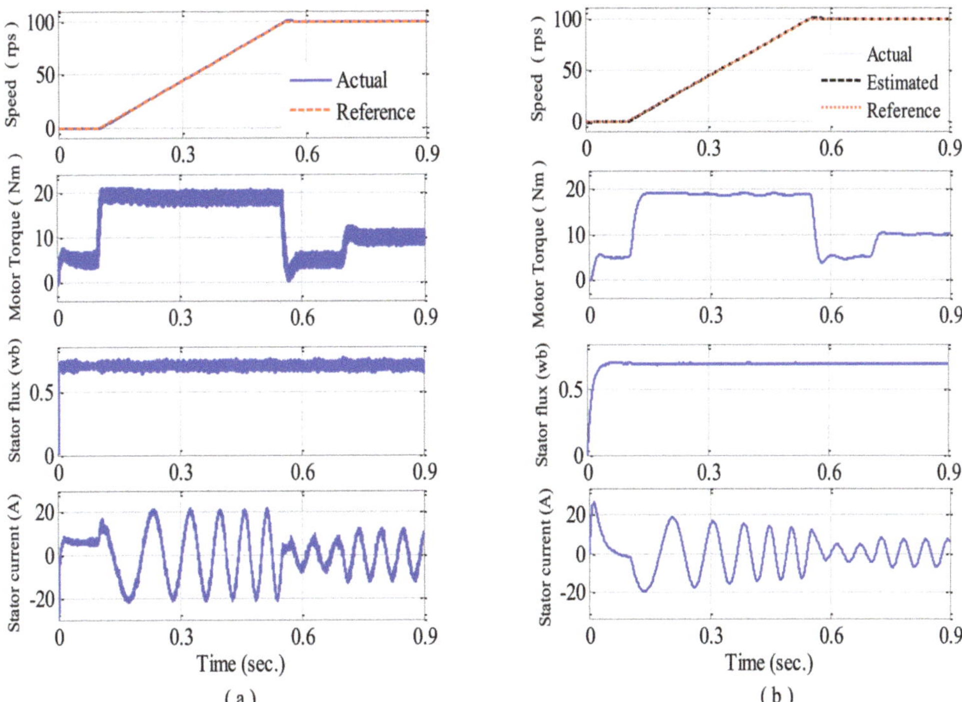

Figure 10. Starting response simulation from standstill to 100 rps for: (**a**) method I of the conventional MPTC; (**b**) method II of the proposed MPTFC.

Curves shown in Figure 10, through top to bottom are velocity, torque, stator flux and stator current, respectively. It is obviously seen that in low-speed operation, proposed MPTFC works well and exhibits high robustness regarding load disturbance. It can also be observed that method I has similar dynamic performance, but it produces much higher flux, current and torque ripples.

A more extensive steady-state torque, current and stator flux waveforms with (50%) of rated torque (10 Nm) is shown in Figure 11. The ripples of torque, current and stator flux in method II are shown to be much smaller than in method I, demonstrating the efficiency of suggested approach of MPTFC.

Conventional and suggested drive responses are shown in Figure 12a,b throughout speed reversals. It could be said that the motor accelerates rapidly from (50 rps) to (−50 rps) for three full cycles of forward and reverse speed to test the drive performance in low and reverse speed operation. The load torque is held constant at (5 Nm) during the whole test period. The conventional MPTC works at reverse speed operation while it suffers from unwanted ripples in the flux, torque and the stator currents when compared to the proposed method, as shown in Figure 12a. Moreover, an acceptable error between actual, estimated velocity and measured one is shown in Figure 12b. Furthermore, the error between the estimated stator current and actual ones through our proposed drive is shown in Figure 12b, which validates the proposed observer's operation. The actual and estimate values of stator and rotor fluxes throughout the reverse speed operation are also shown in Figure 12b. Figure 12c shows the comparison of the response of the stator currents and stator flux with the operation of speed reversal at (50 rps) between the conventional MPTC and the proposed MPTFC. The figure proves that the response of the flux and stator currents with the proposed MPTFC is better than those of the conventional MPTC.

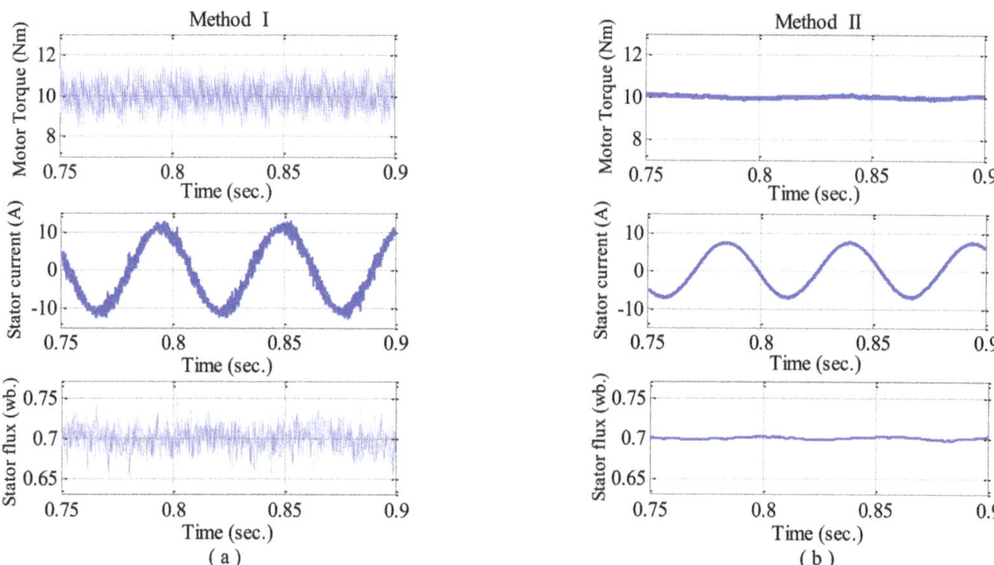

Figure 11. Steady-state response simulation at 100 rps for: (**a**) method I of the conventional MPTC; (**b**) method II of the proposed MPTFC.

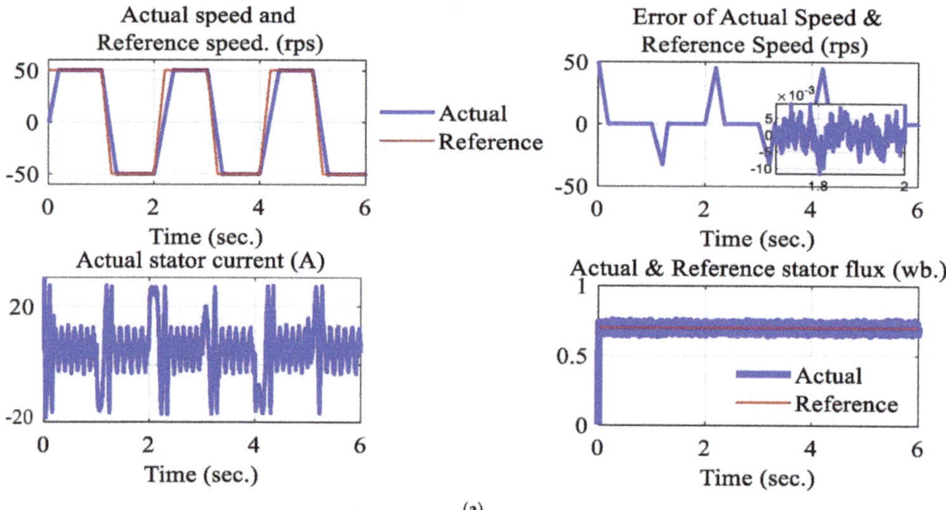

(**a**)

Figure 12. *Cont.*

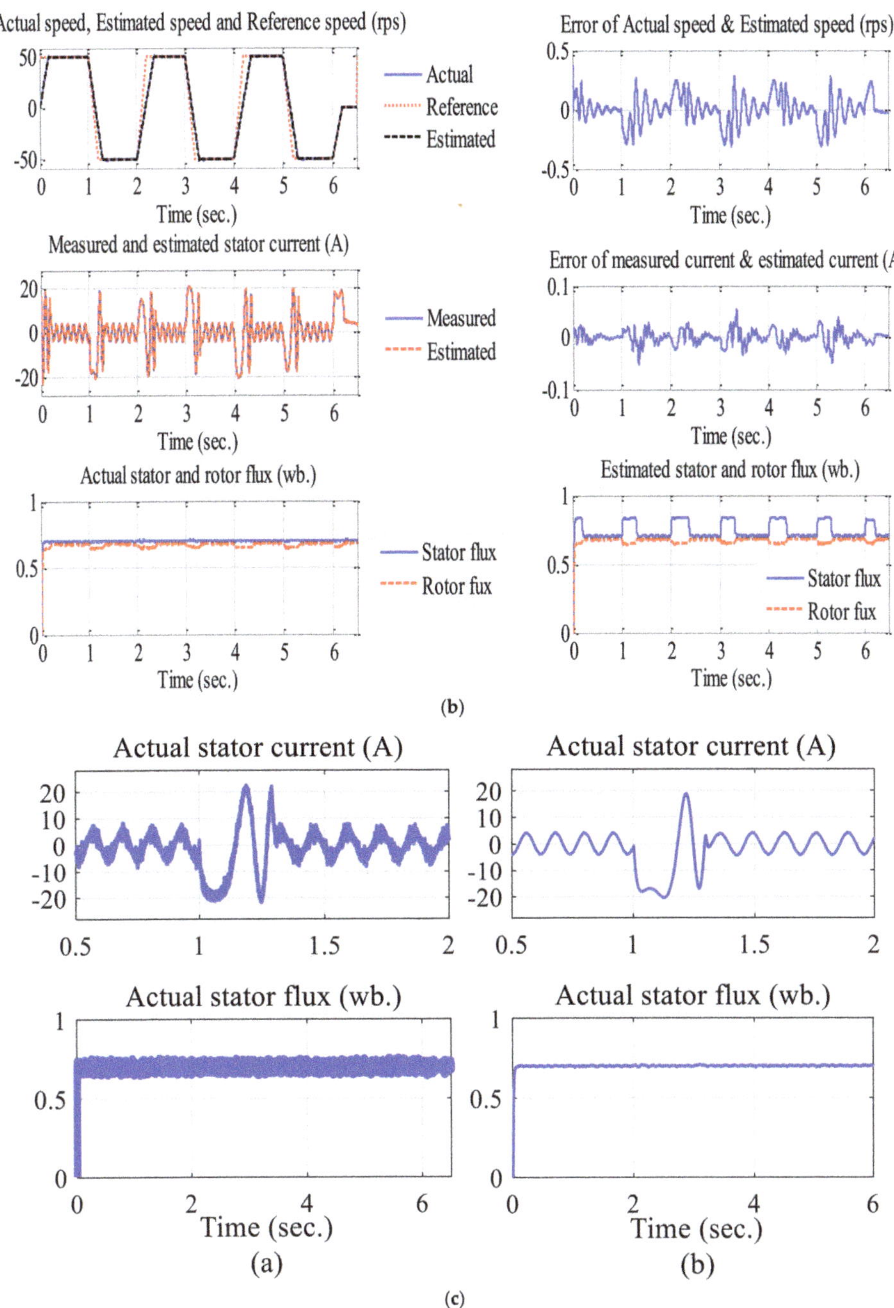

Figure 12. Performance with the operation of speed reversal at 50 rps: (**a**) Simulation responses of speed reversal at 50 rps for conventional MPTC; (**b**) Simulation responses of speed reversal at 50 rps for proposed MPTFC; and (**c**). Comparison of the response of the stator currents and stator flux with the operation of speed reversal at 50 rps between; (**a**) method I of the conventional MPTC; (**b**) method II of the proposed MPTFC.

Furthermore, the drive responses of the conventional and proposed method under step change of the load disturbance are shown in Figure 13a,b at a very low speed operating range of (10 rps). A load torque of (5 Nm) is applied at starting, then load torque is increased to (10 Nm) at (t = 2 s) then decreased to (3 Nm) at (t = 4 s) and finally back to (5 Nm). Figure 13a shows a similar dynamic performance of MPTC at load disturbance; nevertheless, it produces greatly higher flux, current, and torque ripples when comparing it to the proposed MPTFC. Actual values of stator and rotor fluxes are very robust in the suggested drive, as they were compensated by the estimated value of the stator flux, as shown in Figure 13b. Moreover, both speed and current error can be seen as a rather pleasant feature, demonstrating high robustness against external load disturbance of the proposed MPTFC. It is also obvious that the both conventional and proposed method currents in Figure 13a,b seem to be quite sinusoidal in shape. Moreover, Figure 13c displays the comparison of the response of the stator currents, stator flux, and torque with the operation of load disturbance at (10 rps) between the conventional MPTC and the proposed MPTFC. The figure verifies that the response of the flux, stator currents and torque with the proposed MPTFC is better than those of the conventional MPTC.

In the second case, the IM being controlled under the suggested MPTFC with applying of LMC, where stator flux's reference is adjusted to optimal or appropriate flux $\overset{\rightarrow A}{\Psi}_{s,k}$ which is determined in (46) to obtain the highest efficiency or lowest losses. Figure 14a indicates total steady-state losses of IM; the conventional method is represented by dashed lines, while the proposed method is represented by solid lines at various values of speed. Overall loss operated throughout suggested method is certainly smaller than that of the conventional method with constant flux at each motor speed. It is obvious that, as can be seen in Figure 14b, the rotor flux where the losses are minimal depends upon load torque. In most other words, rotor flux that provides maximum IM efficiency stands as a function of load torque.

Optimal rotor flux for maximum operating efficiency that calculated in (45), is drawn in Figure 15 for various driving conditions. It is influenced by motor speed since the core-loss is taken into account. This implies that rotor flux drops to decrease the core-loss, which raises through the excitation frequency further than the other loses at high speeds. Corresponding optimal stator flux for minimum losses that calculated in (46), is drawn in Figure 15b.

IM steady-state efficiency with constant rotor flux is plotted with dash lines and also solid lines represents optimal suggested method, that shown in Figure 15c. It clearly shows that suggested method here achieves higher efficiency over large range of load torques owing to adjusting rotor flux rendering to torque. Thus, it can be said that the proposed system efficiency with LMC has improved significantly in excess of the classical one.

To investigate dynamic performance, load torque will varied as in case of load disturbance in Figure 13 and IM speed is kept constant with very low amount (10 rps). Stator flux's reference is held at its rated value (0.71 wb) for MPTC and suggested estimated value for MPTFC without the LMC. For the suggested MPTFC with LMC, stator flux's reference is adjusted on line through optimal value, which is also presented in Figure 16.

The steady-state optimal flux, which achieves minimum steady-state losses, varies with the operating conditions. To achieve balance among copper and iron losses, it rises with torque and decreases as rotor speed increases.

In the proposed MPTFC with LMC, the optimal stator flux, that calculated in Equation (46), achieves minimum losses, and varies with the operating conditions. To achieve balance among copper and iron losses, it rises with the torque and decreases as rotor speed increases. It can also be seen from Figure 16, in which a comparison of the three control procedures is demonstrated, that proposed MPTFC (without and with LMC) exhibits better dynamic behavior compared to MPTC technique. In addition, it should be noted that losses are effectively minimized during light loading when following the LMC approach and consequently, IM drive efficiency is improved, as clearly seen in Figure 16.

To clarify the speed response throughout low-speed region, that considered among the most major aspects of this study, actual speeds is compared for MPTC and suggested MPTFC (without and with LMC) approaches in Figure 17. Step change to reference speed at (t = 0.05 s) from (0 rps) to (10 rps) is applied with constant load torque(5 Nm). Figure 17 shows a step response of the conventional MPTC and the proposed one of MPTFC with and without considering LMC. To clarify these values more precisely, the over shot, rise time, and settling time were calculated in Table 2. The results show that the overshot of proposed one of MPTFC with LMC is better than those of the conventional one and the proposed MPTFC without LMC. Moreover, the figure shows an increasing in the rise time and settling time of the MPTFC with LMC. Moreover, in general, the speed response of the three schemes is stable as shown in time domine analysis of Figure 17.

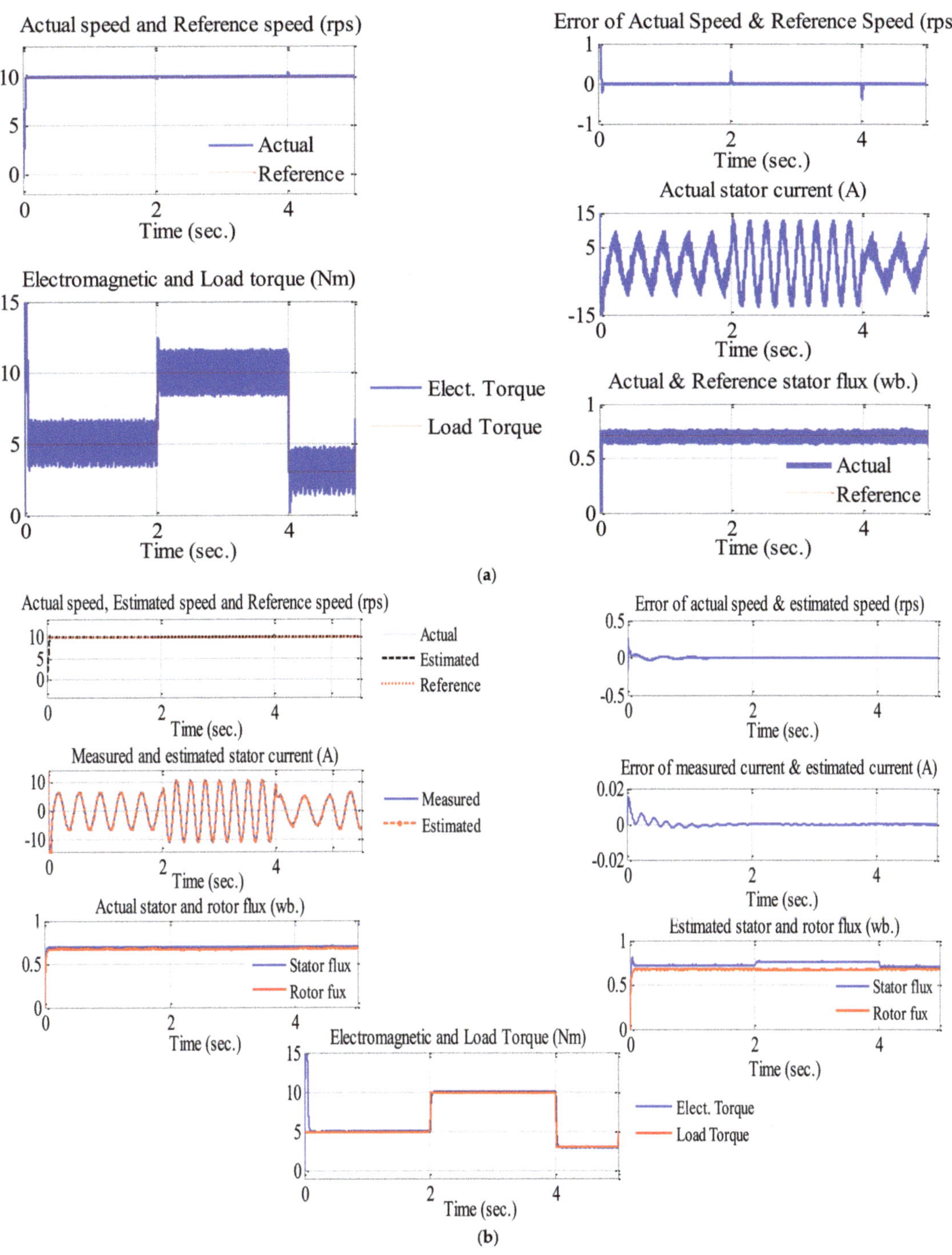

Figure 13. *Cont.*

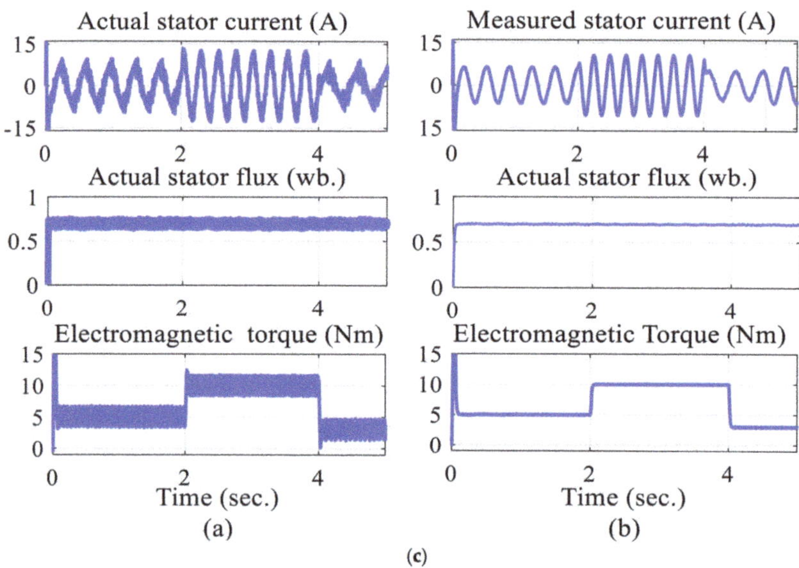

Figure 13. Drive responses of the conventional and proposed method under step change of the load disturbance (**a**) Simulation responses of load disturbance at 10 rps for conventional MPTC; (**b**) Simulation responses of load disturbance at 10 rps for proposed MPTFC. (**c**) Comparison of the response of the stator currents and stator flux with the operation of load disturbance at 10 rps between; (**a**) method I of the conventional MPTC; (**b**) method II of the proposed MPTFC.

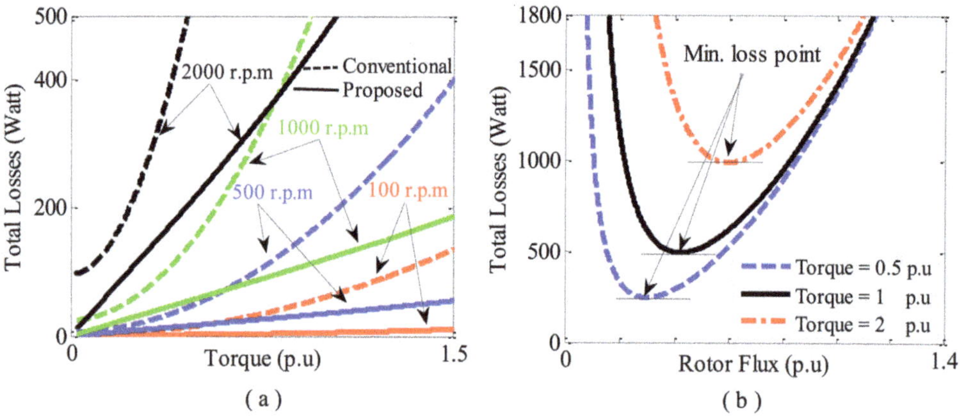

Figure 14. Steady-state overall losses versus: (**a**) load-torque with different rotor speeds; (the dash line is stand to the conventional method while the solid line is for the proposed one); (**b**) rotor flux with different load-torque.

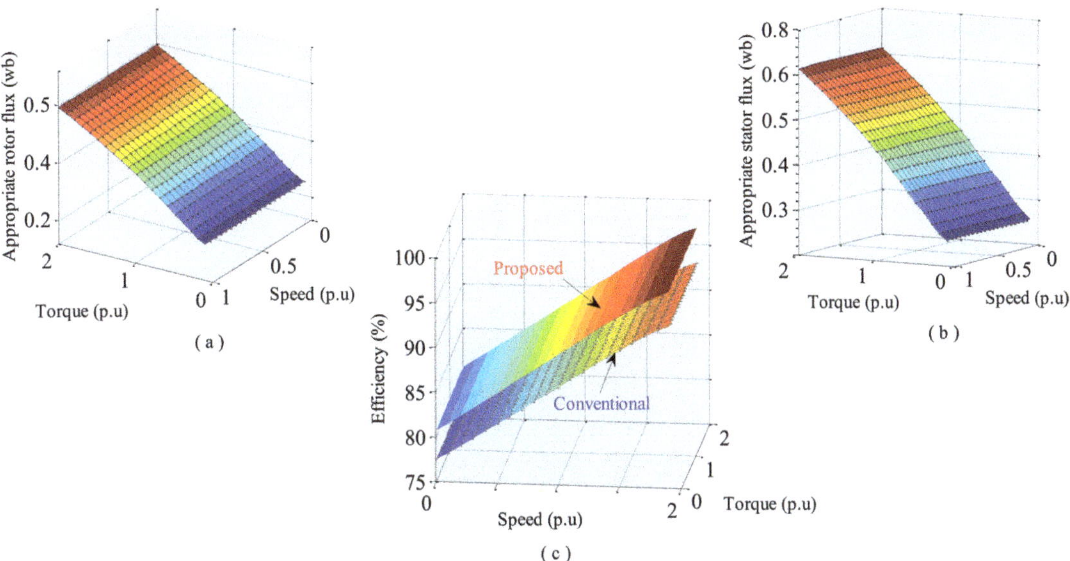

Figure 15. Steady-state IM efficiency: (**a**) appropriate rotor flux map; (**b**) appropriate stator flux map; (**c**) efficiency map.

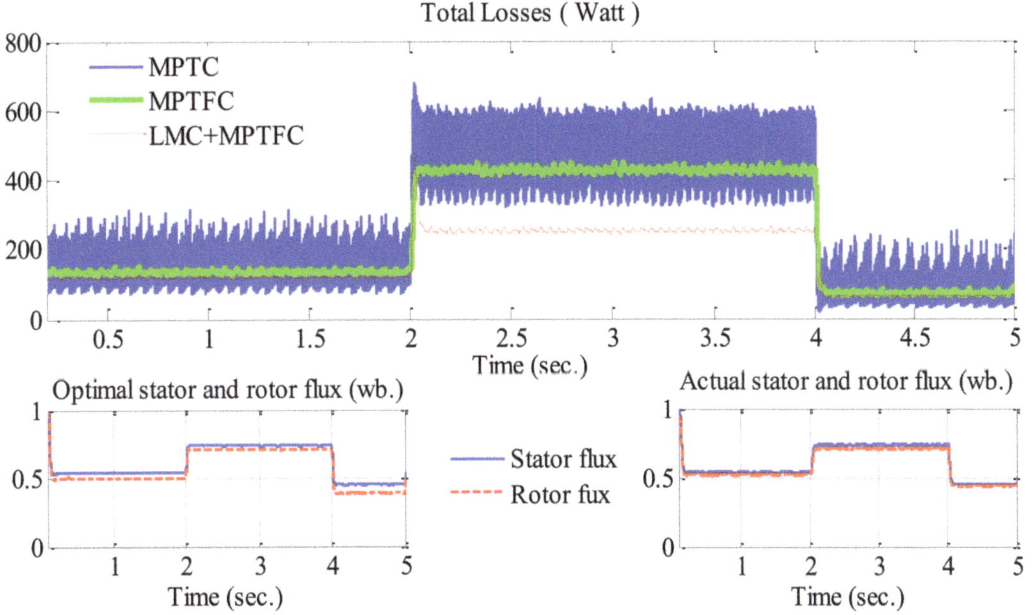

Figure 16. Total IM losses through three control procedures with actual and optimal fluxes.

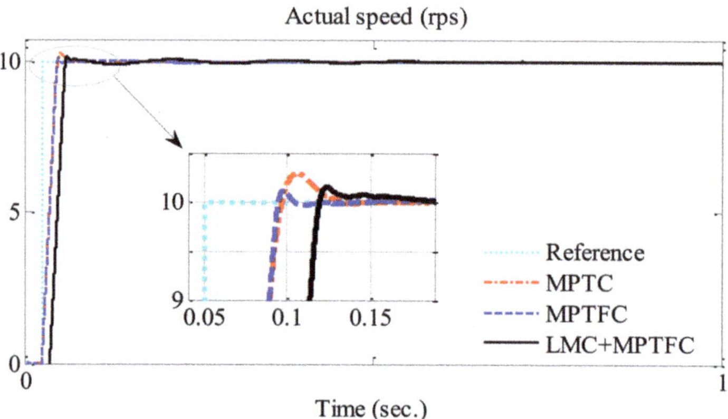

Figure 17. Actual speed response of MPTC and MPTFC (without and with LMC).

Table 2. Values of the over shot, rise time, and settling time for MPTC and suggested MPTFC (without and with LMC) approaches.

Method	Over Shot (%)	Rise Time	Settling Time
MPTC	2.8	0.0399	0.0952
MPTFC	1.1	0.0394	0.0541
MPTFC + LMC	1.5	0.0639	0.1183

7. Conclusions

Paper proposed a new MPTFC solution for 2−level inverter-fed IM drives to address the torque-ripple-phenomenon issue in MPTC. A novel design for sensorless observer centered on AFOO has been proposed. Stator flux estimation in the robust suggested observer provides a strong torque response during very low-speed operations. By introducing the recommended approach, much better steady-state stability in terms of flux/torque ripples is being observed without disturbing the dynamic response. In our proposed MPTFC, a closed feedback-loop is added to the suggested drive; consequently, uncertainties (error of estimated speed, unbalance current measurement and variance of parameters) are being compensated. Therefore, the proposed prediction approach was conducted more accurately. Moreover, the loss-minimization-criterion (LMC) is imposed that minimizes IM losses, particularly at low-speed and light loading, thus improving drive efficiency. In addition to minimizing the losses during steady-state operations, the comparative results affirm the proposed MPTFC effectiveness in achieving fewer oscillations. The proposed MPTFC approach improves IM drive robustness significantly. Finally, it should be noted that the experimental validations of the proposed control system should be considered in future work, which is not considered because of the lack of laboratory equipment and apparatus and the conditions of COVID-19.

Author Contributions: Conceptualization, A.G.M.A.A. and A.A.Z.D.; methodology, A.G.M.A.A. and A.A.Z.D.; software, A.G.M.A.A., H.R. and A.A.Z.D.; validation, H.R., A.G.M.A.A. and A.A.Z.D.; formal analysis, A.G.M.A.A. and A.A.Z.D.; investigation, A.A.Z.D.; resources, A.G.M.A.A., H.R. and A.A.Z.D.; data curation, A.G.M.A.A. and A.A.Z.D.; writing—original draft preparation, A.G.M.A.A. and A.A.Z.D.; writing—review and editing, A.G.M.A.A., H.R. and A.A.Z.D.; visualization, A.G.M.A.A., H.R. and A.A.Z.D.; supervision, A.A.Z.D.; project administration, A.A.Z.D.; funding acquisition, A.A.Z.D. and H.R. All authors have read and agreed to the published version of the manuscript.

Funding: This research received no external funding.

Institutional Review Board Statement: Not applicable.

Informed Consent Statement: Not applicable.

Data Availability Statement: All data are contained within the article.

Conflicts of Interest: The authors declare no conflict of interest.

Appendix A

In this section, the performance of the suggested AFO observer with feedback gains is analyzed to prove the robustness of the estimated speed algorithm. In addition to calculating the observer feedback gains to achieve high performance over a wide range of speed.

The suggested estimate speed in (20) could be rewritten in rotary reference $(d - q)$ frame as:

$$\hat{\omega}_r = \left(K_{P\omega} + \frac{K_{I\omega}}{s} \right) \left(\vec{i}_{qs} - \hat{i}_{qs} \right) \hat{\Psi}_{ds} \tag{A1}$$

A closed-loop schematic diagram of estimate speed is seen as Figure A1:

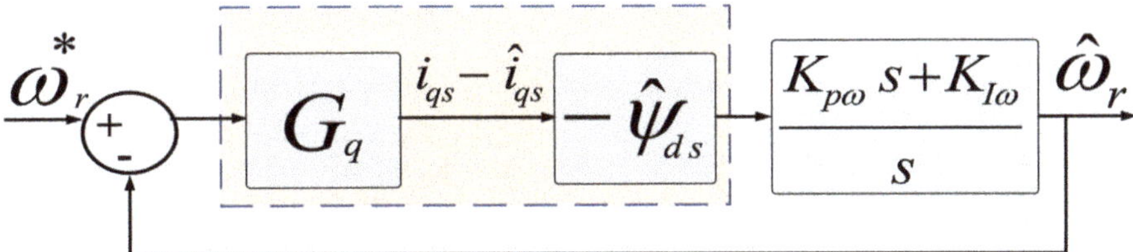

Figure A1. Proposed speed estimation algorithm.

It is potential to derive the open-loop transfer function (OLTF) between both estimated and actual rotor speed from Figure A1.

$$G_{op} = -\hat{\Psi}_{ds} \left(K_{P\omega} + \frac{K_{I\omega}}{s} \right) G_q(s) \tag{A2}$$

Feedback gains can indeed be obtained on the basis of calculated speed stability as:

$$\begin{cases} h_1 = \sigma\, L_s \left[\frac{-R_s}{\sigma\, L_s} - \frac{(1-\sigma)R_r}{\sigma\, L_r} + K\frac{R_r}{L_r} + \frac{R_r L_m^2}{L_r^2} \right] \\ h_2 = -K\,\sigma\, L_s\hat{\omega}_r \\ h_3 = \frac{R_r L_m}{L_r} \\ h_4 = 0 \end{cases} \tag{A3}$$

where, $\Delta\omega_r = \omega_r^* - \hat{\omega}_r$ and $G_q(s)$ is designated as:

$$G_q(s) = \frac{i_{qs} - \hat{i}_{qs}}{\Delta\omega_r} = \frac{s^3 + q_2 s^2 + q_1 s + q_0}{B} \tag{A4}$$

$$B = s^4 + b_3\, s^3 + b_2 s^2 + b_1 s + b_0 \tag{A5}$$

$$q_2 - \frac{L_r(R_s + h_1) + R_r L_s - h_3 L_m}{\sigma\, L_s\, L_r} = x \tag{A6}$$

$$q_1 = (\omega_e)^2 + \frac{h_1 R_r + R_s\, R_r}{\sigma\, L_s\, L_r} - \frac{h_2\omega_r}{\sigma\, L_s} = (\omega_e)^2 + y \tag{A7}$$

$$q_0 = \frac{L_r(R_s + h_1) + R_r L_s - h_3 L_m}{\sigma\, L_s\, L_r}(\omega_e)^2 - \left[\frac{h_2 R_r}{\sigma\, L_s\, L_r} + \frac{h_1 + R_s}{\sigma\, L_s}\omega_r \right]\omega_e \\ = x(\omega_e)^2 + z\omega_e \tag{A8}$$

All zeros of the OLTF must be placed within the left plane to ensure the stability of the speed estimation. With (A2) and (A4) and using the Routh-Hurwitz criterion, where Routh-Table is shown in Table A1, it is potential to obtain the required and adequate conditions for estimate speed stability as [41]:

$$\begin{cases} x > 0 \\ xy - z\omega_e > 0 \\ x(\omega_e)^2 + z\omega_e > 0 \end{cases} \tag{A9}$$

Feedback gains H values are difficult to acquire because values of x, y and z in (A8) are complicated. Deserting the second state in (A9) for easy analysis, the gains of H in (A3) can be provided. To satisfy the conditions in (A9), in this study, the conditions of (A10) are achieved:

$$\begin{cases} x > 0 \\ y > 0 \\ z = 0 \end{cases} \tag{A10}$$

(a) Based on ($z = 0$), the relation among h_1 and h_2 can be optimized:

$$h_2 = \frac{-L_r(h_1 + R_s)}{R_r}\omega_r \tag{A11}$$

(b) Based on ($y > 0$), a stability range of h_1 for speed estimate can be achieved:

$$h_1 > -R_s \tag{A12}$$

(c) Based on ($x > 0$), relation among h_1 and h_2 can be optimized:

$$h_1 > \frac{L_m}{L_r}h_3 \tag{A13}$$

Finally, it is potential to get the required and adequate conditions for stable estimation of speed:

$$\begin{cases} h_2 = \frac{-L_r(h_1+R_s)}{R_r}\,\omega_r \\ h_1 > \frac{L_m}{L_r}h_3 > -R_s \end{cases} \tag{A14}$$

If gains of feedback H satisfy (A14), speed estimation stability can be assured at all IM speed ranges. It could be seen through (A14), that h_4 does not affect the speed estimate stability. For simplicity, feedback gains being given in our study as:

$$\begin{cases} h_1 = h_3 = 0.048 \\ h_2 = \frac{-L_r(h_1+R_s)}{R_r}\,\omega_r \\ h_4 = 0 \end{cases} \tag{A15}$$

As an 4th−order system, the observer has 2−pairs of conjugate poles. To guarantee the observer's stability, each pole must be in the left-hand s-plane side. Observer pole placement with (A15) could be seen in Figure A2a, where the observer's dynamic performance appears to be strong. Via (A2) and (A15), the OLTF zeros with a full load is shown in Figure A2b ($-40\,\pi \le \omega_e \le 40\,\pi$).

It is observed that there are no unstable zeros in the low-speed area, even in regenerating mode. Moreover, velocity estimation can indeed be stable within full load throughout all velocity range.

Table A1. Routh-Table.

s^3	1	q_1
s^2	q_2	q_0
s^1	$q_1 - \frac{q_0}{q_2}$	0
s^0	q_0	0

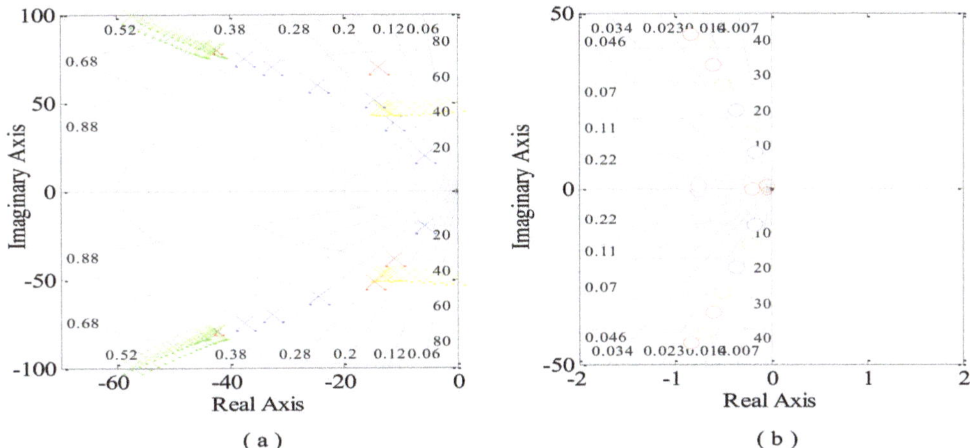

Figure A2. S-plane (**a**) observer poles placement (**b**) transfer function zeros of estimated speed.

References

1. Schauder, C. Adaptive speed identification for vector control of induction motors without rotational transducers. In *Conference Record of the IEEE Industry Applications Society Annual Meeting*; IEEE: Piscataway, NJ, USA; pp. 493–499.
2. Diab, A.A.Z. Novel robust simultaneous estimation of stator and rotor resistances and rotor speed to improve induction motor efficiency. *Int. J. Power Electron.* **2017**, *8*, 267–287. [CrossRef]
3. Kumar, R.; Das, S.; Chattopadhyay, A.K. Comparative assessment of two different model reference adaptive system schemes for speed-sensorless control of induction motor drives. *IET Electr. Power Appl.* **2016**, *10*, 141–154. [CrossRef]
4. Blasco-Gimenez, R.; Asher, G.; Sumner, M.; Bradley, K. Dynamic performance limitations for MRAS based sensorless induction motor drives. Part 1: Stability analysis for the closed loop drive. *IEE Proc. Electr. Power Appl.* **1996**, *143*, 113–122. [CrossRef]
5. Kubota, H.; Matsuse, K.; Nakano, T. DSP-based speed adaptive flux observer of induction motor. *IEEE Trans. Ind. Appl.* **1993**, *29*, 344–348. [CrossRef]
6. Verghese, G.C.; Sanders, S.R. Observers for flux estimation in induction machines. *IEEE Trans. Ind. Electron.* **1988**, *35*, 85–94. [CrossRef]
7. Harnefors, L. Design and analysis of general rotor-flux-oriented vector control systems. *IEEE Trans. Ind. Electron.* **2001**, *48*, 383–390. [CrossRef]
8. Casadei, D.; Profumo, F.; Serra, G.; Tani, A. FOC and DTC: Two viable schemes for induction motors torque control. *IEEE Trans. Power Electron.* **2002**, *17*, 779–787. [CrossRef]
9. Diab, A.A.Z.; Selim, S.; Elnaghi, B.E. Particle swarm optimization based vector control of permanent magnet synchronous motor drive. In *IEEE NW Russia Young Researchers in Electrical and Electronic Engineering Conference (EIConRusNW)*; IEEE: Piscataway, NJ, USA, 2016; pp. 740–746.
10. Alsofyani, I.M.; Idris, N.R.N. Simple flux regulation for improving state estimation at very low and zero speed of a speed sensorless direct torque control of an induction motor. *IEEE Trans. Power Electron.* **2015**, *31*, 3027–3035. [CrossRef]
11. Geyer, T.; Papafotiou, G.; Morari, M. Model predictive direct torque control—Part I: Concept, algorithm, and analysis. *IEEE Trans. Ind. Electron.* **2008**, *56*, 1894–1905. [CrossRef]
12. Rodríguez, J.; Kennel, R.M.; Espinoza, J.R.; Trincado, M.; Silva, C.A.; Rojas, C.A. High-performance control strategies for electrical drives: An experimental assessment. *IEEE Trans. Ind. Electron.* **2011**, *59*, 812–820. [CrossRef]
13. Rojas, C.A.; Rodriguez, J.; Villarroel, F.; Espinoza, J.R.; Silva, C.A.; Trincado, M. Predictive torque and flux control without weighting factors. *IEEE Trans. Ind. Electron.* **2012**, *60*, 681–690. [CrossRef]
14. Diab, A.A.Z.; Kotin, D.A.; Pankratov, V.V. Speed control of sensorless induction motor drive based on model predictive control. In *14th International Conference of Young Specialists on Micro/Nanotechnologies and Electron Devices*; IEEE: Piscataway, NJ, USA, 2013; pp. 269–274.
15. Nacusse, M.A.; Romero, M.; Haimovich, H.; Seron, M.M. DTFC versus MPC for induction motor control reconfiguration after inverter faults. In *Conference on Control and Fault-Tolerant Systems (SysTol)*; IEEE, Piscataway, NJ, USA, 2010; pp. 759–764.
16. Wang, T.; Zhu, J. Finite-control-set model predictive direct torque control with extended set of voltage space vectors. In 2017. In *20th International Conference on Electrical Machines and Systems (ICEMS)*; IEEE: Piscataway, NJ, USA, 2017; pp. 1–6.

17. Zhang, Y.; Wang, Q.; Liu, W. Direct torque control strategy of induction motors based on predictive control and synthetic vector duty ratio control. In *International Conference on Artificial Intelligence and Computational Intelligence*; IEEE: Piscataway, NJ, USA, 2010; Volume 2, pp. 96–101.

18. Lunardi, A.; Sguarezi, A. Experimental results for predictive direct torque control for a squirrel cage induction motor. In *Brazilian Power Electronics Conference (COBEP)*; IEEE: Piscataway, NJ, USA, 2017; pp. 1–5.

19. Kumar, P.; Rathore, V.; Yadav, K. Calculation of Torque Ripple and Derating Factor of a Symmetrical Six-Phase Induction Motor (SSPIM) Under the Loss of Phase Conditions. In Nanoelectronics. In *Circuits and Communication Systems*; Springer: Berlin/Heidelberg, Germany, 2021; pp. 441–453.

20. ML, P.; Eshwar, K.; Thippiripati, V.K. A modified duty-modulated predictive current control for permanent magnet synchronous motor drive. *IET Electr. Power Appl.* **2021**, *15*, 25–38.

21. Jing, K.; Liu, C.; Lin, X. Discrete Dynamical Predictive Control on Current Vector for Three-Phase PWM Rectifier. *J. Electr. Eng. Technol.* **2021**, *16*, 267–276. [CrossRef]

22. Lakhimsetty, S.; Somasekhar, V.T. An Efficient Predictive Current Control Strategy for a Four-Level Open-End Winding Induction Motor Drive. *IEEE Trans. Power Electron.* **2019**, *35*, 6198–6207. [CrossRef]

23. Xue, Y.; Meng, D.; Yin, S.; Han, W.; Yan, X.; Shu, Z.; Diao, L. Vector-based model predictive hysteresis current control for asynchronous motor. *IEEE Trans. Ind. Electron.* **2019**, *66*, 8703–8712. [CrossRef]

24. Miranda, H.; Cortés, P.; Yuz, J.I.; Rodríguez, J. Predictive torque control of induction machines based on state-space models. *IEEE Trans. Ind. Electron.* **2009**, *56*, 1916–1924. [CrossRef]

25. Diab, A.Z.; Kotin, D.; Anosov, V.; Pankratov, V. A comparative study of speed control based on MPC and PI-controller for Indirect Field oriented control of induction motor drive. In *12th International Conference on Actual Problems of Electronics Instrument Engineering (APEIE)*; IEEE: Piscataway, NJ, USA, 2014; pp. 728–732.

26. Rodriguez, J.; Cortes, P. *Predictive Control of Power Converters and Electrical Drives*; John Wiley & Sons: Hoboken, NJ, USA, 2012; Volume 40.

27. Davari, S.A.; Khaburi, D.A.; Kennel, R. An improved FCS–MPC algorithm for an induction motor with an imposed optimized weighting factor. *IEEE Trans. Power Electron.* **2011**, *27*, 1540–1551. [CrossRef]

28. Zhang, Y.; Yang, H. Model-predictive flux control of induction motor drives with switching instant optimization. *J. IEEE Trans. Energy Convers.* **2015**, *30*, 1113–1122. [CrossRef]

29. Zhang, Y.; Yang, H.; Xia, B. Model-predictive control of induction motor drives: Torque control versus flux control. *IEEE Trans. Ind. Appl.* **2016**, *52*, 4050–4060. [CrossRef]

30. Karamanakos, P.; Geyer, T. Model predictive torque and flux control minimizing current distortions. *IEEE Trans. Power Electron.* **2018**, *34*, 2007–2012. [CrossRef]

31. Geyer, T. Algebraic tuning guidelines for model predictive torque and flux control. *IEEE Trans. Ind. Appl.* **2018**, *54*, 4464–4475. [CrossRef]

32. Geyer, T. *Model Predictive Control of High Power Converters and Industrial Drives*; John Wiley & Sons: Hoboken, NJ, USA, 2016.

33. El Fadili, A.; Giri, F.; El Magri, A.; Lajouad, R.; Chaoui, F.Z. Towards a global control strategy for induction motor: Speed regulation, flux optimization and power factor correction. *Int. J. Electr. Power Energy Syst.* **2012**, *43*, 230–244. [CrossRef]

34. Diab, A.Z.; Vdovin, V.; Kotin, D.; Anosov, V.; Pankratov, V. Cascade model predictive vector control of induction motor drive. In *12th International Conference on Actual Problems of Electronics Instrument Engineering (APEIE)*; IEEE: Piscataway, NJ, USA, 2014; pp. 669–674.

35. Borisevich, A.; Schullerus, G. Energy efficient control of an induction machine under torque step changes. *IEEE Transactions Energy Convers.* **2016**, *31*, 1295–1303. [CrossRef]

36. Ta, C.-M.; Hori, Y. Convergence improvement of efficiency-optimization control of induction motor drives. *IEEE Trans. Ind. Appl.* **2001**, *37*, 1746–1753.

37. Tang, Y. Flux Controlled Motor Management. U.S. Patent 20100090629A1, 14 June 2011.

38. Diab, A.A.Z. Robust simultaneous estimation of stator and rotor resistances and rotor speed for predictive maintenance of sensorless induction motor drives. *Int. J. Power Energy Convers.* **2017**, *8*, 411–434. [CrossRef]

39. Diab, A.A.Z. Implementation of a novel full-order observer for speed sensorless vector control of induction motor drives. *Electr. Eng.* **2017**, *99*, 907–921. [CrossRef]

40. Matsuse, K.; Katsuta, S.; Tsukakoshi, M.; Ohta, M.; Huang, L.-p. Fast rotor flux control of direct-field-oriented induction motor operating at maximum efficiency using adaptive rotor flux observer. In *IAS'95. Conference Record of the 1995 IEEE Industry Applications Conference Thirtieth IAS Annual Meeting*; IEEE: Piscataway, NJ, USA, 1995; Volume 1, pp. 327–334.

41. Sun, W.; Gao, J.; Yu, Y.; Wang, G.; Xu, D. Robustness improvement of speed estimation in speed-sensorless induction motor drives. *IEEE Trans. Ind. Appl.* **2015**, *52*, 2525–2536. [CrossRef]

42. Kubota, K.; Matsuse, K. Compensation for core loss of adaptive flux observer-based field-oriented induction motor drives. In Proceedings of the 1992 International Conference on Industrial Electronics, Control, Instrumentation, and Automation; IEEE: Piscataway, NJ, USA, 1992; pp. 67–71.

43. Hu, J.; Wu, B. New integration algorithms for estimating motor flux over a wide speed range. *IEEE Trans. Power Electron.* **1998**, *13*, 969–977.

44. Yang, G.; Chin, T.-H. Adaptive-speed identification scheme for a vector-controlled speed sensorless inverter-induction motor drive. *IEEE Trans. Ind. Appl.* **1993**, *29*, 820–825. [CrossRef]
45. Aziz, A.G.M.A.; Diab, A.A.Z.; Sattar, M.A.E. Speed sensorless vector controlled induction motor drive based stator and rotor resistances estimation taking core losses into account. In *2017 Nineteenth International Middle East Power Systems Conference (MEPCON)*; IEEE: Piscataway, NJ, USA, 2017; pp. 1059–1068.
46. Astrom, K.; Wittenmark, B. *Computer Controlled Systems, Englewood Cliffs, NJ 07632*; Prentice-Hall International, Inc.: New York, NY, USA, 1990.
47. Yuz, J.I.; Goodwin, G.C. On sampled-data models for nonlinear systems. *IEEE Trans. Autom. Control* **2005**, *50*, 1477–1489. [CrossRef]

*mathematics*

MDPI

Finite Element Based Overall Optimization of Switched Reluctance Motor Using Multi-Objective Genetic Algorithm (NSGA-II)

Mohamed El-Nemr [1,2,*], Mohamed Afifi [1], Hegazy Rezk [3,4] and Mohamed Ibrahim [5,6,7]

[1] Electromagnetic Energy Conversion Laboratory, Tanta University, Tanta 31527, Egypt; emec@f-eng.tanta.edu.eg
[2] Electrical Power and Machines Engineering Department, Faculty of Engineering, Tanta University, Tanta 31527, Egypt
[3] College of Engineering at Wadi Addawaser, Prince Sattam Bin Abdulaziz University, Wadi Aldawaser 11991, Saudi Arabia; hr.hussien@psau.edu.sa
[4] Electrical Engineering Department, Faculty of Engineering, Minia University, Minia 61111, Egypt
[5] Department of Electromechanical, Systems and Metal Engineering, Ghent University, 9000 Ghent, Belgium; m.nabil@ugent.be
[6] FlandersMake@UGent—Corelab EEDT-MP, 3001 Leuven, Belgium
[7] Electrical Engineering Department, Kafrelshiekh University, Kafrelshiekh 33511, Egypt
* Correspondence: melnemr@f-eng.tanta.edu.eg; Tel.: +20-111-3535-272

Citation: El-Nemr, M.K.; Afifi, M.M.; Rezk, H.; Ibrahim, M.N. Finite Element Based Overall Optimization of Switched Reluctance Motor Using Multi-Objective Genetic Algorithm (NSGA-II). *Mathematics* **2021**, *9*, 576. https://doi.org/10.3390/math9050576

Academic Editor: David Greiner

Received: 29 January 2021
Accepted: 23 February 2021
Published: 8 March 2021

Publisher's Note: MDPI stays neutral with regard to jurisdictional claims in published maps and institutional affiliations.

Abstract: The design of switched reluctance motor (SRM) is considered a complex problem to be solved using conventional design techniques. This is due to the large number of design parameters that should be considered during the design process. Therefore, optimization techniques are necessary to obtain an optimal design of SRM. This paper presents an optimal design methodology for SRM using the non-dominated sorting genetic algorithm (NSGA-II) optimization technique. Several dimensions of SRM are considered in the proposed design procedure including stator diameter, bore diameter, axial length, pole arcs and pole lengths, back iron length, shaft diameter as well as the air gap length. The multi-objective design scheme includes three objective functions to be achieved, that is, maximum average torque, maximum efficiency and minimum iron weight of the machine. Meanwhile, finite element analysis (FEA) is used during the optimization process to calculate the values of the objective functions. In this paper, two designs for SRMs with 8/6 and 6/4 configurations are presented. Simulation results show that the obtained SRM design parameters allow better average torque and efficiency with lower iron weight. Eventually, the integration of NSGA-II and FEA provides an effective approach to obtain the optimal design of SRM.

Keywords: optimal design; switched reluctance machine; NSGA-II optimization; finite element analysis

1. Introduction

The switched reluctance motor (SRM) is the type of motor that has saliency in both stator and rotor without permanent magnets or windings on rotor [1]. SRM develops electromagnetic torque based on variation of reluctance values for rotor position change with respect to phases when they are switched on. SRM provides several merits compared to other types of electric machines [2]. For instance, the topology of SRM is simple and very robust. Moreover, the power density, efficiency and torque output of SRM are high over a wide speed range [3–5]. The previous merits have increased the research efforts recently and made SRM preferred for high speed applications [1,6]. However, the torque ripple of SRM is the major problem that results in a high noise and variation. The latter can be improved by both control and design [7]. The control of this machine plays an essential role in the operation and hence it is required to overcome its challenges, which differ depending on the application [8–10].

The SRM construction has a lot of geometric parameters. The design is achieved by specifying all of these parameters values. Since SRM has salient poles in both stator and rotor, a wide range of geometric parameters combinations exist for a certain design objective. Searching for the best design is then not a simple task. Therefore, it is necessary to obtain the geometrical parameters that achieve the required objectives (e.g., maximum torque and efficiency and minimum volume and cost) in the best way (optimum design).

Optimization is a general term used to describe types of problems and solution techniques that are concerned with the best ("optimal") allocation of limited resources in projects. The problems are called optimization problems and the methods optimization methods [11,12]. Initially random values of variables are chosen for a number of solutions then the objective function values are evaluated for all solutions and classified from best to worst. The algorithm produces other solutions (variables) from these classified as the best .

If one objective function is desired to optimize, it is called single objective optimization. If more than one objective functions are desired to optimize simultaneously, it is called multi objective optimization. In this context, a set of solutions is obtained where their objective values form what is referred to as the Pareto front or non-dominated front. For the same Pareto front, all solutions are equally good because there is no way of telling which one is better or worse. In other words, all solutions in the same Pareto front are the optimal solutions (for optimal Pareto front) of the problem in a multi-objective sense [13].

In [14], the optimization of SRM design was made considering a certain ratio between the length of SRM core to the pole arcs of stator and rotor. The stator and rotor pole arcs were then varied between the limits that achieve self-starting not causing negative torque as this reduces the total developed torque. With every variation in stator and rotor pole arcs the objective functions—which were average torque and torque per volume—were calculated. The arc's values were chosen based on a compromise between the average torque and the torque per volume values. However, this work assumed fixed ratios for the lengths and arcs and did not study the other values, which give other designs.

In [15], the design optimization of a switched reluctance motor (SRM) by using a combination of two-dimensional electromagnetic and thermal finite-element analysis, three-dimensional correction factors and computer search techniques were presented.

The sub-problem approximation analysis was initially performed to locate an approximate optimum in the feasible design space, and then the first-order method was used to perform the final search. The core losses were calculated from a 2-D finite element analysis (FEA), based on a pre-calculated Fourier series of the flux density distribution in the SRM with typical phase currents.

In [16], a method of the optimization design with multi-objectives for switched reluctance motors for electric vehicle (EV) applications was proposed. From the requirements of EVs on electric motors in [16], three objective functions were chosen to optimize. They are the average torque, the average torque per copper loss and the average torque per motor core volume. The stator and the rotor pole arc angles are selected as the optimized parameters in this paper. The optimized parameters are only the stator and rotor pole arcs.

In [17], a multi-objective optimization for 16/20 SRM design and control were introduced based on a non-dominated sorting genetic algorithm intended for high volume traction applications. The proposed methodology considers a lot of parameters as variables for optimization process, also it considers the optimal firing angles (on and off angles) as an objective function in addition to frequently used objective functions like average torque, efficiency and torque ripples. The optimization of firing angles has the advantage of achieving minimum size of motor for specific requirements. The firing angles are optimized for this design by trying 100 different combinations of turn-on and turn-off angles to get the highest average torque and efficiency while concurrently minimizing the torque ripple. The proposed optimization framework succeeded to achieve the optimal geometry design for the special application intended for motor to be used.

Hayashi and Miller [18] represented the different flux density waveforms in matrices form and calculated eddy-current losses and hysteresis losses separately which was used in this paper for core losses calculations as will come later.

SRM's geometric parameters have an indirect and non-linear relationship with performance indices, that is, efficiency and average torque. Hence, sensitivity analysis on SRM geometric parameters is usually made as in [19–22] to reduce the complexity of the optimization process. The sensitivity analysis is to study the degree of influence of optimization problem's variables on the objective functions. Most influencing variables are only considered in optimization in order to reduce computational time. However, eliminating some of variables in optimization problem eliminates some of the indirect influence of these variables on the performance indices and makes the optimization limited to the specified objective functions and variables. Therefore, the method presented in this paper enables the optimization of 11 dimensions independently in order to include all possible design candidates which are within search area. Note that only seven dimensions are optimized in this paper as the remaining four are specified by application constraints (outer diameter D_o, axial length L and shaft diameter D_{sh}) or for practical reasons (air gap length g).

Various methods of analyzing SRM include magnetic equivalent circuit (MEC), FEA and regression methods exist [1,23]. In this paper, FEA is used for its accuracy. Multi objective optimization of SRM design is achieved by the non-dominated sorting genetic algorithm method (NSGA-II). The program of optimization is made in Lua script to run from FEMM4.2 software. The FEA is performed each candidate design evaluation. The three objective functions average torque (T_{av}), efficiency (η) and iron weight (W_{iron}) are chosen to be optimized. Numerical methods are used to perform integration and differentiation on flux density waveforms to calculate eddy current losses as demonstrated later on this paper. The results of optimizations are compared and verified.

2. Design of SRM

The design procedure of switched reluctance machine starts with specifying the available dimensions from space constraints, for example, frame size, shaft size and axial length then continues until all other dimensions are obtained. The number of stator and rotor poles are specified at the beginning as well. In the conventional analytic design methods, the inductance in the aligned and unaligned is calculated. Using the values of inductances in both aligned and unaligned positions the average torque is calculated. This step is repeated with modified values of the main dimensions until the requirements are justified. The number of turns per phase is calculated for every modification of dimensions such that the flux density doesn't saturate in stator poles for normal operation. This is demonstrated in Figure 1. Many other characteristics can be calculated such as efficiency, volume, weight and torque ripple. In this section, the SRM variables are discussed and the methods of characteristics calculations are emphasized.

2.1. Pole Selection

The number of stator poles P_s and the number of rotor poles P_r are usually selected based on previous experience with the application requirements and converter configuration to be used. The combinations of stator and rotor poles are of few choices for good overall design of SRM to be used in general; however, special applications may lead the designer to explore more of less frequent combinations to achieve the application requirements. This paper primarily focuses on the popular combination of 6/4 and 8/6 machines. The 6/4 machine has the advantage of using less switches in the converter, two less terminals and less core losses because of less switching losses than 8/6 machine; however, it has the disadvantage of higher torque ripple than the other common combination of (8/6 machine).

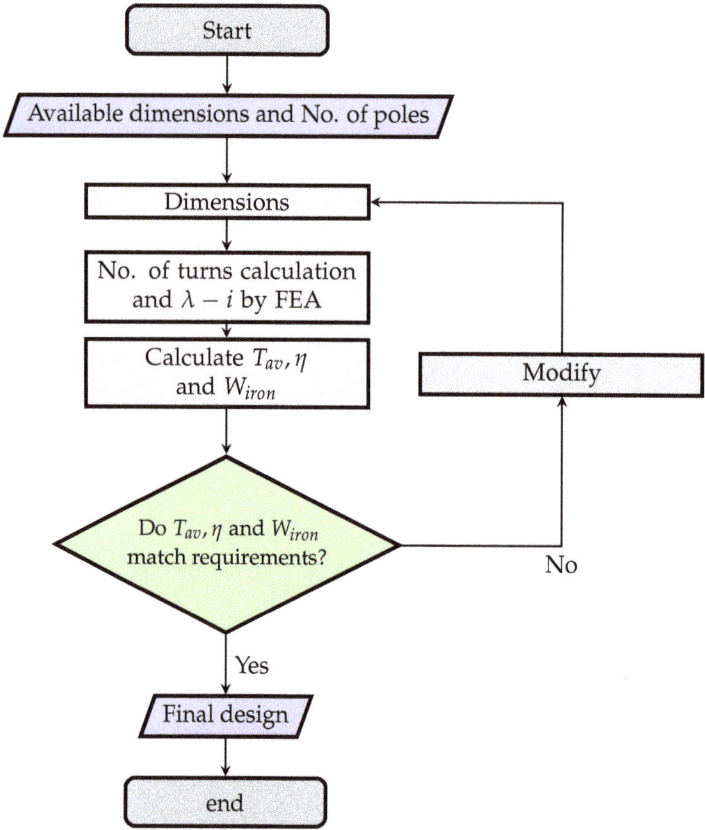

Figure 1. Design process flow chart.

2.2. Rotor and Stator Poles Arcs Selection

Referring to [2], the minimum stator pole arc to achieve self starting :

$$min[\beta_s] = \frac{4\pi}{P_s P_r} \, , rad. \tag{1}$$

The angle between the corners of adjacent rotor poles must be greater than the stator pole arc or there will be an overlap between the stator and rotor poles in the unaligned position. This condition is represented as:

$$\beta_s + \beta_r \leq \frac{2\pi}{P_r}. \tag{2}$$

The implication of this condition not being followed is that the machine will start having a positive inductance rate of change before reaching the minimum value. This causes the unaligned inductance value to be higher and leads to a lower torque generation.

2.3. Main Dimensions

For the dimensions shown in Figure 2 and Table 1, the outer diameter (D_o) is determined by the available space in the application. The shaft diameter (D_{sh}) is obtained from the shaft's standard sizes. The outer diameter(D_o) and (D_{sh}) are fixed and they are not changed while searching for the suitable design since they are space constraints. However, the axial length (L) and the bore diameter (D) can be changed during the design process (Note that the axial length increase is limited by the maximum axial length available in the application). The air gap length(g) can be changed too but here it is fixed of 0.5 mm. The remaining dimensions are to be changed in order to reach the desired value of torque, efficiency $\cdots$ and so forth.

Table 1. SRM dimensions.

Dimension	Unit
outer diameter, D_o	mm
shaft diameter, D_{sh}	mm
axial length, L	mm
bore diameter, D	mm
air gap length, g	mm
stator pole length, h_s	mm
rotor pole length, h_r	mm
stator back iron length, b_s	mm
rotor back iron length, b_r	mm
stator pole arc, β_s	degree
rotor pole arc, β_r	degree
stator poles, P_s	NA
rotor poles, P_r	NA

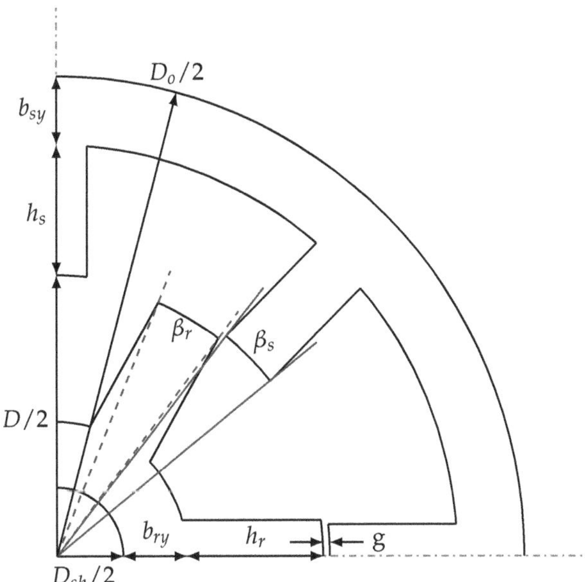

Figure 2. Lamination dimensions considered in optimization process.

2.4. Limits of Variables

The limits of variables depend on the application and the available space. Here, the frame size, shaft diameter, air gap length and axial length are kept constant by making their limits at the same value. The rest of the limits are set by the previous experience. Table 2 shows the maximum and minimum values of all variables.

Table 2. Limits of variables.

Variable	Min	Max	Unit
D_o	130	130	mm
L	100	100	mm
D	44	95	mm
b_{sy}	5	20	mm
b_{ry}	5	20	mm
h_s	7	52	mm
h_r	5	23	mm
D_{sh}	24	24	mm
g	0.5	0.5	mm
β_s	$0.85 \times 720/(P_s P_r)$	$0.6 \times 360/P_s$	degree
β_r	$0.85 \times 720/(P_s P_r)$	$0.6 \times 360/P_s$	degree

2.5. Windings Clearance

The clearance between two adjacent windings is calculated as in [14]. Taking a wedge of $h_{wed} = 4$ mm is required to hold the windings in place, the stator pole arc length t_s at the closest point of the winding to the center of the shaft is given by:

$$t_s = \left(\frac{D}{2}\right)\beta_s + 2h_{wed}, \text{ mm.} \tag{3}$$

Accounting for the wedges that hold the windings in place leads to the calculation of a modified stator pole pitch p_b as:

$$p_b = \frac{\pi(D + 2h_{wed})}{P_s}, \text{ mm.} \tag{4}$$

Assuming a suitable value of allowable current density ($J = 6$ A/mm^2), the area of conductor a_c is calculated. Hence, the wire diameter (d_w) including insulation is obtained from standards wires tables. The maximum height of the winding (h_w) is obtained by subtracting a margin length (h_{wed}) from stator pole height (h_s):

$$h_w = h_s - h_{wed}, \text{mm.} \tag{5}$$

Assuming $K_f = 0.95$ fill factor, the number of layers that can be accommodated in this available winding height is given by:

$$N_v = \frac{h_w K_f}{d_w}. \tag{6}$$

The value of N_v is rounded off to the nearest lower integer. Now the number of horizontal layers required for winding is given by:

$$N_h = \frac{T_{ph}}{N_v}. \tag{7}$$

The space between 2 stator pole tips at the bore is given by:

$$Z = p_b - t_s \text{ mm.} \tag{8}$$

The width of the winding w_t is given by:

$$w_t = d_w \frac{N_h}{K_f} \text{ mm.} \tag{9}$$

The clearance between the windings at the bore is given by:

$$CL = Z - 2w_t \text{ mm.} \tag{10}$$

This value has to be positive and preferably greater than 3 mm. Here it is allowed greater than 0.5 mm.

2.6. Average Torque Calculation

Average torque of SRM is calculated based on the assumptions that flux linkage (λ) vs. current (i) characteristics are available and phase current is kept constant at its maximum value between the unaligned and aligned positions [2].The average torque is the total work done per stroke multiplied by number of strokes of one revolution divided by 2π:

$$T_{av} = \frac{W P_s P_r}{4\pi}, \ N.m \tag{11}$$

$$W = W_{aligned} - W_{unaligned}, \tag{12}$$

where $W_{aligned}$ and $W_{unaligned}$ are the areas under $\lambda - i$ curves at aligned and unaligned positions, respectively. W is the area of energy loop and then calculated as in [22].

2.7. Losses and Efficiency Calculation

The prediction of switched reluctance motor efficiency requires knowledge of losses [18,24]. The calculation of losses in the SRM, especially the assessment of core losses, is a very difficult task mainly because the flux waveforms are non-sinusoidal and the differences in shape of flux density waveforms in the different sector of SRM's magnetic circuit. Furthermore, core losses are also conditioned by the type of control used and rotation speed(ω). For low speeds, the mechanical losses can be neglected. Hence losses may be calculated as:

$$Losses(\omega) = Core\ Loss + Copper\ Loss. \tag{13}$$

Once the losses are obtained the efficiency is calculated as follows :

$$\eta = \frac{\omega T_{av}}{\omega T_{av} + Losses(\omega)}. \tag{14}$$

In this paper, the speed at which efficiency is calculated is the rated speed of 1000 rpm for all SRM designs candidates. Copper losses value depends on the control technique used as it impacts the value of phase current. Considering n is number of phases, R_j is phase dc resistance and I_j is phase current, total copper loss instantaneous value may be calculated by the equation:

$$P_{cu}(t) = \sum_{j=1}^{j=n} I_j^2(t) R_j. \tag{15}$$

The average copper losses can be calculated by equation:

$$P_{cu} = \frac{1}{T} \int_0^T P_{cu}(t) dt, \tag{16}$$

where, T is the period of time for $P_s/2$ strokes. For sake of simplification, we assume no overlap between phases. Since the current of phase is not pure dc. The peak value of it (I_p) is considered for copper losses calculation as a pessimistic prediction. Copper loss is then calculated straight forwardly by the equation:

$$P_{cu} = I_p^2 R_{ph}. \tag{17}$$

2.8. Eddy Currents Losses

Referring to [25] the eddy current losses in SRM can be calculated by the equation:

$$P_e = \frac{e^2}{4k_{cir}\rho_{fe}\delta}\frac{1}{T}\int(\frac{\partial B}{\partial t})^2 dt \ \ w/kg,$$

(18)

where e: sheet thickness in meter, k_{cir}: constant ($1 < k_{cir} < 3$) introduced to account for the fact that paths in the interior of the lamination will have smaller emfs than those near the surface; ρ_{fe}: the electrical resistivity of the ferromagnetic material (in Ωm); δ: density of the ferromagnetic material (in kg/m^3).

From Equation (18), the waveform of flux density (B) for all SRM sectors must be known. Once they are available, P_e is calculated by numerical integration and differentiation. There are a lot of methods to obtain these waveforms and many of them are time consuming. In [18], a mathematical method using matrices is introduced to obtain the waveforms of all the SRM sectors in a systematic manner. The calculation of B waveforms for all sectors is achieved by modulation of triangular pulses. The stator poles waveforms consist only of unipolar triangular pulses, while those of the rotor poles contain both positive and negative pulses. The stator and rotor yoke waveforms have more complicated relationship with the triangular pulses. This method is demonstrated in details in [18] and used here for 8/6 and 6/4 SRMs.

2.9. Hysteresis Losses

Referring to [18], the hysteresis losses can be calculated for various sectors of SRM using the following equations:

$$P_h = P_{sph} + P_{rph} + P_{syh} + P_{ryh}$$

(19)

$$P_{sph} = \frac{\omega}{2\pi}P_s P_r W_{sp} E_h(0, B_{spm})$$

(20)

$$P_{rph} = \frac{\omega}{2\pi}P_r P_s W_{rp}\left[\frac{h_{rph}}{2}E_h(-B_{rpm}, B_{rpm}) + (1 - h_{rph})E_h(0, B_{rpm})\right]$$

(21)

$$P_{syh} = \frac{\omega}{2\pi}P_s P_r N_{ph} W_{sy}\left[\frac{h_{syh}}{2}E_h(-B_{sym}, B_{sym}) + (1 - h_{syh})E_h(B_{sy0}, B_{sym})\right]$$

(22)

$$P_{ryh} = \frac{\omega}{2\pi}P_r^2 N_{ph} W_{ry}\left[\frac{h_{ryh}}{2}E_h(-B_{rym}, B_{rym}) + (1 - h_{ryh})E_h(B_{ry0}, B_{rym})\right],$$

(23)

where :

$$E_h(-B_{max}, B_{max}) = C_h f B_{max}^{(a+bB_{max})}$$

(24)

$$Or \ E_h(-B_{max}, B_{max}) = a B_{max} + b B_{max}^2.$$

(25)

Referring to [14], the second formula is used and the constants a, b are $-4.6445 \times 10^{-3}, 0.01652$ respectively. The rest of symbols are shown in Table 3.

The flux density waveforms depend on the phase current waveform and the speed of the motor. Flux density waveforms calculated in this paper rated the speed of 1000 rpm and control is by a single pulse voltage. Hence, it is expected that the resulted designs will have maximum efficiency at 1000 rpm and rated torque average. Note that phase current has the peak of 6 ampere for all SRMs candidates.

Table 3. Symbols in Equations (19)–(25).

Symbol	Description	Symbol	Description
P_h	Total hysteresis losses	B_{rym}	Rotor yoke maximum flux density
P_{sph}	Stator pole hysteresis losses	B_{spm}	Stator pole maximum flux density
P_{rph}	Rotor pole hysteresis losses	B_{rpm}	Rotor pole maximum flux density
P_{syh}	Stator yoke hysteresis losses	B_{sy0}	Stator yoke initial flux density
P_{ryh}	Rotor yoke hysteresis losses	B_{ry0}	Rotor yoke initial flux density
$E_h(B_1, B_2)$	Hysteresis loss energy per unit weight for a hysteresis loop where flux density changes between B_1 and B_2.	h_{rph}	Normalized count of the flux polarity changes in rotor pole.
h_{syh}	Normalized count of the flux polarity changes in stator yoke.	W_{sp}	Weight of stator pole in kg.
h_{ryh}	Normalized count of the flux polarity changes in rotor yoke.	W_{rp}	Weight of rotor pole in kg.
C_h	Hysteresis losses coefficient.	W_{sy}	Weight of stator yoke in kg.
f	Frequency.	W_{ry}	Weight of rotor yoke in kg.
P_s	Stator poles number.	N_{ph}	No. of phases.
P_r	Rotor poles number.	B_{sym}	Stator yoke maximum flux density
B_{max}	Magnitude of maximum flux density in hysteresis loop		

3. SRM Design Optimization Techniques

The optimization is a search problem that seeks better objectives. One of the most popular techniques is the genetic algorithm. Wherein, better generations are produced by crossover between the best individuals of the previous generation. To make a decision which is the best, the objective of the optimization problem is needed to be defined. There are two types of optimization, single objective and multi-objective optimization. The decision making criteria is then different, in single objective optimization the criteria is to choose the greater value in the maximization problem to be the best (the smaller for minimization problem). At the end, the best value is considered as the optimal solution .

In the optimization of SRM ,the dimensions in Table 1 represent one possible solution (individual). All individuals information are stored in a vector in suitable data structure.

3.1. Constraints

Since there are several SRM dimensions, there must be certain constraints on them to prevent any non-logical values of variables with non-related physical meanings. The dimensions are checked to satisfy the following constraints:

$$D_{sh} + 2b_{ry} + 2h_r + 2g = D \tag{26}$$

$$D + 2b_{sy} + 2h_s = D_o \tag{27}$$

$$\beta_r > \beta_s \tag{28}$$

$$\frac{D}{2}\left(1 - \frac{\beta_r P_r}{2\pi}\right) \geq h_r. \tag{29}$$

It is also needed to specify certain limits to each variable (dimension). The maximum and minimum limits are then added to be constraints for all of the variables. If it is needed to keep a certain dimension fixed, this can be simply achieved by setting both the minimum and maximum limits to the desired value.

3.2. Objective Functions

The objectives of SRM optimization depends on the application and its conditions. For general purpose SRM, average torque and efficiency are needed to be maximized and weight is to be minimized. In some applications other objectives (i.e. torque ripples, acoustic noise, vibrations...etc.) are important.

4. NSGA-II for SRM Design Optimization

Non-dominated sorting genetic algorithm (NSGA-II) is one of the best and most popular techniques, which is used in multi-objective optimization problems. It depends mainly on the concept of dominance to judge the individuals of the same generation. This concept takes all of objective functions in consideration with their direction to minimize or maximize. It is required to check each individual with the rest on the dominance basis. As in [26], assuming X_1, X_2 are two vector individuals, m is the number of objective functions, if $X_1 \prec X_2$ (X_1 dominates X_2) is true, Pareto dominance conditions must all be true and they are:

$$f_j(X_1) \not\triangleright f_j(X_2) \forall j = \{1, ..., m\} \tag{30}$$

$$f_j(X_1) \triangleleft f_j(X_2) \exists j = \{1, ..., m\}. \tag{31}$$

The non-dominated individual is the individual that is not dominated by any of the other individuals in the population of a certain generation. After that, the individuals are sorted in the form of groups depending on their degree of dominance. These groups are called non-dominated sets or simply fronts. The first front is the group consists of the best (non-dominated) individuals. Multi-objective optimization by NSGA-II eliminates the need of weights in multi-objective optimization by a single function ($f = w_1 f_1 + w_2 f_2 + \cdots w_n f_n$). Moreover, it eliminates the conflict between weights (as the summation of weights must equal to 1) and hence a wider search area is covered.

Crowding distance is a criterion used to compare between solutions, which are in the same non-dominated front. The more space there is around a solution, the higher is the crowding distance. Therefore, solutions with a high crowding distance should have a rank better than those with a low crowding distance in order to maintain diversity in the population. Crowding distance is computed in the same manner as mentioned in [26]. Crowding distance is computed for each solution using Equation (32). If solutions of the same non-dominated fronts are numbered with their associate objective functions in lists, crowding distances are calculated as follows:

$$CD_j = CD_j + \frac{f_m^{j+1} - f_m^{j-1}}{f_m^{max} - f_m^{min}}, \tag{32}$$

where j is a solution in the sorted list, f_m is the objective function value of mth objective, f_m^{max} and f_m^{min} are the population-maximum and population-minimum values of mth objective functions.

SRM optimization using NSGA-II requires the setting of variables, objective functions, constraints, population size and number of maximum generations. Population size is preferred to be high in order to enhance the possibility of finding better individuals. Since T_{av} calculation by FEMM4.2 requires the SRM magnetic circuit to be analyzed several times, the computation time must be taken into consideration while deciding the population size. Hence, population size is chosen to be 30 candidates. A maximum generations number is used as a termination condition. It is chosen to be more than 300 generations. The constraints of the SRM design problem are mainly the limits of variables and the clearance between windings as shown in Table 2.

Code Algorithm

The code is made using the Lua programming language. The code is executed using Lua Console in FEMM4.2 software. The choice of the Lua programming language to be used is due to its simplicity and that it is adopted by FEMM4.2, which provides the FEA analysis in good accuracy. The code's algorithm is shown in Figure 3. First, SRM optimization data are entered. These data include the population size, problem variables,

variables limits, objective functions, to specify which objective function to maximize and which to minimize the maximum iterations limit and numbers of rotor and stator poles (P_r and P_s). Then, solutions are initialized by random choice of variables within the search space area. After that, constraints in Equations (26)–(29) are maintained in this step by changing the values of variables resulted. Then, the FEA is accomplished using FEMM4.2 software to calculate average torque, maximum stator and rotor poles flux densities and volume of iron. After that, the results of FEA analysis is used to calculate the remaining objective functions (η and W_i). Then, non-dominated sorting is performed and crowding distance is calculated for all solutions. Next, the selection of the best designs to be used in crossover and mutation. Lastly, termination condition is checked such that if number of iterations exceeds the maximum limit the whole process is finished and the highest rank of all solutions (non-dominated front) is the given in the output of optimization process.

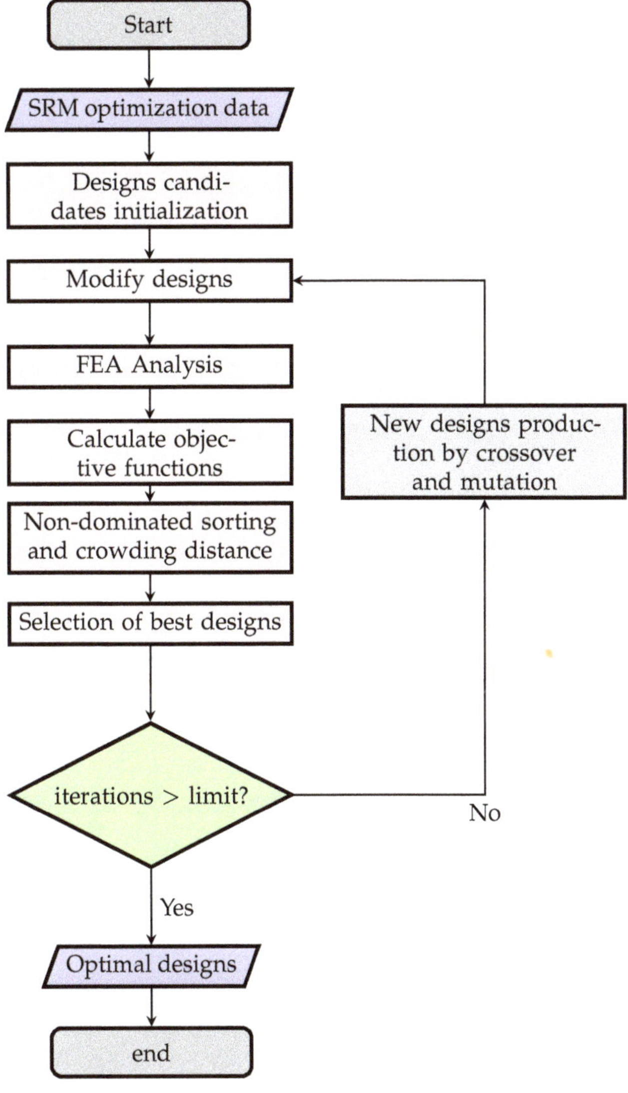

Figure 3. NSGA-II optimization program flowchart.

5. Results and Discussions

Both 6/4 and 8/6 SRMs are optimized using the same technique. Since there are three objective functions, it is difficult to show them all together in one figure to see the progress of the optimization process with generations. Hence, the objective functions are taken in pairs and shown as in Figure 4 for 8/6 SRM and Figure 5 for 6/4 SRM for more than 300 generations.

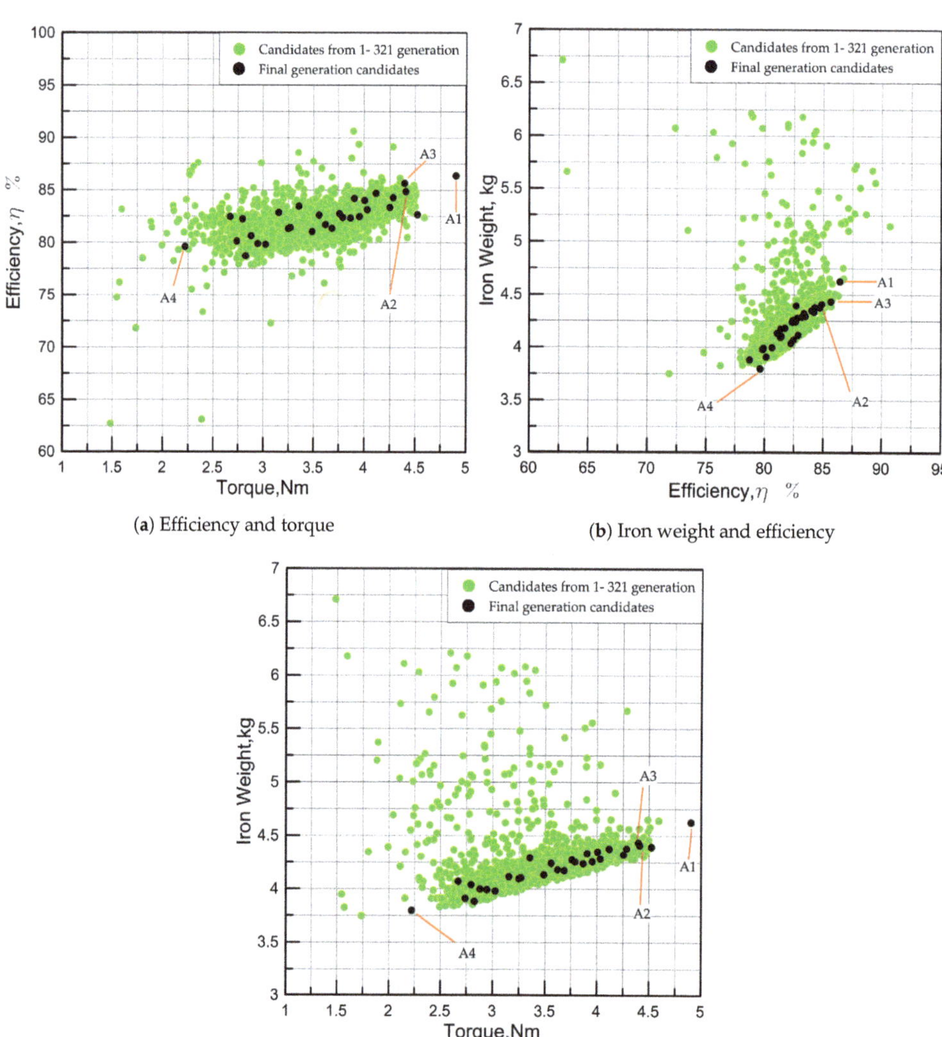

(**a**) Efficiency and torque

(**b**) Iron weight and efficiency

(**c**) Iron weight and torque

Figure 4. Objective functions results for 8/6 SRM.

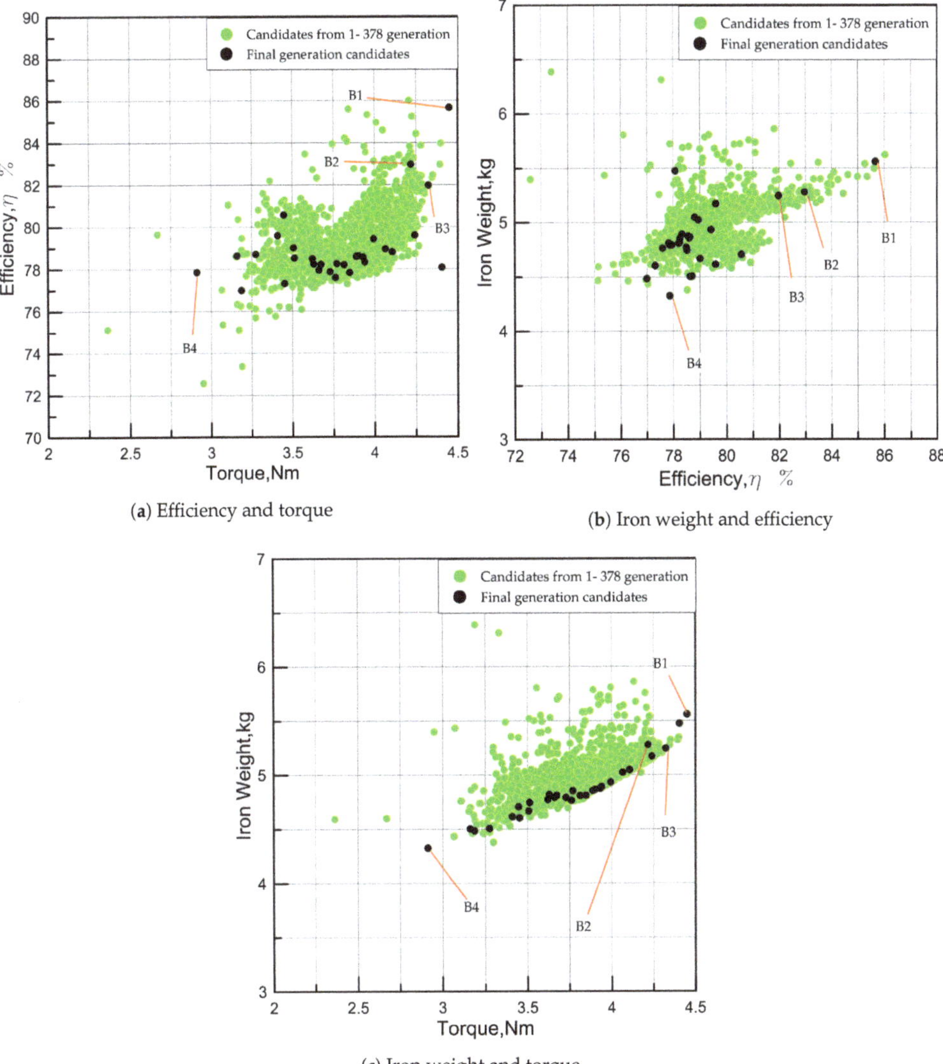

(**a**) Efficiency and torque

(**b**) Iron weight and efficiency

(**c**) Iron weight and torque

Figure 5. Objective functions results for 6/4 SRM.

These figures show the search direction or the optimization progress as more generations are produced. As we intended to maximize both average torque and efficiency and minimize iron weight, it is obvious that the crowded area (which indicates the majority of search) in Figure 4a, for example, exists in the upper right quarter (considering the axes limits). Wherein, higher values for both efficiency and average torque are sought. This means that the optimization program has searched a lot in the area of variables that produce candidates with more average torque and efficiency in the same time. Figure 4b shows the same concept with the difference that the objective functions are iron weight and efficiency. The program tries to minimize iron weight while maximizing efficiency but because of the complexity of the problem and the constraints, results have a unique shape. The program tries to achieve better candidates by searching right or left of the crowded area. The same goes for Figure 4c, replacing the efficiency in Figure 4b with average torque. For 6/4 SRM, the results represented by Figure 5 show the same features of optimization as in 8/6 SRM. Figure 6 shows the candidates of final generation with the three objective functions. The results shown confirm the accuracy of the search direction. Moreover, it indicates the diversification of the method used as it shows variety in objective functions' values.

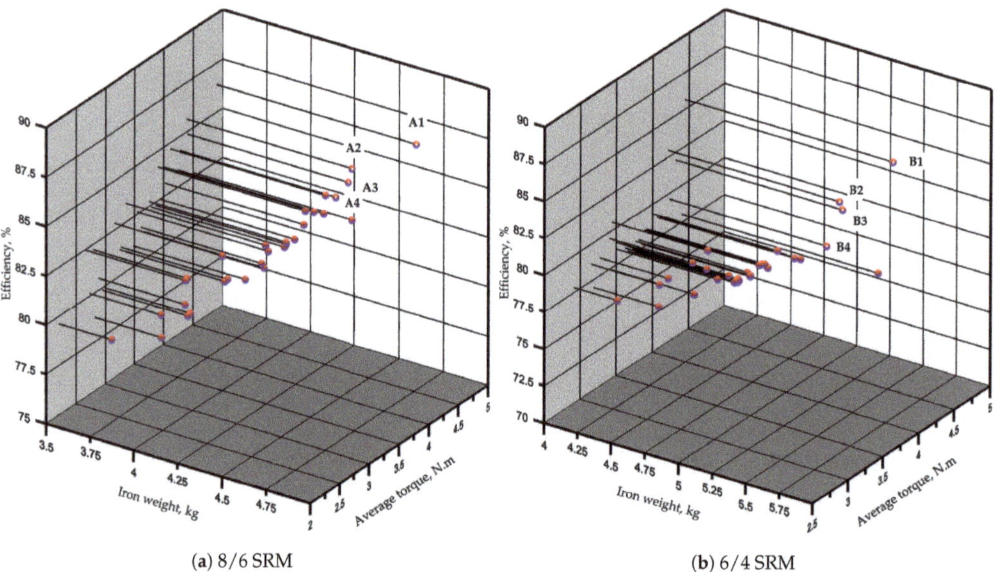

(a) 8/6 SRM   (b) 6/4 SRM

Figure 6. Objective functions 3D representation of the last generation (30 candidates).

The progress of optimization with generations for both 8/6 SRM and 6/4 SRM is shown in Figure 7. It can be seen that average torque and efficiency are maximized as more generations are produced and the iron weight is minimized at the same time.

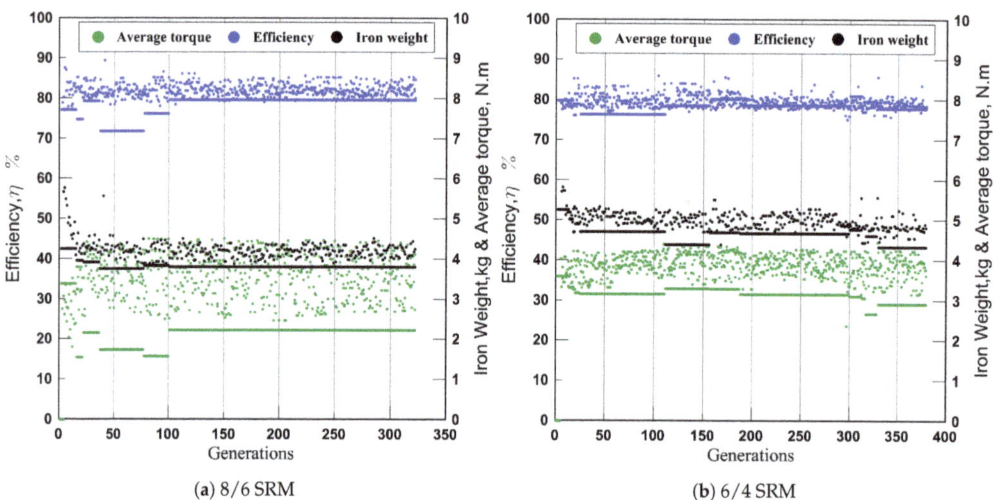

(a) 8/6 SRM   (b) 6/4 SRM

Figure 7. Objective functions progress with number of generations.

Table 4 shows the best selected candidates among the first non-dominated front (also known as rank 1). Four candidates are selected for each configuration (set A for 8/6 SRM and B for 6/4 SRM). A1 and B1 achieve the maximum values of average torque and efficiency among the final generation candidates. A4 and B4 achieve the minimum iron weight. A2, A3, B2 and B3 are compromise designs, which satisfy each objective function in a certain degree. A1 and B1 designs are selected from these candidates for further investigation. All designs of the selected sets in Table 4 are shown in Figures 4–6. The values of objective functions, parameters, dimensions and other details for selected sets are shown in Table 5. From Table 5, it can be seen that the variation of dimensions to produce better designs matches with the SRM design experience, which indicate the accuracy of calculation methods. For example, the difference between aligned inductance (L_a) and unaligned inductance (L_u) is higher in designs of higher average torque. This result matches with design experience as the energy conversion increases with higher difference of flux level between aligned and unaligned positions. This is indicated by Equation (12). Other important observations indicated in Table 5 should be highlighted and they are:

- For the same axial length wider rotor and stator poles result in higher flux. Hence, higher average torque. This is observed by values of β_s, β_r and T_{av} for all selected designs.
- Higher torque densities are achieved by 8/6 configuration (A) due to their higher number of phases and, hence, average torque.
- Efficiencies values are calculated for all candidates at 1000 rpm, which is a relatively low speed. Hence, there is not much of a difference between iron losses of 8/6 (A) and 6/4 (B) configurations. However, B1 design is proven to give higher efficiency values than that of A1 for a wide range of speeds as will be provided later. The same can be indicated for all designs of sets A and B by all meaning that designs in set B (6/4 SRM) would have higher efficiency values than those in set A (8/6 SRM) for higher values of speeds.
- Number of turns and inductance values are higher in B (6/4 SRM) due to wider stator and rotor poles as the same flux density is assumed for both.

Table 4. Candidates in first non-dominated front (rank 1).

Candidate	Configuration	Average Torque (N.m)	Efficiency (%)	Iron Weight (kg)
A1	8/6	4.90	86.4	4.63
A2	-	4.41	84.86	4.41
A3	-	4.4	85.65	4.44
A4	-	2.22	79.6	3.8
B1	6/4	4.45	85.66	5.56
B2	-	4.22	82.97	5.28
B3	-	4.32	81.89	5.24
B4	-	2.91	77.86	4.33

Further investigations are made on the selected designs A1 and B1. The torque is shown in Figure 8 for constant phase current from an unaligned position to aligned positions for selected optimal designs. Note that the peak torque for B1 (6/4 SRM) is higher than that of A1 (8/6 SRM) due to the higher difference between aligned and unaligned inductances in 6/4 SRM. However, the average value of A1 (8/6 SRM) torque is higher than that of B1 (6/4 SRM) due to the increased number of phases in 8/6 SRM configurations.

Table 5. Parameters and objective functions values of the selected optimal designs.

Value	A1	A2	A3	A4	B1	B2	B3	B4	Unit
T_{av}	4.9	4.41	4.4	2.22	4.45	4.22	4.32	2.91	N.m
η	86.4	84.86	85.65	79.6	85.66	82.97	81.89	77.86	%
Iron weight	4.63	4.41	4.44	3.8	5.56	5.28	5.24	4.33	kg
Torque density	8315	7848	7780	4589	6282	6238	6473	5283	N.m/m^3
T_{ph}	254	250	250	204	270	276	280	256	Turns
L_a	94.6	68	68.6	37.65	181.8	112.3	112.8	83.7	mH
L_u	13.15	14.2	14.5	9.75	18.8	18.9	18.8	17.2	mH
R_{ph}	1.69	1.66	1.66	1.36	1.781	1.46	1.48	1.35	Ω
D_o	130	130	130	130	130	130	130	130	mm
L	100	100	100	100	100	100	100	100	mm
D	73.77	64.32	63.83	54.88	72.35	70.76	72.26	57.86	mm
b_{sy}	5.14	5.18	5.16	5.15	6.58	5.92	5.84	5.25	mm
b_{ry}	7.28	6.17	6.24	5.9	11.03	9.8	10.4	8.34	mm
h_s	22.97	27.66	27.91	32.4	22.23	23.7	23.02	30.81	mm
h_r	17.1	13.49	13.18	9.02	12.63	13.08	13.2	8.07	mm
D_{sh}	24	24	24	24	24	24	24	24	mm
g	0.5	0.5	0.5	0.5	0.5	0.5	0.5	0.5	mm
β_s	16.6	17.26	17.5	13.76	31.5	30.66	30.31	25.85	degree
β_r	18.3	19.14	19.68	21.68	33.9	32.3	30.61	29.58	degree

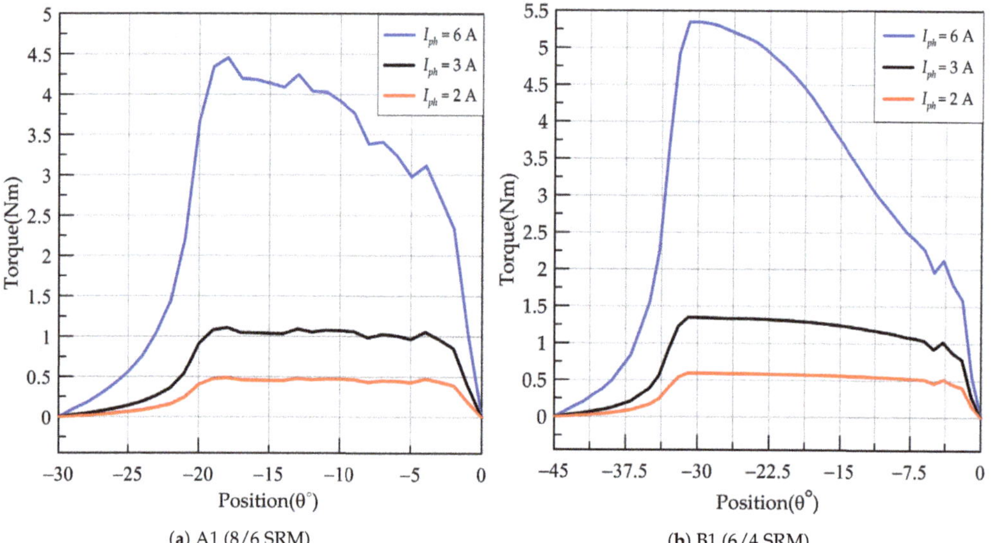

(**a**) A1 (8/6 SRM)

(**b**) B1 (6/4 SRM)

Figure 8. Developed torque of one phase for different excitation levels.

Figure 9 shows the magnetic flux of both selected designs. It can be seen that the value of flux density in the stator pole is about 1.8 T, which represents the knee point of B-H curve for industrial steel used. The stator yoke flux density is obviously higher than stator flux density for 6/4 SRM. This is because the stator yoke thickness does not impact the objective functions strongly, that is, it can be neglected. Hence, the optimization program tends to decrease it to a minimum to get less weight of iron. To achieve a good overall SRM design, other objective functions must be added such as torque ripples, acoustic noise … and so forth. When these functions are added, the program will not reduce the stator yoke thickness to a minimum as it influences other objective functions negatively (increases acoustic noise for example).

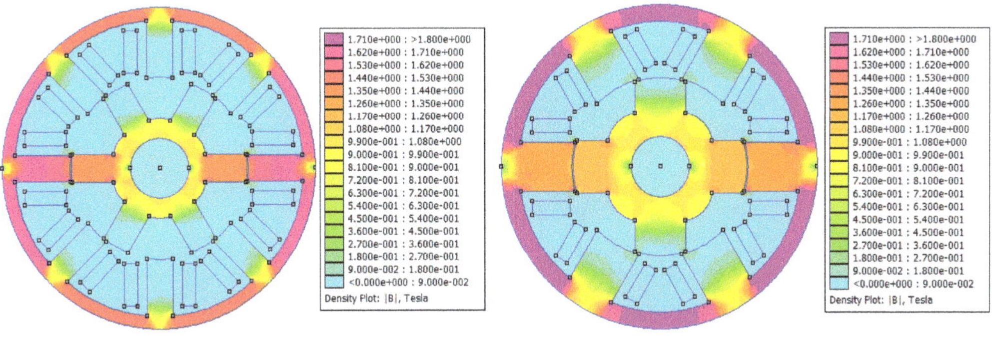

(**a**) A1 (8/6 SRM)   (**b**) B1 (6/4 SRM)

Figure 9. Flux density of selected optimal designs in an aligned position.

In Figure 10, the efficiency of optimal selected designs for a range of speeds up to 10th of rated speed is shown. It can be seen that the values of efficiency for selected optimal designs are almost identical to the value of speed before 1300 rpm. This result is due to the lower core losses in this region as shown in Figure 11. The program was given the rated speed of 1000 rpm to calculate core losses and efficiency and then seek better values at same speed. It can also be seen that 6/4 SRM has a better efficiency profile than 8/6 SRM over a wide speed range, which is expected due to higher core loss values as demonstrated in [14]. SRM with 8/6 configuration has a higher number of poles than 6/4 SRM and hence flux changes are higher, which leads to core losses. Figure 11 shows the core losses of 8/6 SRM is much higher than that of 6/4 SRM. Figure 12 shows a comparison between calculated and FEA waveforms of flux density (*B*) of all sectors for the chosen optimal designs. It can be seen that the results of the calculation are very close to FEA waveforms. The calculation method is used to produce flux density (*B*) waveforms and to then calculate eddy currents losses as introduced in Section 2.8.

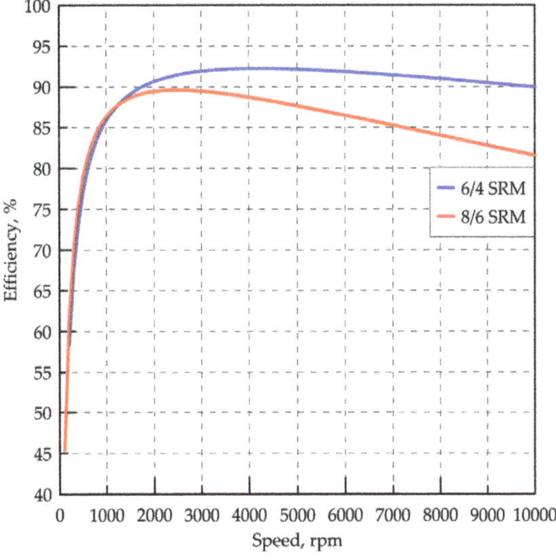

Figure 10. Efficiency of selected optimal designs A1 (8/6 SRM) and B1 (6/4 SRM) at different speeds.

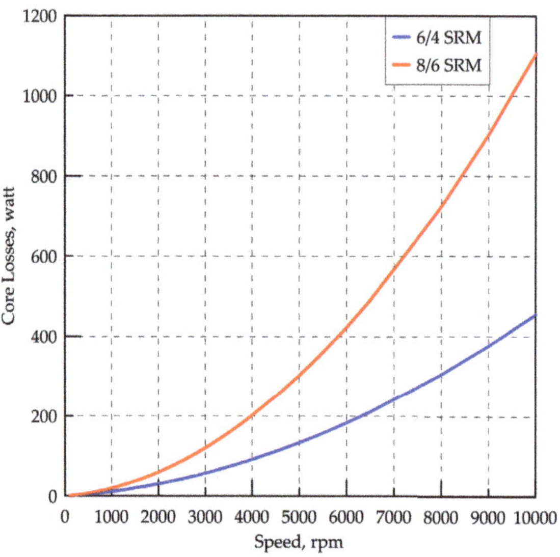

Figure 11. Core losses of selected optimal designs A1 (8/6 SRM) and B1 (6/4 SRM) at different speeds.

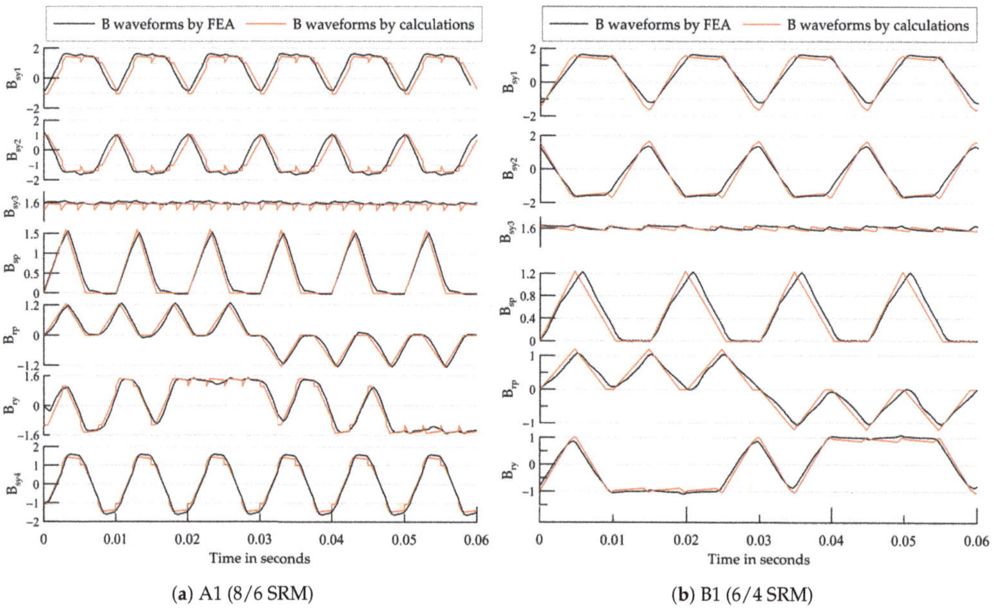

(a) A1 (8/6 SRM)

(b) B1 (6/4 SRM)

Figure 12. The flux density waveforms in all sectors of optimal SRM designs A1 (8/6 SRM) and B1 (6/4 SRM) for one revolution at 1000 rpm.

The proposed techniques in [16,17,20,27] are considered for the evaluation of the methodology presented by this paper, as they studied the same SRM configurations or multi objective optimization. The methodology presented in this paper achieves high accuracy in analysis due to the use of FEA. The optimization technique performance has also shown successful progress towards the optimal. In [17], the same technique is used for a specific application. The values of efficiency and torque density are higher than what is presented in this paper. This is due to the generality of the approach in this paper. The approaches in [20,27] are general, however, they use mathematical models in analysis to reduce computational time, which makes the process more complex and needs more analytic work before optimization starts. For torque density values, [27] achieves 1200–1580 N.m/m^3 while [17] reached 15,950–17,030 N.m/m^3. In this paper, torque densities of 7780–8315 N.m/m^3 and 6238–6473 N.m/m^3 are achieved for 8/6 and 6/6 configurations, respectively. Efficiency values are 75%–80%, 86%–91% and 80%–85% in [17,20,27], respectively, while in this paper efficiency values of 80%–86% are achieved for most of the design candidates. The approach proposed in this paper has shown success in optimization, as objective function values indicate. Also, it can be used for almost any application if the suitable objective functions are added.

6. Conclusions

This paper introduces a method of SRM design optimization by genetic algorithm. Non-dominated sorting multi-objective genetic algorithm (NSGA-II) is used for its high performance and intensification in optimization problems. Since the NSGA-II optimization technique provides optimal set of solutions (non-dominated front), the final decision is left to the designer to choose the most convenient design of optimal non-dominated front to be picked. FEA analysis is adopted in optimization process as it provides high accuracy. Core losses are calculated numerically based on flux density waveforms and hysteresis loops. Three objective functions T_{av} ,η and W_{iron} were chosen to be optimized. The results show the variation of variables to optimize these objectives only, regardless of other important considerations like torque ripples, acoustic noise and mechanical vibrations. This proves the success of the optimization program as a framework to optimize the specified objective functions. However, further work is required to include more objective functions and to use this framework for a specific applications.

Author Contributions: Funding acquisition, M.I. and H.R.; Methodology, M.E.-N.; Resources, M.E.-N.; Validation, M.I. and H.R.; Writing—original draft, M.A.; Writing—review & editing, M.E.-N. All authors have read and agreed to the published version of the manuscript.

Funding: This research received no external funding.

Data Availability Statement: Data available on request from the authors.

Conflicts of Interest: The authors declare no conflict of interest.

References

1. Li, S.; Zhang, S.; Habetler, T.G.; Harley, R.G. Modeling, Design Optimization and Applications of Switched Reluctance Machines—A Review. *IEEE Trans.Ind. Appl.* **2019**, *55*, 2660–2681. [CrossRef]
2. Krishnan, R. *Switched Reluctance Motor Drives, Modeling, Simulation, Analysis, Design, and Applications*; CRC Press: Boca Raton, FL, USA, 2001.
3. Besbes, M.; Gasbi, M.; Hoang, E.; Lecrivian, M.; Grioni, B.; Plasse, C. SRM design for starter-alternator system. *Proc. Int. Conf. Electr. Mach.* **2000**, *6*, 1931–1935.
4. Sugiura, M.; Ishihara, Y.; Ishikawa, H.; Naitoh, H. Improvement of Efficiency by Stepped-Skewing Rotor for Switched Reluctance Motors. In Proceedings of the 2014 International Power Electronics Conference, Hiroshima, Japan, 18–21 May 2014; pp. 1135–1140.
5. Gao, J.; Sun, H.; He, L. Optimization design of Switched Reluctance Motor based on Particle Swarm Optimization. In Proceedings of the 2011 International Conference on Electrical Machines and Systems (ICEMS), Beijing, China, 20–23 August 2011; pp. 1–5.
6. Lukman, G.F.; Hieu, P.T.; Jeong, K.-I.; Ahn, J.-W. Characteristics Analysis and Comparison of High-Speed 4/2 and Hybrid 4/4 Poles Switched Reluctance Motor. *Machines* **2018**, *6*, 4. [CrossRef]
7. Besmi, M.R. Geometry Design of Switched Reluctance Motor to Reduce the Torque Ripple by Finite Element Method and Sensitive Analysis.*J. Electr. Power Energy Convers. Syst.* **2016** , *1*, 23–31.
8. Mousavi-Aghdam, S.R.; Feyzi, M.R.; Bianchi, N.; Morandin, M. Design and Analysis of a Novel High Torque Stator-Segmented SRM. *IEEE Trans. Ind. Electron.* **2015**, *63*, 1458–1466. [CrossRef]
9. Hu, Y.; Ding, W.; Wang, T.; Li, S.; Yang, S. Investigation on a Multi-Mode Switched Reluctance Motor: Design, Optimization, Electromagnetic Analysis and Experiment. *IEEE Trans. Ind. Electron.* **2017**, *64*, 9886–9895. [CrossRef]
10. Bogusz, P.; Korkosz, M.; Prokop, J. A three-phase switched reluctance motor for a high-speed drive. *Electr. Eng. Electron.* **2016**. [CrossRef]

11. Kreyszig, E.; Kreyszig, H.; Norminton, E.J. Unconstrained Optimization.Linear Programming. In *Advanced Engineering Mathematics*, 10th ed.; John Wiley Inc.: Chichester, UK, 2011; pp. 950–969.
12. Mirjalili, S.; Dong, J.S. *Multi-Objective Optimization using Artificial Intelligence Techniques*; Springer: New York, NY, USA, 2019.
13. Minella, G.; Ruiz, R. A Review and Evaluation of Multiobjective Algorithms for the Flowshop Scheduling Problem. *Informs J. Comput.* **2008**, *20*, 451–471. [CrossRef]
14. Vijayraghavan, P. *Design of Switched Reluctance Motors and Development of a Universal Controller for Switched Reluctance and Permanent Magnet Brush-Less DC Motor Drives*; Virginia Tech: Blacksburg, VA, USA, 2001.
15. Wu, W.; Dunlop, J.B.; Collocott, S.J.; Kalan, B.A. Design Optimization of a Switched Reluctance Motor by Electromagnetic and Thermal Finite-Element Analysis. *IEEE Trans. Magn.* **2003**, *39*, 3334–3336. [CrossRef]
16. Xue, X.D.; Cheng, K.W.E.; Cheung, N.C. Multi-objective optimization design of in-wheel switched reluctance motors in electric vehicles. *IEEE Trans. Ind. Electron.* **2010**, *57*, 2980–2987. [CrossRef]
17. Anvari, B.; Toliyat, H.A.; Fahimi, B. Simultaneous Optimization of Geometry and Firing Angles for In-Wheel Switched Reluctance Motor Drive. *IEEE Trans. Transp. Electrif.* **2017**, *4*, 322–329. [CrossRef]
18. Hayashi, Y.; Miller, T.J.E. A New Approach to Calculating Core Losses in the SRM. *IEEE Trans. Ind. Appl.* **1995**, *31*, 5. [CrossRef]
19. Cui, X.; Sun, J.; Gan, C.; Gu, C.; Zhang, A.Z. Optimal Design of Saturated Switched Reluctance Machine for Low Speed Electric Vehicles by Subset Quasi-Orthogonal Algorithm. *IEEE Access* **2019**, *7*. [CrossRef]
20. Ma, C.; Qu, L. Multiobjective optimization of switched reluctance motors based on design of experiments and particle swarm optimization. *IEEE Trans. Energy Convers.* **2015**, *30*, 1144–1153. [CrossRef]
21. Zhu, Y.; Yang, C.; Yue, Y.; Wei, W.; Zhao, C. Design and optimization of an In-wheel switched reluctance motor for electric vehicles. *IET Intell. Transp. Syst.* **2019**, *13*, 175–182.
22. El-Nemr, M.K.; AI-Khazendar, M.A.; Rashad, E.M.; Hassanin, M.A. Modeling and Steady-State Analysis of Stand- Alone Switched Reluctance Generators. *IEEE Power Eng. Soc. Meet.* **2003**, *3*, 1894–1899.
23. Mirzaeian, B.; Moallem, M.; Tahani, V.; Lucas, C. Multiobjective optimization method based on a genetic algorithm for switched reluctance motor design. *IEEE Trans. Magn.* **2002**, *38*, 1524–1527. [CrossRef]
24. Ibrahim, M.N.; Sergeant, P. Prediction of Eddy Current Losses in Cooling Tubes of Direct Cooled Windings in Electric Machines. *Mathematics* **2019**, *7*, 1096. [CrossRef]
25. Torrent, M.; Andrada, P.; Blanque, B.; Martinez, E.; Perat, J.I.; Sanchez, J.A. Method for estimating core losses in switched reluctance motors. *Eur. Trans. Electr. Power* **2011**, *21*, 757–771. [CrossRef]
26. Emmerich, M.T.M.; Deutz, A.H. A tutorial on multiobjective optimization: fundamentals and evolutionary methods. *Nat. Comput.* **2018**, *17*, 585–609. [CrossRef] [PubMed]
27. Li, S.; Zhang, S.; Jiang, C.; Habetler, T.G.; Harley, R.G. A fast control-integrated and multiphysics based multiobjective design optimization of switched reluctance machines. In Proceedings of the 2017 IEEE Energy Conversion Congress and Exposition (ECCE), Cincinnati, OH, USA, 1–5 October 2017; pp. 730–737.

Article

Comparison of Flux-Switching and Interior Permanent Magnet Synchronous Generators for Direct-Driven Wind Applications Based on Nelder–Mead Optimal Designing

Vladimir Prakht [1,*], Vladimir Dmitrievskii [1], Vadim Kazakbaev [1] and Ekaterina Andriushchenko [2]

[1] Department of Electrical Engineering and Electric Technology Systems, Ural Federal University, 620002 Yekaterinburg, Russia; vladimir.dmitrievsky@urfu.ru (V.D.); vadim.kazakbaev@urfu.ru (V.K.)

[2] Department of Electrical Power Engineering and Mechatronics, Tallinn University of Technology, 19086 Tallinn, Estonia; ekandr@taltech.ee

* Correspondence: va.prakht@urfu.ru; Tel.: +7-343-375-45-07

Abstract: The permanent magnet flux-switching machine (PMFSM) is one of the most promising machines with magnets inserted into the stator. To determine in which applications the use of PMFSM is promising, it is essential to compare the PMFSM with machines of other types. This study provides a theoretical comparison of the PMFSM with a conventional interior permanent magnet synchronous machine (IPMSM) in the gearless generator of a low-power wind turbine (332 rpm, 51.4 Nm). To provide a fair comparison, both machines are optimized using the Nelder–Mead algorithm. The minimized optimization objectives are the required power of frequency converter, cost of active materials, torque ripple and losses of a generator averaged over the working profile of the wind turbine. In order to reduce the computational time, the substituting profile method is applied. Based on the results of the calculations, the advantages and disadvantages of the considered machines were revealed: the IPMSM has significantly lower losses and higher efficiency than the PMFSM, and the PMFSM requires much less rare-earth magnets and copper and is, therefore, cheaper in mass production.

Keywords: direct-drive; electric machine analysis computing; interior permanent magnet machine; mathematical model; optimal-design; permanent magnet flux-switching machine; wind generator

Citation: Prakht, V.; Dmitrievskii, V.; Kazakbaev, V.; Andriushchenko, E. Comparison of Flux-Switching and Interior Permanent Magnet Synchronous Generators for Direct-Driven Wind Applications Based on Nelder–Mead Optimal Designing. *Mathematics* **2021**, *9*, 732. https://doi.org/10.3390/math9070732

Academic Editor: Jinfeng Liu

Received: 17 December 2020
Accepted: 25 March 2021
Published: 29 March 2021

1. Introduction

Interior permanent magnet synchronous machines (IPMSMs) have been widely used in gearless generators of low-power wind turbines [1,2]. IPMSMs are used in direct-driven wind turbines with power ratings from a fraction of kW to over MW [3–5]. Alternatively, permanent magnet flux-switching machines (PMFSMs) can be used in this application, which has several advantages over IPMSM, such as simpler and more reliable rotor and higher specific torque at the same mass of magnets [6]. PMFSMs can also be used both in low-power wind turbines [7] and in wind turbines with a power rating of more than MW [8]. It is important for electrical generator designers to understand why permanent magnet flux-switching generators (PMFSGs), with their obvious benefits, are not as widely used in this application as the traditional machines with permanent magnets on the rotor.

There is a large volume of published studies that compares PMFSMs with synchronous machines (SMs). For instance, in [9] a comparison of PMFSMs (44 kW, 350 N·m, 1200 ÷ 6000 rpm) and IPMSMs with V-shaped magnetic poles (48 kW, 382 N·m, 1200 ÷ 6000 rpm) for traction applications is presented. A PMFSM has 12 teeth on the stator, 10 teeth on the rotor, and an IPMSM has eight poles on the rotor and 48 teeth on the stator and a distributed winding with the number of slots per pole and phase $q = 2$. Both machines have an external diameter of 269 mm and the stack length of 84 mm. It has been revealed that a PMFSM has higher efficiency (losses are 1.1 times lower) and the torque ripple is 2.25 times lower. However, the mass of rare-earth magnets used in a PMFSM is higher by 2.3 times. Additionally, the power factor at maximum output torque and based speed is higher for IPMSMs (0.92) than for PMFSMs (0.76). This is a key reason why PMFSMs produce lower power and torque when used with the same inverter. Thus, it has been shown that IPMSMs have lower costs and higher torques [9]. However, this study is not entirely reliable since a PMFSM is designed without optimization, while an IPMSM is optimized a priori.

In the case of traction applications of maximum torque 30 N·m and based speed 1200 ÷ 6700 rpm, the PMFSM has been compared with different SMs in [10–12]. In particular, the IPMSM with V-shaped magnetic poles, the surface permanent magnet synchronous machine (SPMSM) and the IPMSM with flux concentration have been considered. In these studies, the PMFSM has 12 teeth on the stator, 10 teeth on the rotor, and the SM has 10 poles on the rotor and 12 teeth on the stator. All machines have a concentrated winding, their external diameter is 134 mm, and their stack length is 90 mm. The authors have shown that the PMFSM has a torque ripple 1.4 times higher than the IPMSM with flux concentration, and 3.5 and 2.2 times higher than the SPMSM and IPMSM with V-shaped magnetic poles, respectively. At the same time, the cost of the magnets used in the PMFSM compared to the cost of the magnets used in the IPMSM with flux concentration, SPMSM and IPMSM with V-shaped magnetic poles is higher by 3, 2.4 and 2.4 times, respectively. On the other hand, authors have reported that the PMFSM has a wider constant power-speed range, while the IPMSM with flux concentration and the SPMSM rapidly lose power with increasing speed [12]. This conclusion contradicts other studies [13,14], where the IPMSM with flux concentration and the SPMSM have had a wider constant power-speed range in traction applications. Moreover, the studies [10–12] have a lack of data regarding the efficiency and power factor. Therefore, the discussed comparison of the machines is concluded to be incomplete.

Y. Pang et al. (2007) have compared the gearless traction PMFSM (100 W, 400 rpm) with the IPMSM with V-shaped magnetic poles. Overall, the maximum torque capability of the PMFSM is about 10% higher than that of the IPMSM [15]. Nevertheless, the paper does not contain any information regarding the efficiency of the machines, cost of magnetic materials, their power factor and torque ripple.

Besides, an SPMSM with 14 poles and 12 slots on the stator has been compared analytically with a PMFSM with 14 teeth on the rotor and 12 slots on the stator [16]. Both machines have an external diameter of 480.6 mm and the stack length 120 mm. It has been shown that the PMFSM has a higher torque and less overheating. However, the paper has a lack of data regarding the efficiency and power factor. Moreover, the comparison is carried out without the design optimization of the machines.

Previously mentioned studies have been dedicated to the comparison of PMFSMs with SMs working as a motor. However, there is little published data on the comparison of PMFSMs with SMs working as a generator.

A study has been carried out in the comparison of a high-speed PMFS generator (PMFSG) and SPMS generator (SPMSG) [17]. Both machines have an external diameter of 540 mm and the stack length of 80 mm. The PMFSG has 12 and 10 slots on the stator and rotor, respectively. The SPMSG has 20 poles and 24 slots on the stator and a concentrated winding with the number of slots per pole and phase $q = 2/5$. The air gap in the PMFSG is 2 mm, and in the SPMSG—1 mm. To avoid mechanical damaging and ensure the reliability of the SPMSG, a thick retaining ring made from carbon fiber is required, which further increases the complexity and the cost of the rotor manufacturing. Moreover, the use of the retaining ring increases the effective air gap; subsequently, magnets on the rotor are required to be thicker. The study has revealed that the PMFSM has lower copper losses at the same torque. Nevertheless, the study has a lack of data regarding the efficiency and power factor, as well as the torque ripple and mass of magnetic materials.

PMFSGs and SPMSGs have been compared theoretically as well [6]. Three-phase machines have been considered in gearless wind generator application (15 rpm, 1900 kN·m). The PMFSG has 48 teeth on the stator and 56 teeth on the rotor, and the SPMSG has 225 teeth on the stator and 150 poles on the rotor. The generators have approximately the same external radius of 2.6 m and the stack length of 1.9 m. It has been shown that the PMFSG has higher efficiency, but considerably lower power factor (0.69, while the SPMSG has 0.96). Therefore, the PMFSG requires a frequency converter with 40% more power than the SPMSG. Additionally, the PMFSG requires 1.27 times more usage of magnetic material than the SPMSM. The authors have chosen magnets with residual induction (1.3 T) for the PMFSG and cheaper magnets with less residual induction (1.23 T) for the SPMSG. Still, the paper does not contain any information regarding the torque ripple and magnetic losses, which can be significantly higher in the PMFSG. Besides, a rated voltage of 400 V has been chosen, which is hardly possible for generators with a power of more than 2 MW. Particularly, the literature has shown a successful application of a rated voltage of 600–1000 V [18,19], since there is no industrial 2.5 kA transistor and a voltage of 400 V.

In [20], 12-phase PMFSGs and SPMSGs in gearless wind generator application (16 kW, 500 rpm) are considered. The PMFSM has 24 teeth on the stator and 22 teeth on the rotor. The SPMSG has the best characteristics, with 48 teeth on the stator and 44 poles on the rotor. The generators have the same external radius of 163 mm and the stack length of 185 mm. The study has demonstrated that both machines have a low torque ripple (less than 1.5%). At the same time, the magnetic material used in the PMFSG has a 1.15 times bigger mass. Besides, the rated efficiency, torque and power output of the SPMSG are higher.

Nevertheless, previous studies that cover the comparison of PMFSGs and interior permanent magnet synchronous generators (IPMSGs) for wind generator application [6,20] did not use any formal optimization techniques (e.g., genetic algorithms or the Nelder–Mead algorithm) for the machine design. Applying the optimization methods for the PMFSG and the IPMSG will help to better estimate the potential of these generators and their advantages and disadvantages. Therefore, this research aims to optimize PMFSGs and SPMSGs and provide their comparison.

Various approaches have been described in the literature for optimizing PMFSGs and IPMSGs. In [3], procedures for the manual optimization of efficiency and torque ripple are described for a 4 kW IPMSG. In [5], a multicriteria optimization of an IPMSG with a power rating of 0.6 kW using the Taguchi method is described. Efficiency, EMF THD (total harmonic distortion of back electromotive force), and EMF amplitude were chosen as optimization criteria. In [21], the optimization of the average torque of the PMFSG with a power rating of 5 kW using the random optimization is described. In [22], a multi-criteria optimization of a PMFSG with a power rating of 1.5 kW was described using the particle swarm optimization method. The generator weight and EMF THD were chosen as the optimization criteria. The optimization of PMFSGs with rare earth [7] and ferrite magnets [23] using the Nelder–Mead method was also described. In [7,23], the objective optimization function was selected to reduce losses, torque ripple, and the required power of the semiconductor inverter. In [24], a comparison was made of PMFSGs with rare-earth and ferrite magnets developed in [7,23]. However, no comparison has been made between PMFSGs and IPMSGs with rare-earth magnets when using the Nelder–Mead method.

This paper considers three-phase PMFSGs and IPMSGs with V-shaped magnetic poles in gearless wind generator application (332 rpm, 1784 W). The PMFSG has 24 teeth on the stator and 22 teeth on the rotor. The IPMSG has 24 teeth on the stator and 20 poles on the rotor. The generators have the same external radius of 80 mm and the stack length of 100 mm. The machines are designed and optimized using the approach described in the previous papers [7,23] based on the Nelder–Mead algorithm with the optimization method, the mathematical models of the machines, and the substituting profile method. An important advantage of the Nelder–Mead method over other methods that are often used to optimize electrical machines [25,26] is the significant savings in computational time [23].

Using substituting profiles instead of initial ones reduces the calculation efforts, which is extremely significant for the optimization of electric machines. The optimization objectives are to minimize the power required of the frequency converter, the cost of active materials, the torque ripple, and the average generator losses for the working profile of the wind turbine. Additionally, optimization is carried out using a substituting two-mode load profile that considerably reduces the computational time.

Section 2 outlines the general design parameters of the PMFSG and IPMSG, such as the number of stator slots and the number of poles. Section 3 describes the application of the substituting load profile method to reduce the computational time in design optimization. Section 4 describes the optimization criteria and procedures for the PMFSG and IPMSG. Initial geometry parameters are also listed in this section. Section 5 compares the calculated characteristics of the PMFSG and the IPMSG before optimization. Section 6 compares the calculated characteristics of the PMFSG and the IPMSG after optimization, and also lists the optimized geometry parameters. Section 7 summarizes the general findings of the comparative study.

2. General Design Parameters of PMFSG and IPMSG

Figure 1 shows an IPMSG with V-shaped magnetic poles (a) and a PMFSG (b). The stator of the IPMSG has 24 teeth with half-open slots. The rotor is made of electrical steel laminations and has V-shaped slots where rare-earth magnets are inserted. The number of poles is $2p = 20$, and each pole includes two magnets, where p is the number of pole pairs in the rotor. Concentrated stator winding has the number of slots per pole and phase $q = 4/10$.

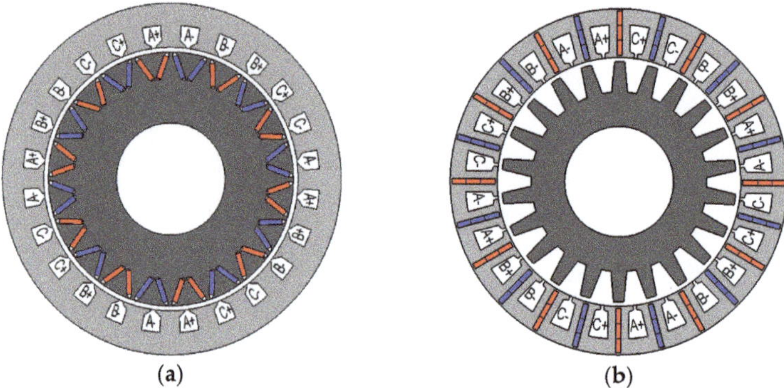

(a)   (b)

Figure 1. Sketch of the machine geometry: (**a**) interior permanent magnet synchronous generator (IPMSG); (**b**) permanent magnet flux-switching generator (PMFSG).

The supply frequency is $f_{e,IPM} = p \times f_m/60$ Hz, where f_m is the mechanical rotational speed, rpm. The stator of the PMFSG has 24 teeth with half-open slots, while the magnets are inserted in each stator tooth. The magnets located in the nearest slots are magnetized in opposite directions. The stator integrity is ensured by thin ribs at the outer and inner stator surfaces. Each second stator tooth is wound. Each magnet is divided into three insulated parts to reduce the eddy currents losses in the magnets. The toothed rotor is made of electrical steel laminations and has no magnets and winding. The supply frequency $f_{e,FSM} = Z_r \times f_m/60$, where $Z_r = 22$ is the number of rotor teeth. Consequently, the required supply frequency is 2.2 times higher for the PMFSG than for the IPMSG at the same mechanical rotational frequency.

3. Two-Mode Substituting Load Profile of Wind Turbine

In this study, minimizing the loss averaged over the load profile of the wind generator is one of the objectives of the optimization because it results in a higher efficiency machine over the entire load range. In order to reduce the calculation time of the losses without considerable effect on the accuracy, this study applied an approach for designing a gearless generator for wind turbines using a two-modes substituting load profile. This approach has been described in detail while used for the design of a PMFSG with ferrite magnets [7,23].

The authors considered PMFSGs and IPMSGs designed for the wind turbine that operated at the wind speed range from 4 to 12 m/s [27]. Table 1 provides the data on the rotational speed n_i, mechanical power $P_{mech\,i}$, and torque T_i of the turbine at different operating points (modes). Data on the wind speed V_i and statistical probability of an i-th mode p_i are also presented.

Table 1. Initial nine-modes profile of the wind generator. n_i: rotational speed; $P_{mech\,i}$: mechanical power; T_i: torque; V_i: wind speed; p_i statistical probability of an i-th mode.

Mode, i	V_i, MPS	n_i, rpm	$P_{mech\,i}$, W	T_i, N·m	p_i
1	4	111	82	7.02	0.134
2	5	140	142	9.69	0.144
3	6	163	237	13.9	0.146
4	7	196	362	17.6	0.138
5	8	221	542	23.4	0.124
6	9	247	761	29.4	0.107
7	10	276	1038	35.9	0.087
8	11	308	1383	42.9	0.069
9	12	332	1784	51.5	0.051

An annual wind speed distribution was approximated by the one-parameter Rayleigh distribution [28], while the average speed was taken as 7 m/s. As long as the considered wind speeds of the modes were integers, the probabilities of the modes were assumed to be equal to the Rayleigh distribution density at the given wind speed. The values p_i were formed by normalizing the probabilities of the nine modes, in a way that they got summed up as 1. The torque provided in Table 1 was interpolated by a cubic polynomial as a function of the mechanical power P_{mech} using the least-squares method ("cftool", Matlab toolbox) with the adjusted R^2 equal to 0.9997 and SSE (Error Sum of Squares) equal to 0.3642 (Figure 2). The interpolated dependence is as follows:

$$T = 4.73 \cdot 10^{-9} P_{mech}^3 - 1.904 \cdot 10^{-5} P_{mech}^2 + 0.04571 \cdot P_{mech} + 3.571. \tag{1}$$

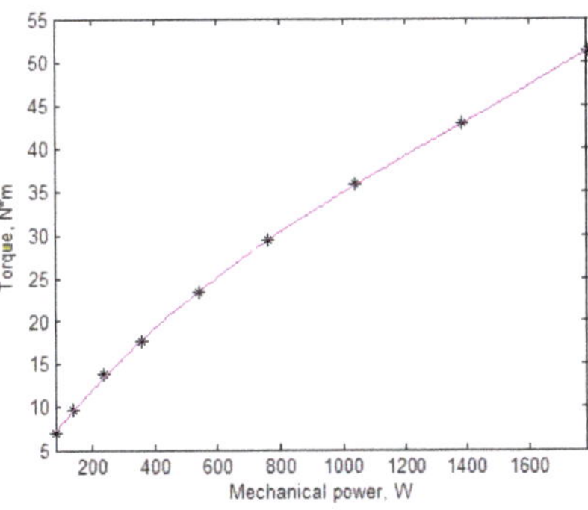

Figure 2. Loading torque profile of the wind turbine. Individual points of the load profile from Table 1 are marked with asterisks. The polynomial interpolation dependence is shown with the solid line.

In Figure 2, the obtained relation between the torque and mechanical power is presented. This interpolated dependence allows one to consider various mode characteristics (the torque, the losses, the torque ripple, etc.) as a function of the only argument P_{mech}. Then, the substituting load profile was created in a way that the means $< P_{mech} >, < P_{mech}^2 > \ldots < P_{mech}^n >$ were coincident with the means of the original profile [7]. Operator $< \ldots >$ denotes a (weighted) averaged over all points of a selected load profile. Therefore, the means of the polynomial interpolated values over the original profile and substituting load profile were coincident. Additionally, the loading mode with the highest power was included in the substituting mode. The loading mode determined the required converter power used as one of the optimization objectives. The rest of the modes of the substituting profile were characterized by two parameters: probability and mechanical power, which determined the point on the turbine's maximum power curve (Figure 2). Generally, m-mode substituting load profile, $(m-1)$ modes can be adjusted, and the polynomial degree becomes $2 \cdot (m-1)$. For instance, when the values can accurately be interpolated by quadric polynomial, the two-mode substituting profile is sufficient. The substituting two-mode load profile of the wind turbine is presented in Table 2. The proposed two-mode load profile significantly decreased the computational time without considerable effect on the accuracy of calculations [7].

Table 2. Substituting two-mode profile.

Mode, i	n_i, rpm	$P_{mech\,i}$, W	T_i, N·m	p_i
1	194	362	17.8	0.873
2	332	1784	51.4	0.127

4. PMFSG and IPMSG Optimization Models

4.1. Optimization Objectives

The mathematical models of both machines, based on solving 2D magnetostatic boundary problems, were used for the optimization. To reduce the computation time, the model took into account the motors' instantaneous symmetry and the periodicity of the electromagnetic processes. The optimization model of the PMFSG consisted of three objectives that were joined into one objective function [7]:

$$F = K_1 \cdot K_2{}^2 \cdot K_3{}^{0.5} \tag{2}$$

where K_1, K_2, and K_3 are the optimization objectives. Particularly, $K_1 = <P_{loss}>$ represents the losses averaged over the substituting load profile and annual energy production. K_2 is the required power of the frequency converter:

$$K_2 = \sqrt{3} \cdot I_{ampl,rated} \cdot V_{DC,rated}/2 \tag{3}$$

where $I_{ampl,rated}$ and $V_{DC,rated}$ are the current amplitude and the DC voltage of the frequency converter required at the rated conditions ($T = 100\%$, $n = 100\%$), respectively. At these conditions, K_1 and K_2 reach their peak. The objective K_2 is coincident with the apparent power when the current and voltage are sinusoidal and symmetric. Moreover, K_2 includes $V_{DC,rated}$, which is the converter limitation. It is important to notice that $V_{DC,rated}$ should not include the losses. The reason is that if the active power decreases, K_2 would be reduced and could be considered as a gain in the objective function. K_3 represents the cost of magnetic material:

$$K_3 = L \cdot (h_{mag} + 0.001 \text{ m}) \cdot l_{mag} \tag{4}$$

where $l_{mag} = R_1 - R_3 - 2 \cdot \Delta_3$ is the radial length of the magnets, h_{mag} is their thickness, R_1 is the outer stator radius; R_3 is the inner stator radius; Δ_3 is the thickness of the outer and inner stator ribs holding the magnets (see Figure 3) and L is the stack length. K_3 is to decrease the cost of the magnets and similar to the volume, i.e., equal to the product of three dimensions of the magnets, but the magnet thickness is increased by 0.001 m (meters) in (4) to take into account that thin magnets cost more than thick ones [29].

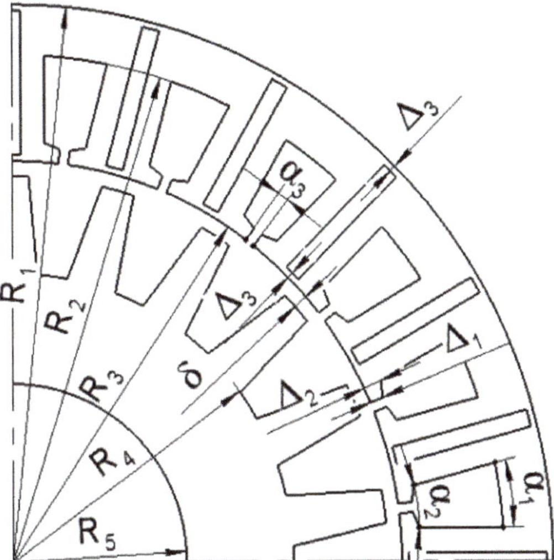

Figure 3. PMFSG parameters.

The power of K_2 and K_3 in (2) represents the percentage of the decrease in K_1 as valuable as the decrease in K_2 and K_3 by 1%. Particularly, the change in K_2 by 1% is as valuable as the change in K_1 by 2%, and the change in K_3 by 1% is as valuable as the change in K_1 by 0.5%. Basically, the higher the exponent of the objective, the more valuable the objective is.

It is worth mentioning that although the torque ripple of the initial design of the IPMSM was not very high, it was much higher than that of PMFSG. Therefore, an additional element K_4 equal to the torque ripple averaged over the substituting profile was added in (2) to construct the objective function used for the IPMSM optimization:

$$F = K_1 \cdot K_2^2 \cdot K_3^{0.5} \cdot K_4^{0.25} \tag{5}$$

where K_4 is the relative torque ripple averaged over the substituting profile. The factors "1", "2", "0.5", and "0.25" were not calculated by any formal method. The values simply indicate the approximate relative importance of each of the K1–K4 optimization objectives, according to the authors' experience in the design of similar machines.

It is important to highlight that functions (2) and (5) had noise due to round errors coming from FEM (Finite Element Method) calculations and differences in the grid for multiple calls of the objective function. Therefore, the Nelder–Mead method was chosen since it is able to deal with objective functions with noise [30].

The Nelder–Mead algorithm is well known [31] and is included in the basic MATLAB software package (function "fminsearch"). The unconstrained one-criterion Nelder–Mead method is applied in this work to optimize the PMFSG design. After the initial simplex was built, the Nelder–Mead algorithm as it is described in [31] was applied. The reflection, expansion, and contraction coefficients are 1, 2, and 1/2, respectively. The used generator model based on the FEM is described in [7].

4.2. Optimization Parameters of PMFSG

Figure 3 shows the main dimensions of the PMFSG. Additionally, Table 3 shows the geometric parameters fixed during optimization, while Table 4 provides the geometric parameters varied during the optimization and their initial values.

Table 3. Fixed parameters of the PMFSG. t_s: the angular tooth pitch of the stator.

Parameter	Value	Units
Outer radius of the stator R_1	80	mm
Air gap δ	0.35	mm
Rotor slot depth $(R_3 - \delta - R_4)$	9	mm
Rotor yoke thickness $(R_4 - R_5)$	5	mm
Stator ribs thickness Δ_3	500	μm
Remanent flux density of the magnets	1.2	T
Angular size of the stator slot opening α_3	0.162	t_s
Rotor and stator core length L	100	mm

Table 4. Initial values of the PMFSG parameters varied during the optimization. t_r: the angular tooth pitch of the rotor.

Parameter	Value	Units
Radius of the stator slot bottom R_2	74	mm
Inner stator radius R_3	61.7	mm
Angular size of the stator slot α_1	0.45	t_s
Angular size of the stator slot α_2	0.5	t_s
Magnet's thickness	2	mm
Angular size of the rotor tooth surface facing the air gap	0.265	t_r
Current angle in the rated mode	0.1	electrical radians

In order to reduce the reactive power, the current angle was varied. The current angle was supposed to be proportional to the torque. The zero current angle corresponds to the maximum torque when the current does not influence the steel saturation.

In Tables 3 and 4 and below, t_s is the angular tooth pitch of the stator; t_r is the angular tooth pitch of the rotor.

4.3. Optimization Parameters of IPMSG with V-Shaped Magnetic Poles

Figure 4 shows the main dimensions of the stator and rotor of the IPMSG. Table 5 shows the parameters fixed during the optimization and Table 6 reports the parameters varied during the optimization and their initial values. It is important to mention that during the optimization, the slot opening remained the same and was determined by the winding technology. The geometry of the rotor was created by the following procedure. Firstly, point 1 was set at the angle α in the circle with the radius R' remoted from the outer rotor boundary. D and d_2 defined the position of point 2. Point 5 was located in a distance of d_5 from the outer rotor boundary and on the same radius as point 4. Segment 1–4 (the magnet thickness) was perpendicular to segment 1–2, whose length was chosen to obtain the fixed values d. Point 7 was placed at the same circle with point 5, and the angular distance between these points was fixed (see Figure 4b).

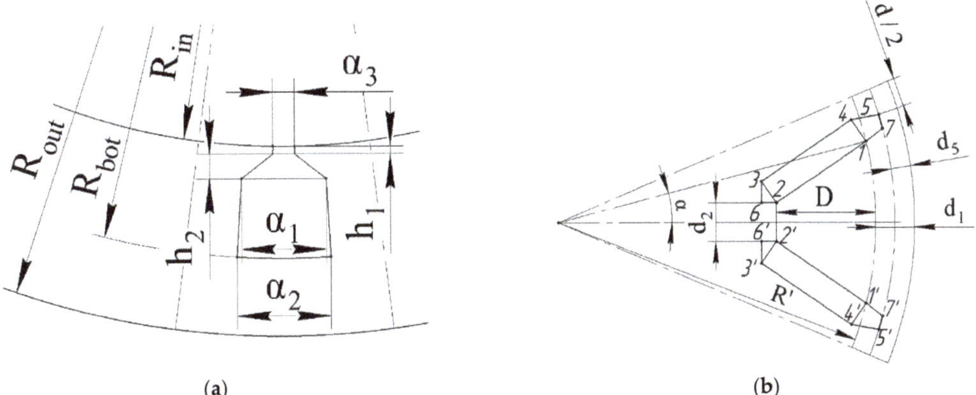

Figure 4. IPMSG parameters: (**a**) Stator parameters; (**b**) Rotor parameters.

Table 5. Fixed parameters of the interior permanent magnet synchronous machine (IPMSM).

Parameter	Value	Units
Outer stator radius R_{out}	80	mm
Parameter of the stator slot h_1	0.7	mm
Parameter of the stator slot h_2	2.3	mm
Stator slot thickness α_3	2	mm
Parameter of the rotor d	2	mm
Air gap δ	0.35	mm
Parameter of the rotor d_1	2	mm
Thickness of the rotor ribs d_5	1	mm
Angular distance between points 5 and 7	0.02	degrees

Table 6. Initial values of the IPMSM parameters varied during the optimization.

Parameter	Value	Units
Radius of the stator slot bottom R_{bot}	74	mm
Stator inner radius R_{in}	60	mm
Stator tooth thickness α_1	7.1	degrees
Stator tooth thickness α_2	7.1	degrees
Parameter of the rotor D	10	mm
Parameter of the rotor α	5.73	degrees
Current angle at the rated mode	0.3	electrical radians

5. PMFSG and IPMSG Comparison before Optimization

The initial geometry and magnetic flux density at the rated mode of the IPMSG and PMFSG are shown in Figure 5a,b, respectively. The calculation of the cost of active materials was based on the assumption that the steel price was USD 1/kg, the copper price was USD 7/kg, and the permanent magnet price was USD 126.6/kg [29,32]. Table 7 provides the main characteristics of the PMFSG and IPMSG initial designs.

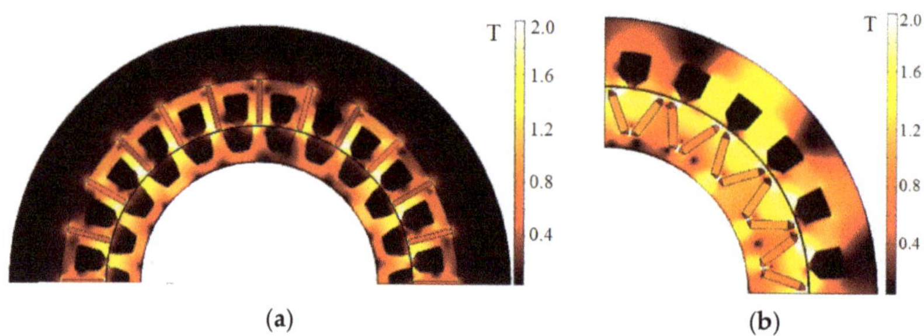

Figure 5. Initial design and magnetic flux density at the rated mode: (**a**) PMFSG; (**b**) IPMSG.

Table 7. Characteristics of the PMFSG and IPMSG initial designs.

Specification	PMFSG	IPMSG	Units
Rated mechanical power on the shaft (332 rpm, torque 51.4 N·m)	1784	1784	W
Rated active output power	1314	1338	W
Required converter power	1973	1867	V·A
Active output power at low load (194 rpm, moment 17.8 N·m)	305	305	W
Efficiency at rated speed	73.3	75.0	%
Efficiency at low load	83.7	84.4	%
Average efficiency	79.4	79.8	%
Average losses K_1	112	109	W
Torque ripple at rated mode	1.6	9.8	%
Torque ripple at low load	3.5	15.7	%
Cogging torque	0.6	2.7	N·m
Mass of magnets	0.61	0.79	kg
Mass of copper	1	0.81	kg
Mass of stator core	3.88	5.43	kg
Mass of rotor core	2.17	3.34	kg
Active material mass	7.66	10.38	kg
Cost of active materials	90.3	114.5	$

6. Optimization Results and PMFSG and IPMSG Comparison after Optimization

Figures 6–8 show the changes in the efficiencies, the required converter powers, and the objective function value during the process of the optimization. The left figures (a) are for flux-switching machines (FSMs). The right figures (b) are for interior permanent magnets (IPMs). Objective function (2) was used for the PMFSG optimization. Objective function (5) was used for the IPMSG optimization. It is seen that the optimization goes through the very bad solution with very high converter power and very low efficiency and objective function value. However, there are trends of the efficiency growth and the decline in the required converter power. The maximum efficiencies of both FSMs and in both modes are achieved approximately at the middle of the optimizations. These efficiencies are a little bit lower in the final optimization point because the optimization is a compromise between the optimization objectives.

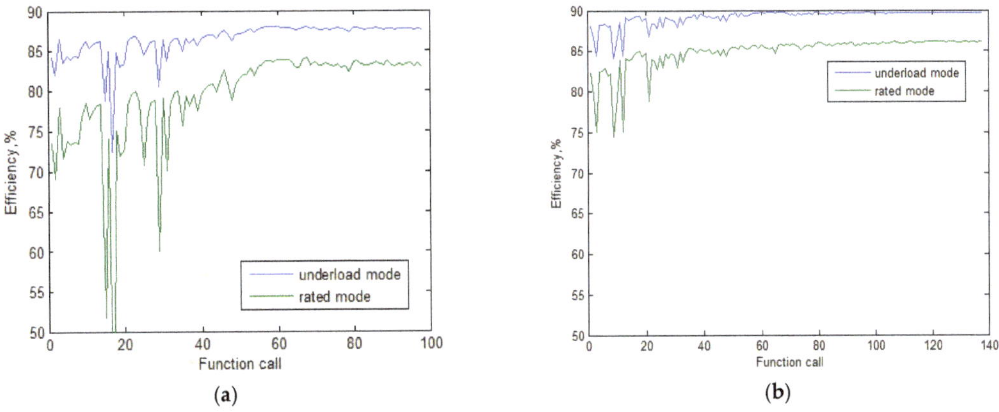

Figure 6. The change in efficiency during optimization. (**a**) PMFSG; (**b**) IPMSG.

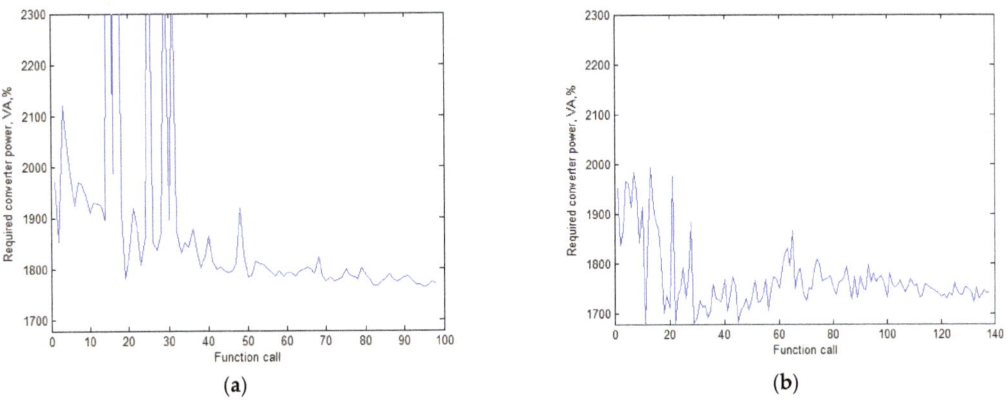

Figure 7. The change in the required converter power during optimization. (**a**) PMFSG; (**b**) IPMSG.

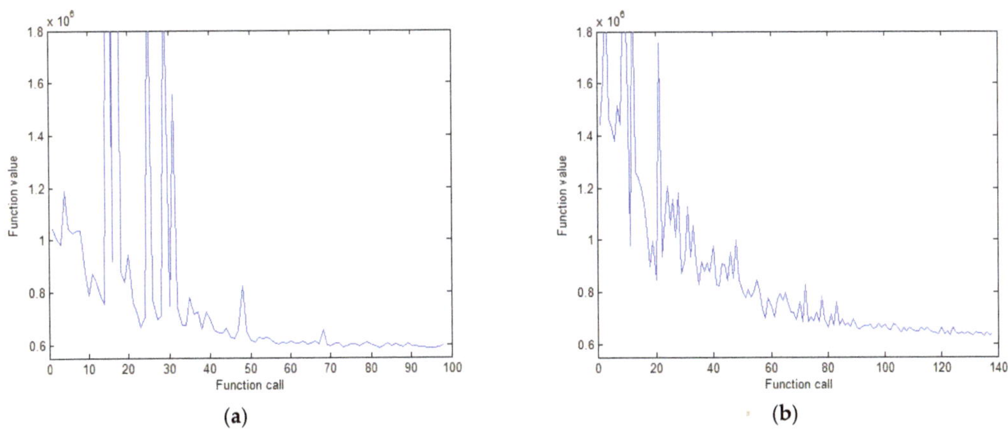

Figure 8. The change in the objective function call during optimization. (**a**) PMFSG; (**b**) IPMSG.

The optimized designs of IPMSGs and PMFSGs and the magnetic flux density at the rated mode are shown in Figure 6a,b, respectively. The optimized values of the varied parameters are given in Tables 8 and 9. Although the stator inner radius of the PMFSG is greater than that of the IPMSM by 19%, the moment of inertia of the active materials of the PMFSG rotor is more than thrice as low. It is because the PMFSG rotor core is thin and has thick teeth.

Table 8. Optimized values of the varied parameters of the PMFSG.

Parameter	Value	Units
Radius of the stator slot bottom R_2, mm	77.2	mm
Inner stator radius R_3, mm	63.3	mm
Angular size of the stator slot α_1	0.28	t_s
Angular size of the stator slot α_2	0.558	t_s
Magnet's thickness, mm	2.41	mm
Angular size of the rotor tooth surface facing the air gap	0.324	t_r
Current angle in the rated mode	0.086	electrical radians

Table 9. Optimized values of the varied parameters of the IPMSM.

Parameter	Value	Units
Radius of the stator slot bottom R_{bot}	76.5	mm
Stator inner radius R_{in}	53.2	mm
Stator tooth thickness α_1	6.95	degrees
Stator tooth thickness α_2	7.45	degrees
Parameter of the rotor D	11.6	mm
Parameter of the rotor α	5	degrees
Current angle at the rated mode	0.31	electrical radians

Table 10 reports the main characteristics of PMFSGs and IPMSGs after optimization. During the optimization, the torque ripple of both motors reduces significantly. The optimized values of the varied parameters are given in Tables 8 and 9. Figure 6a presents the dependence of total losses on mechanical power for both machines in the optimized condition. Figure 6b shows the electric losses in different parts of the machines at the rated conditions. It can be clearly seen that the most considerable losses for both machines were the copper losses. At the same time, the copper losses, as well as the rotor steel losses, in the IPMSG were noticeably lower.

Table 10. Characteristics of the optimized PMFSG and IPMSG.

Specification	PMFSG	IPMSG	Units
Rated mechanical power on the shaft (332 rpm, 51.4 N·m)	1784	1784	W
Rated active output power	1488	1538	W
Required converter power	1973	1867	V·A
Active output power at low load (194 rpm, moment 17.8 N·m)	318	325	W
Efficiency at rated speed	83.4	86.2	%
Efficiency at low load	87.8	89.7	%
Average efficiency	85.4	87.7	%
Average losses K_1	79.0	66.5	W
Torque ripple at rated mode	1.3	3.9	%
Torque ripple at low load	2.4	1.8	%
Cogging torque	0.4	0.3	N·m
Copper losses at rated mode	262.5	221.9	W
Stator core losses at rated mode	22.9	21.8	W
Rotor core losses at rated mode	10.3	2.3	W
Magnets Losses at rated mode	0.5	0.4	W
Mass of the magnets	0.674	0.924	kg

Table 10. *Cont.*

Mass of the copper	1.06	2.18	kg
Mass of the stator core	3.26	5.25	kg
Mass of the rotor core	2.46	2.87	kg
Momentum of inertia of the rotor active materials	7.4	24.2	g·m^2
Active material mass	7.45	11.2	kg
Cost of active materials	98.5	140.4	$
Objective function F	$5.90 \cdot 10^5$	$6.37 \cdot 10^5$	$\text{W}^3\text{·m}^{1.5}$

What stands out in Table 8 that the average losses in the IPMSG are 1.2 times lower than in the PMFSG. The active output power of the IPMSG is higher both at the rated mode and at the low load mode of the two-point profile. Moreover, the IPMSG required a frequency converter with less power than the PMFSG. Talking about the advantages of the PMFSG, it is important to highlight that the torque ripple at the rated mode in the PMFSG was three times lower than in the IPMSG, and the mass of the magnets is lower by 1.4 times. The optimized geometry and magnetic flux density at the rated mode of the IPMSG and PMFSG are shown in Figure 9a,b, respectively.

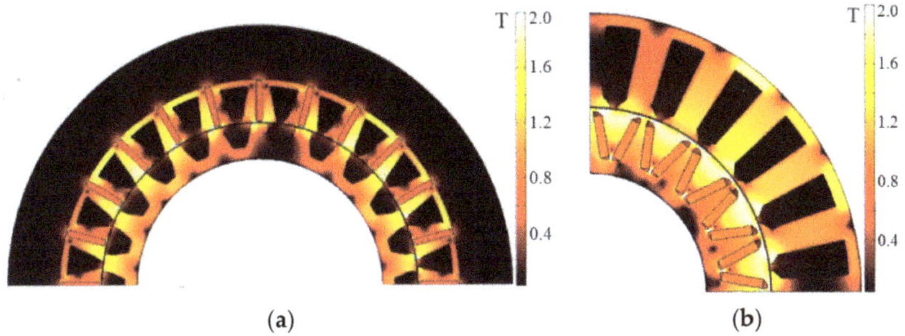

Figure 9. Optimized design and magnetic flux density at the rated mode: (a) PMFSG; (b) IPMSG.

The losses in the PMFSG (blue asterisks) and the IPMSG (red asterisks) at nine modes of the profile presented in Table 1 are shown in Figure 10a. The losses in the PMFSG are higher than those in the IPMSG in a wide range of powers.

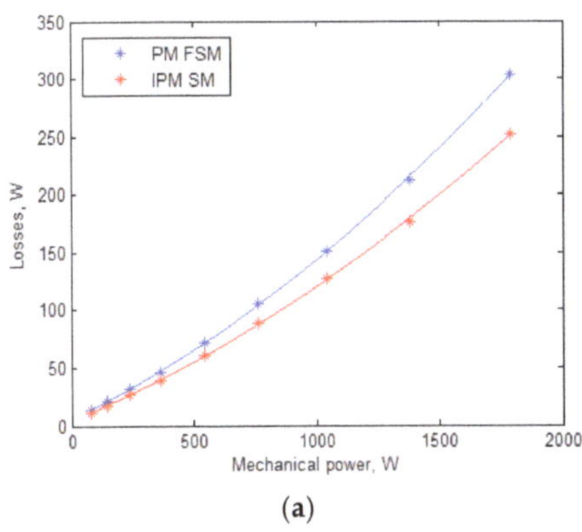

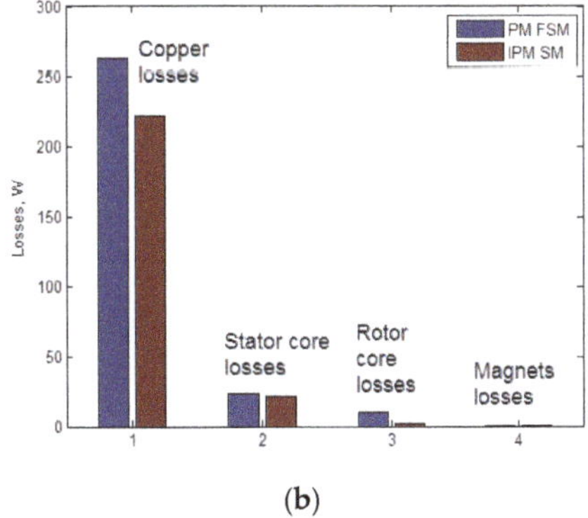

Figure 10. Comparative dependencies: (a) dependence of total losses on mechanical power of PMFSG and IPMSG; (b) electric losses of PMFSG and IPMSG at the rated conditions.

The two-profile approach is developed for the cases when the dependence of these losses on the mechanical power can be approximated by a quadric polynomial [7]. Such approximations shown in Figure 10a with continuous lines are rather accurate ($R^2 = 0.9998$ for both cases), which justifies the accuracy of substituting the initial nine-point profile with the two-point one. Figure 11 shows a comparison of the efficiency and torque ripple of the optimized generators at all points of the considered working profile.

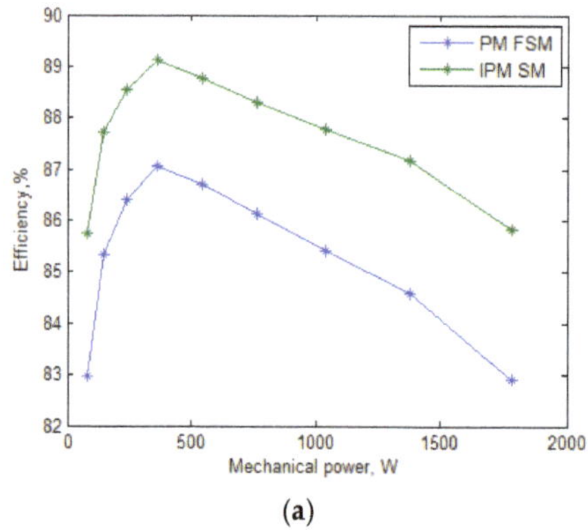

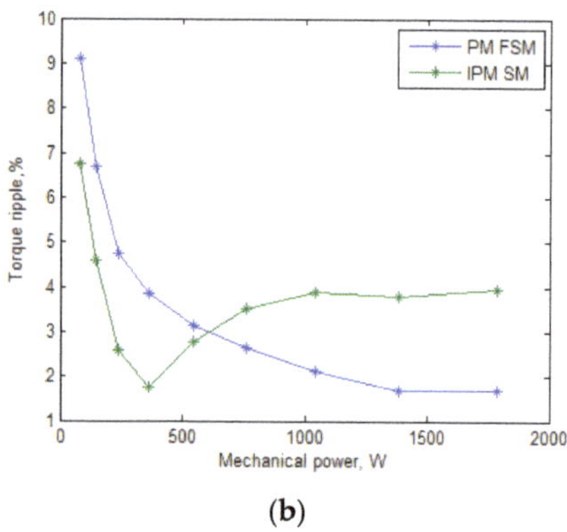

(a) (b)

Figure 11. Comparative dependencies on mechanical power: (**a**) efficiency (**b**) torque ripple.

The torque ripple in the IPMSG was quite low, namely, below 4%. Another revealed advantage of the PMFSG was a lower cost of rare-earth magnets. Particularly, the PMFSG required 1.4 times less rare-earth magnetic material than the IPMSG. Considering that the mining of rare-earth materials and their manufacturing have a harmful effect on the environment, this advantage of the PMFSG is highly valuable for mass production [33]. Furthermore, Table 8 shows that the demand for copper in the PMFSG was twice lower than for the IPMSG. Consequently, the cost of active materials for the PMFSG was 1.4 times lower than for the IPMSG, mainly due to the reduction in the magnets mass.

7. Conclusions

The aim of this study was to theoretically compare the PMFSG with a conventional IPMSG in a direct-driven low-power wind application. To ensure a fair comparison, both machines were first optimized using the Nelder–Mead algorithm. The optimization objective functions were selected so as to take into account the required power of frequency converter, the evaluation of the active materials cost, and losses in generators averaged over the working profile of the wind turbine. Additionally, the torque ripple was included in the optimization function of the IPMSG, since this parameter was significantly higher in the IPMSG than in the PMFSG. To reduce the computational time, the substituting profile method was applied. As a result, the strengths and weaknesses of the machines were revealed.

This study shows that the PMFSG requires 1.4 times less rare earth-earth magnet material than the IPMSG. Considering that the mining of rare-earth materials and their manufacturing have a harmful effect on the environment, this advantage of the PMFSG is highly valuable for mass production. In addition, the PMFSG torque ripple is less. However, the PMFSG requires a slightly higher power rating of the semiconductor inverter, and the average losses in the IPMSG are 1.2 times lower than in the PMFSG.

The results of this comparative study can be used by developers of direct-driven wind generators when selecting the type of electric machine suitable for a specific wind power application.

Author Contributions: Conceptual approach, V.P. and V.D.; data curation E.A. and V.K.; software V.D. and V.P.; calculations and modeling, V.P and V.D. and V.K.; writing of original draft, E.A.,

V.P., V.D. and V.K.; visualization, E.A., V.D. and V.K.; review and editing, E.A., V.P., V.D. and V.K. All authors have read and agreed to the published version of the manuscript.

Funding: This research received no external funding.

Informed Consent Statement: Not applicable.

Data Availability Statement: Data is contained within the article.

Acknowledgments: The authors thank the editors and reviewers for careful reading, and constructive comments.

Conflicts of Interest: The authors declare no conflict of interest.

References

1. Saleh, S.; Khan, M.; Rahman, M. Steady-state performance analysis and modelling of directly driven interior permanent magnet wind generators. *IET Renew. Power Gener.* **2011**, *5*, 137. [CrossRef]
2. Andriushchenko, E.; Kallaste, A.; Vaimann, T.; Rassolkin, A.; Heidari, H.; Demidova, G.L. Simulation of Wind Turbine Vibrations. In Proceedings of the 2020 27th International Workshop on Electric Drives: MPEI Department of Electric Drives 90th Anniversary (IWED), Moscow, Russia, 27–30 January 2020; pp. 1–4.
3. Ahsanullah, K.; Dutta, R.; Rahman, M.F.; Dutta, R. Analysis of Low-Speed IPMMs With Distributed and Fractional Slot Concentrated Windings for Wind Energy Applications. *IEEE Trans. Magn.* **2017**, *53*, 1–10. [CrossRef]
4. Chen, H.; Qu, R.; Li, J.; Li, D. Demagnetization Performance of a 7 MW Interior Permanent Magnet Wind Generator with Fractional-slot Concentrated Windings. *IEEE Trans. Magn.* **2015**, *51*, 1. [CrossRef]
5. Karimpour, S.R.; Besmi, M.R.; Mirimani, S.M. Optimal design and verification of interior permanent magnet synchronous generator based on FEA and Taguchi method. *Int. Trans. Electr. Energy Syst.* **2020**, *30*, 12597. [CrossRef]
6. Lehr, M.; Dietz, D.; Binder, A. Electromagnetic design of a permanent magnet Flux-Switching-Machine as a direct-driven 3 MW wind power generator. In Proceedings of the 2018 IEEE International Conference on Industrial Technology (ICIT), Lyon, France, 20–22 February 2018; pp. 383–388.
7. Dmitrievskii, V.; Prakht, V.; Kazakbaev, V. Design Optimization of a Permanent-Magnet Flux-Switching Generator for Direct-Drive Wind Turbines. *Energies* **2019**, *12*, 3636. [CrossRef]
8. Ditmanson, C.; Hein, P.; Kolb, S.; Molck, J.; Bernet, S. A New Modular Flux-Switching Permanent-Magnet Drive for Large Wind Turbines. *IEEE Trans. Ind. Appl.* **2014**, *50*, 3787–3794. [CrossRef]
9. Su, P.; Hua, W.; Zhang, G.; Chen, Z.; Cheng, M. Analysis and evaluation of novel rotor permanent magnet flux-switching machine for EV and HEV applications. *IET Electr. Power Appl.* **2017**, *11*, 1610–1618. [CrossRef]
10. Fasolo, A. Multi Polar Direct Drive Permanent Magnet Synchronous Machines for Renewable Energy. Ph.D. Thesis, University of Padova, Padova, Italy, 2013. Available online: http://paduaresearch.cab.unipd.it/5818 (accessed on 19 March 2021).
11. Fasolo, A.; Alberti, L.; Bianchi, N. Performance comparison between switching-flux and IPM machine with rare earth and ferrite PMs. In Proceedings of the 2012 XXth International Conference on Electrical Machines, Marseille, France, 2–5 September 2012; pp. 731–737.
12. Fasolo, A.; Alberti, L.; Bianchi, N. Performance Comparison Between Switching-Flux and IPM Machines with Rare-Earth and Ferrite PMs. *IEEE Trans. Ind. Appl.* **2014**, *50*, 3708–3716. [CrossRef]
13. Howe, D.; Chen, Y.S.; Zhu, Z.Q. Online optimal flux-weakening control of permanent-magnet brushless AC drives. *IEEE Trans. Ind. Appl.* **2000**, *36*, 1661–1668. [CrossRef]
14. Pellegrino, G.-M.L.; Vagati, A.; Guglielmi, P.; Boazzo, B. Performance Comparison Between Surface-Mounted and Interior PM Motor Drives for Electric Vehicle Application. *IEEE Trans. Ind. Electron.* **2012**, *59*, 803–811. [CrossRef]
15. Pang, Y.; Zhu, Z.Q.; Howe, D.; Iwasaki, S.; Deodhar, R.; Pride, A. Comparative study of flux-switching and interior perma-nent magnet machines. In Proceedings of the 2007 International Conference on Electrical Machines and Systems (ICEMS), Seoul, Korea, 1–2 November 2007; pp. 757–762. Available online: https://ieeexplore.ieee.org/document/4412185 (accessed on 19 March 2021).
16. Tang, Y.; Motoasca, E.; Paulides, J.J.; Lomonova, E.A. Comparison of flux-switching machines and permanent magnet synchronous machines in an in-wheel traction application. *COMPEL Int. J. Comput. Math. Electr. Electron. Eng.* **2012**, *32*, 153–165. [CrossRef]
17. Thomas, A.; Zhu, Z.; Jewell, G. Comparison of flux switching and surface mounted permanent magnet generators for high-speed applications. *IET Electr. Syst. Transp.* **2011**, *1*, 111. [CrossRef]
18. Isfahani, A.H.; Boroujerdi, A.H.-S.; Hasanzadeh, S. Multi-objective design optimization of a large-scale directdrive permanent magnet generator for wind energy conversion systems. *Front. Energy* **2014**, *8*, 182–191. [CrossRef]
19. Ibrahim, H.A. *Wind Turbines*; IntechOpen: London, UK, 2011; p. 652.
20. Shao, L.; Hua, W.; Soulard, J.; Zhu, Z.-Q.; Wu, Z.; Cheng, M. Electromagnetic Performance Comparison Between 12-Phase Switched Flux and Surface-Mounted PM Machines for Direct-Drive Wind Power Generation. *IEEE Trans. Ind. Appl.* **2020**, *56*, 1408–1422. [CrossRef]
21. Ojeda, J.; Simoes, M.G.; Li, G.; Gabsi, M. Design of a Flux-Switching Electrical Generator for Wind Turbine Systems. *IEEE Trans. Ind. Appl.* **2012**, *48*, 1808–1816. [CrossRef]

22. Meo, S.; Zohoori, A.; Vahedi, A. Optimal design of permanent magnet flux switching generator for wind applications via artificial neural network and multi-objective particle swarm optimization hybrid approach. *Energy Convers. Manag.* **2016**, *110*, 230–239. [CrossRef]
23. Prakht, V.; Dmitrievskii, V.; Kazakbaev, V. Optimal Design of Gearless Flux-Switching Generator with Ferrite Permanent Magnets. *Mathematics* **2020**, *8*, 206. [CrossRef]
24. Prakht, V.; Dmitrievskii, V.; Kazakbaev, V.; Ibrahim, M.N. Comparison between rare-earth and ferrite permanent magnet flux-switching generators for gearless wind turbines. *Energy Rep.* **2020**, *6*, 1365–1369. [CrossRef]
25. Cupertino, F.; Pellegrino, G.; Gerada, C. Design of Synchronous Reluctance Motors with Multiobjective Optimization Algorithms. *IEEE Trans. Ind. Appl.* **2014**, *50*, 3617–3627. [CrossRef]
26. Krasopoulos, C.T.; Beniakar, M.E.; Kladas, A.G. Robust Optimization of High Speed PM Motor Design. *IEEE Trans. Magn.* **2017**, *53*, 1. [CrossRef]
27. Anandavel, P.; Rajambal, K.; Chellamuthu, C. Power Optimization in a Grid-Connected Wind Energy Conversion System. In Proceedings of the 2005 International Conference on Power Electronics and Drives Systems, Kuala Lumpur, Malaysia, 28 November–1 December 2005; pp. 1617–1621.
28. Pishgar-Komleh, S.; Keyhani, A.; Sefeedpari, P. Wind speed and power density analysis based on Weibull and Rayleigh distributions (a case study: Firouzkooh county of Iran). *Renew. Sustain. Energy Rev.* **2015**, *42*, 313–322. [CrossRef]
29. ChenYang NdFeB Magnets. Price List of Standard Block Magnets. Available online: http://www.ndfebmagnets.de/CY-PriceList-NdFeB-Block.pdf (accessed on 19 March 2021).
30. Audet, C.; Tribes, C. Mesh-based Nelder–Mead algorithm for inequality constrained optimization. *Comput. Optim. Appl.* **2018**, *71*, 331–352. [CrossRef]
31. Nelder, J.A.; Mead, R. A Simplex Method for Function Minimization. *Comput. J.* **1965**, *7*, 308–313. [CrossRef]
32. Prakht, V.; Dmitrievskii, V.; Kazakbaev, V.; Oshurbekov, S. Comparison of High-Speed Single-Phase Flux Reversal Motor and Hybrid Switched Reluctance Motor. In Proceedings of the 2019 20th International Symposium on Power Electronics (Ee), Novi Sad, Serbia, 21–26 October 2019; pp. 1–5.
33. De Lima, I.B.; Filho, W.L. *Rare Earths Industry*; Elsevier BV: Amsterdam, The Netherlands, 2016; p. 434.

mathematics

Article

An Enhanced DC-Link Voltage Response for Wind-Driven Doubly Fed Induction Generator Using Adaptive Fuzzy Extended State Observer and Sliding Mode Control

Mohammed Mazen Alhato [1], Mohamed N. Ibrahim [2,3,4,*], Hegazy Rezk [5,6] and Soufiene Bouallègue [1,7]

1   Research Laboratory in Automatic Control (LARA), National Engineering School of Tunis, BP 37, Le Belvédère, Tunis 1002, Tunisia; mohammed.alhato@enit.utm.tn (M.M.A.); soufiene.bouallegue@issig.rnu.tn (S.B.)
2   Department of Electromechanical, Systems and Metal Engineering, Ghent University, 9000 Ghent, Belgium
3   FlandersMake@UGent—Corelab EEDT-MP, 3001 Leuven, Belgium
4   Electrical Engineering Department, Kafrelsheikh University, Kafrelsheikh 33511, Egypt
5   College of Engineering at Wadi Addawaser, Prince Sattam Bin Abdulaziz University, Al-Kharj 11911, Saudi Arabia; hr.hussien@psau.edu.sa
6   Electrical Engineering Department, Faculty of Engineering, Minia University, Minia 61517, Egypt
7   High Institute of Industrial Systems of Gabès (ISSIG), University of Gabès, Omar Ibn Khattab, Gabès 6029, Tunisia
*   Correspondence: m.nabil@ugent.be

Citation: Alhato, M.M.; Ibrahim, M.N.; Rezk, H.; Bouallègue, S. An Enhanced DC-Link Voltage Response for Wind-Driven Doubly Fed Induction Generator Using Adaptive Fuzzy Extended State Observer and Sliding Mode Control. *Mathematics* 2021, 9, 963. https://doi.org/10.3390/math9090963

Academic Editor: Nicu Bizon

Received: 3 March 2021
Accepted: 23 April 2021
Published: 25 April 2021

Publisher's Note: MDPI stays neutral with regard to jurisdictional claims in published maps and institutional affiliations.

Abstract: This paper presents an enhancement method to improve the performance of the DC-link voltage loop regulation in a Doubly-Fed Induction Generator (DFIG)- based wind energy converter. An intelligent, combined control approach based on a metaheuristics-tuned Second-Order Sliding Mode (SOSM) controller and an adaptive fuzzy-scheduled Extended State Observer (ESO) is proposed and successfully applied. The proposed fuzzy gains-scheduling mechanism is performed to adaptively tune and update the bandwidth of the ESO while disturbances occur. Besides common time-domain performance indexes, bounded limitations on the effective parameters of the designed Super Twisting (STA)-based SOSM controllers are set thanks to the Lyapunov theory and used as nonlinear constraints for the formulated hard optimization control problem. A set of advanced metaheuristics, such as Thermal Exchange Optimization (TEO), Particle Swarm Optimization (PSO), Genetic Algorithm (GA), Harmony Search Algorithm (HSA), Water Cycle Algorithm (WCA), and Grasshopper Optimization Algorithm (GOA), is considered to solve the constrained optimization problem. Demonstrative simulation results are carried out to show the superiority and effectiveness of the proposed control scheme in terms of grid disturbances rejection, closed-loop tracking performance, and robustness against the chattering phenomenon. Several comparisons to our related works, i.e., approaches based on TEO-tuned PI controller, TEO-tuned STA-SOSM controller, and STA-SOSM controller-based linear observer, are presented and discussed.

Keywords: doubly fed induction generator; DC-link voltage regulation; second-order sliding mode control; extended state observer; fuzzy gain scheduling; advanced metaheuristics

1. Introduction

Increasing demand for energy, decreasing conventional fossil-fuel energy sources, and environmental concerns are driving forces toward renewable energy sources [1]. Wind energy as a renewable source turns is becoming an important, promising source [2]. It has been the most growing source in terms of installed capacity. However, the dense researches that are executed in the wind area market create various wind energy topologies. One of the common configurations is the Doubly Fed Induction Generators (DFIGs) equipped with variable speed Wind Turbines (WTs) [3]. This configuration is widely adopted due to its significant merits, such as the independent control of active and reactive powers, low

Mathematics 2021, 9, 963. https://doi.org/10.3390/math9090963   https://www.mdpi.com/journal/mathematics

converter costs, and mechanical stress reduction [4]. The DFIG has rotor windings that are connected to the grid through power converters which are composed of the Rotor Side Converter (RSC) and the Grid Side Converter (GSC).

The classical control diagram of DFIG-based wind energy conversion systems is mainly constructed based on a vector control strategy in which the well-known Proportional-Integral (PI) regulators are commonly used [5]. At the GSC control part, the performance of the DC-link voltage dynamics depends on the proper tuning of the PI regulator gains and the system parameters, such as resistances and inductances [6]. Therefore, the performance of the DC-link voltage may degrade due to the deviation of the real system's parameters from its nominal values. Besides, the DFIGs are very sensible to grid disturbances because of the direct link of stator windings to the electrical net. Especially, the dip voltage action is not accepted in wind conversion systems since it leads to over-voltages and over-currents in the rotor windings. Moreover, such a phenomenon generates much oscillation in the DC-link voltage as well as in the stator active-reactive powers' dynamics, which leads to disconnect the WT and stop the power generation [7].

Hence, there is a need to develop a robust control strategy that can deal with parameters' mismatch, uncertainties, and external disturbances. To this end, the Sliding Mode Control (SMC) approaches were proposed. In the DFIG-based WT framework, a conventional SMC approach was adopted to regulate the DC-link voltage loop in [7,8]. However, the chattering issue stays a grave challenge when implementing the SMC method, which is highly undesirable in the DFIG systems. To tackle this problem, a new SMC law was developed to enhance the tracking performance and the robustness regarding operational uncertainties [9]. In [10], the fast exponential reaching law was performed and contrasted with the conventional one. In these studies, the obtained results indicated better performances in the DC-link voltage maintenance against parameters' variations and grid voltage disturbances in comparison with the classical PI controller. However, the DC-link voltage ripples introduced by the SMC strategies still result in poor performance.

To enhance the performance of the DC-link voltage dynamics in the presence of disturbances, feed-forward compensation methods were introduced. Several sensor-less-based DC-link voltage control methods were proposed. However, the implementation of these methods was adopted only for the AC-DC converters. The load current was only considered as a constant DC-disturbance term as shown in [11–13]. On the contrary, the load current that represented the DC rotor had severe ripples due to the RSC control and switching behaviors of the IGBTs in the DFIG system. Therefore, the effect of load current on the dynamics of the DC-link voltage should be studied [14]. To this end, an Extended State Observers (ESO) combined with other controllers is proposed to improve the performance of the voltage control loop. The ESO is used to offer estimations of internal states and external disturbances with minimal data about the system. The ESO considers the lumped disturbances, i.e., parameters' mismatch and unmodeled dynamics, as an extended state, which is estimated and compensated in the control law.

Nevertheless, in the design of an ESO, the selection of the bandwidth parameter is a challenging task that affects the closed-loop system performance. Generally, the larger the ESO's bandwidth, the more accurate the estimation of states will be achieved. On the other hand, the increase in such a bandwidth may lead to noise vulnerability and loss of robustness. The design and tuning of an observer are a trade-off between the estimation performance and noise action [15]. Usually, the bandwidth of the ESO is selected to be about 5 to 15 times the DC-link voltage controller's bandwidth. This ensures that the estimated state dynamics have a fast-tracking performance when the actual state dynamics change. Also, to guarantee that the ESO does not affect the current controller's performance, the selected bandwidth of the ESO should prevent overlapping with the current controller's bandwidth [16]. To this end, the bandwidth of the designed ESO is selected to be between 1/200 and 15/100 of the switching frequency. However, to achieve a fast response of the ESO dynamics during the perturbation, this paper introduces a novel adaptive fuzzy method to tune the value of the ESO's bandwidth within the limited region. Fuzzy gains-scheduling mechanism is introduced to adjust the appropriate values of the observer's bandwidth and further better the performance of the DC-link voltage control scheme. The concept utilizes a fuzzy logic inference as adaptive supervisors to tune the bandwidth gain in real-time.

Further, in this paper, a Second Order Sliding Mode (SOSM) controller based on the well-known Super Twisting Algorithm (STA) is firstly developed to regulate the DC-link voltage. The total disturbance in the DC-link voltage dynamics is theoretically investigated and the proposed fuzzy tuned-ESO is adopted to estimate it. Thus, a combined control law consisting of SOSM controller and disturbance compensation through the fuzzy gains-scheduling based-ESO is elaborated for the DC-link voltage regulation loop. Moreover, this paper investigates the stability conditions using the Lyapunov theory of nonlinear systems to be used as operational constraints for the formulated optimization-based control problem consisting of the tuning of all effective parameters of the designed ESO-based SOSM controller. Such a hard and constrained optimization problem is solved thanks to advanced competitive global metaheuristics, such as Thermal Exchange Optimization (TEO) [17,18], Water Cycle Algorithm (WCA) [19], Particle Swarm Optimization (PSO) [20], Genetic Algorithm (GA) [21], Grasshopper Optimization Algorithm (GOA) [22] and Harmony Search Algorithm (HSA) [23]. The main advantages of the designed adaptive controller are: (1) it presents a fast time-domain response and high robustness of the closed-loop DC-link voltage loop under external disturbances, (2) it clearly reduces the chattering phenomenon that is introduced by the SMC strategies, and (3) it ensures high tracking performance through a selection of the optimal and adaptive gains for the proposed control method.

The rest of the paper is arranged as follows: An elaborated dynamical model of the GSC component is described in Section 2. Section 3 investigates the second-order sliding mode controller design based on constrained optimization methods. The established uncertainties and stability conditions in the sense of Lyapunov for nonlinear systems are served as operational constraints for the reformulated optimization problem. Section 4 discusses the design of the fuzzy tuned-ESO-based DC-link voltage control loop. Comparative simulation results, gained on a 1.5 MW DFIG, are investigated in Section 5. Finally, Section 6 states the conclusion and future works.

2. Modeling of the DFIG Based Wind Energy Converter

The GSC component is joined to the electrical net via an L or LCL filter. However, for a better harmonics reduction of the grid currents, architecture with an LCL filter is adopted [3,24]. The mathematical modeling of the GSC system is defined in the d-q synchronous reference frame as given by Equation (1). In these dynamics, L_T states the sum of the converter and grid side inductances. Indeed, at the fundamental frequency, the LCL filter can be considered as an L-one with an inductance equal to the sum of the LCL-filter inductors [3]:

$$\begin{cases} L_T \frac{di_{dg}}{dt} = -R_T i_{dg} + \omega_g L_T i_{qg} + V_{dg} - V_{df} \\ L_T \frac{di_{qg}}{dt} = -R_T i_{qg} - \omega_g L_T i_{dg} + V_{qg} - V_{qf} \\ \frac{dV_{dc}}{dt} = \frac{1}{C_{dc}} \left(\frac{3}{2} \frac{V_{dg}}{V_{dc}} i_{gd} - i_{rdc} \right) \end{cases} \tag{1}$$

where C_{dc} is the capacitance of the DC-link circuit and i_{rdc} is the current between the DFIG rotor and the DC-link component.

3. Design of Second-Order Sliding Mode Controller

The main aim of the GSC control is to keep the DC-link voltage constant [5,7]. To perform this goal, Figure 1 presents the proposed control scheme. In such a design setup, the direct grid current is controlled thanks to the STA-based SOSM controller to maintain the DC-link voltage constant. The quadrature grid current can be used to regulate the reactive power's flow between the GSC and the grid.

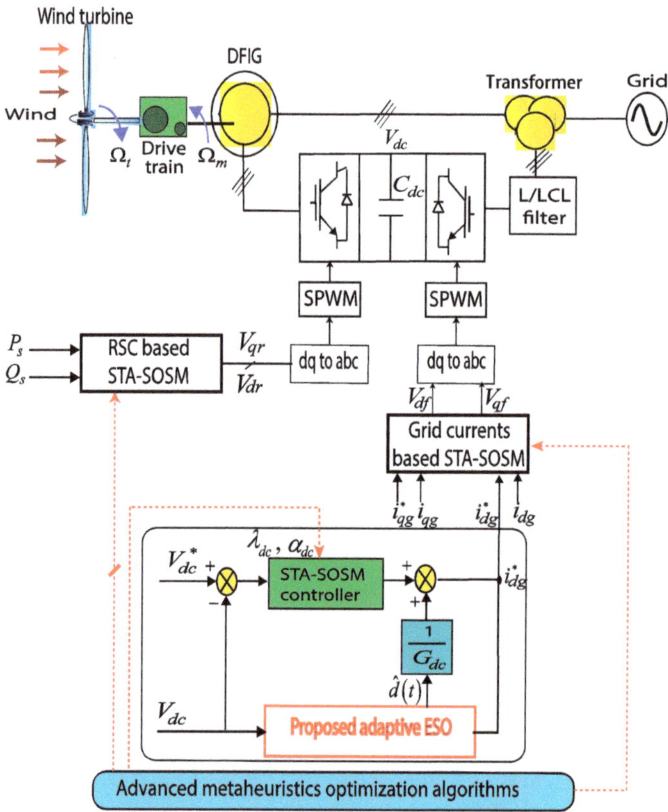

Figure 1. General block diagram for the proposed control scheme method.

3.1. Controller Design

The DC-link voltage dynamics (1) can be rearranged under the following form [7,25]:

$$\frac{dV_{dc}}{dt} = G_{dc}i_{dg} - \frac{1}{C_{dc}}i_{rdc} = G_{dc}i_{dg} + \eta_{dc} \tag{2}$$

where η_{dc} represents the uncertainties, parametric mismatch, and external disturbances of the DC-link dynamics, and it is considered to be a bounded parameter and $G_{dc} = \frac{1}{C_{dc}}\frac{3}{2}\frac{V_{dg}}{V_{dc}}$.

Let us specify the dynamics of the DC-link voltage error as [7,25]:

$$s_{V_{DC}} = V_{dc} - V_{dc}^* \tag{3}$$

The time derivative of the sliding surface $s_{V_{DC}}$ is given as follows:

$$\frac{ds_{V_{dc}}}{dt} = G_{dc}i_{dg} + \eta_{dc} - \frac{dV_{dc}^*}{dt} = u_{dc} + \eta_{dc} \tag{4}$$

where u_{dc} is the new control input as follows:

$$u_{dc} = G_{dc}i_{dg} - \frac{dV_{dc}^*}{dt} \tag{5}$$

While using the concept of the super twisting algorithm, a SOSM control law is designed to regulate the DC-link voltage as follows [15,25]:

$$\begin{cases} u_{dc} = -\lambda_{dc}\left|s_{V_{dc}}\right|^{0.5}\mathrm{sgn}\left(s_{V_{dc}}\right) + y_{dc} \\ \frac{dy_{dc}}{dt} = -\alpha_{dc}\mathrm{sgn}\left(s_{V_{dc}}\right) \end{cases} \tag{6}$$

where λ_{dc} and α_{dc} are parameters to be designed.

Finally, the reference direct grid current i_{dg}^{*} can be derived from Equations (5) and (6) as follows:

$$\begin{cases} i_{dg}^{*} = \frac{1}{G_{dc}}\left(-\lambda_{dc}\left|s_{V_{dc}}\right|^{0.5}\mathrm{sgn}\left(s_{V_{dc}}\right) + y_{dc} + \frac{dV_{dc}^{*}}{dt}\right) \\ \frac{dy_{dc}}{dt} = -\alpha_{dc}\mathrm{sgn}\left(s_{V_{dc}}\right) \end{cases} \tag{7}$$

3.2. Operational Constraints

The selection of the control gains of the SOSM is a significant issue where the best gains should guarantee good dynamic performances, stability, and robustness of the DFIG. The tuning of the control parameter by the conventional trial-and-error strategy is a time-consuming approach. Moreover, other classical methods such as the use of Hurwitz stability criterion for the linearized model of the DC-link voltage dynamics can attain the stability of the control law, but without securing that, the captured gains are the best parameters [26]. To improve the chosen of the control coefficients of the SOSM controller, advanced optimization algorithms are employed to process the tuning problem. In this instance, the restrictions that are forced on the control parameters by the Lyapunov theory are considered as inequality constraints for the optimization problem. This results in an improvement in tracking execution and chattering reduction.

The Lyapunov stability theory of the DFIG dynamics generates nonlinear limitations on the decision variables of the given control problem. By using the control law of Equation (7) and taking into consideration the derivative of the sliding surface as stated in Equation (4), the dynamics of $s_{V_{dc}}$ will be defined as:

$$\begin{cases} \frac{ds_{V_{dc}}}{dt} = -\lambda_{dc}\left|s_{V_{dc}}\right|^{0.5}\mathrm{sgn}\left(s_{V_{dc}}\right) + y_{dc} + \eta_{dc} \\ \frac{dy_{dc}}{dt} = -\alpha_{dc}\mathrm{sgn}\left(s_{V_{dc}}\right) \end{cases} \tag{8}$$

Let us consider that the perturbation term η_{dc} of the DC-link voltage dynamics (2) is globally bounded and described as follows [27,28]:

$$\left|\eta_{dc}\right| \le \Psi_{dc}\left|s_{V_{dc}}\right|^{0.5} \; ; \Psi_{dc} \ge 0 \tag{9}$$

To draw sufficient conditions on the robust stability of the studied DC-link voltage dynamics, let us define the following quadratic Lyapunov function:

$$V_{dc}\left(s_{V_{dc}}, y_{dc}\right) = \zeta_{dc}^{\mathrm{T}}L_{dc}\zeta_{dc} \tag{10}$$

where $\zeta_{dc} = [\zeta_1\zeta_2]^{\mathrm{T}} = \left[\left|s_{V_{dc}}\right|^{0.5}\mathrm{sgn}\left(s_{V_{dc}}\right)y_{dc}\right]^{\mathrm{T}}$ and L_{dc} is a symmetric positive definite matrix, which takes the following form:

$$L_{dc} = \frac{1}{2}\begin{bmatrix} 4\alpha_{dc} + \lambda_{dc}^{2} & -\lambda_{dc} \\ -\lambda_{dc} & 2 \end{bmatrix} \tag{11}$$

Then, the time derivative of the candidate Lyapunov function $V_{dc}\left(s_{V_{dc}}, y_{dc}\right)$ along the trajectories of the system (6) is calculated as:

$$\frac{dV_{dc}}{dt}\left(s_{V_{dc}}, y_{dc}\right) = -\frac{1}{\left|\zeta_1\right|}\zeta_{dc}^{\mathrm{T}}M^{\mathrm{T}}\zeta_{dc} + \frac{\eta_{dc}}{\left|\zeta_1\right|}N^{\mathrm{T}}\zeta_{dc} \tag{12}$$

where $M = \frac{\lambda_{dc}}{2}\begin{bmatrix} 2\alpha_{dc} + \lambda_{dc}^2 & -\lambda_{dc} \\ -\lambda_{dc} & 1 \end{bmatrix}$ and $N^{\mathrm{T}} = \left[\left(2\alpha_{dc} + \frac{\lambda_{dc}^2}{2}\right) \quad -\frac{\lambda_{dc}}{2}\right]$.

Using the bounds on the perturbations of Equation (9), it can be demonstrated that:

$$\frac{dV_{dc}}{dt}\left(s_{V_{dc}}, y_{dc}\right) = -\frac{1}{|\zeta_1|}\zeta_{dc}^{\mathrm{T}} Q_{dc} \zeta_{dc} \tag{13}$$

where Q_{dc} is a symmetric matrix expressed as:

$$Q_{dc} = \frac{\lambda_{dc}}{2}\begin{bmatrix} 2\alpha_{dc} + \lambda_{dc}^2 - \left(4\frac{\alpha_{dc}}{\lambda_{dc}} + \lambda_{dc}\right)\Psi_{dc} & -(\lambda_{dc} + \Psi_{dc}) \\ -(\lambda_{dc} + \Psi_{dc}) & 1 \end{bmatrix} \tag{14}$$

when this matrix is positive definite $Q_{dc} = Q_{dc}^{\mathrm{T}} > 0$, the stability condition of the dynamics (2) is satisfied in the sense of Lyapunov until $\dot{V}_{dc}\left(s_{V_{dc}}, y_{dc}\right) < 0$ as given in [29].

To have Q_{dc} as a positive definite matrix, the range values of λ_{dc} and α_{dc} should be adjusted as follows:

$$\begin{cases} \lambda_{dc} > 2\Psi_{dc} = \lambda_{dc}^{\min} \\ \alpha_{dc} > \lambda_{dc}\frac{5\lambda_{dc}\Psi_{dc} + 4\Psi_{dc}^2}{2(\lambda_{dc} - 2\Psi_{dc})} = \alpha_{dc}^{\min} \end{cases} \tag{15}$$

To enhance the performance of the DFIG system's control, the optimization theory is adopted to select and tune the effective parameters of the SOSM controllers for the DC-link voltage, stator active/reactive powers, and grid current loops. Hence, the tuning problem related with the STA-SOSM controllers is completely formulated as a constrained optimization problem. Several time-domain performance metrics, i.e., maximum overshoot, steady-state error, rise and/or settling times, as well as the established stability and robustness conditions of Equation (15), are included as inequality constraints for the optimization problem defined as follows:

$$\begin{cases} \text{minimize } J_m(x, t), \ m \in \{IAE, ISE, ITAE, ITSE\} \\ x = \left[\lambda_{Q_s}, \alpha_{Q_s}, \lambda_{dc}, \alpha_{dc}, \lambda_{i_{dg}}, \alpha_{i_{dg}}\right]^{T} \in S \subseteq \mathbb{R}_+^6 \\ \text{subject to :} \\ g_1(x, t) = \delta_{P_s} - \delta_{P_s}^{\max} \leq 0 \\ g_2(x, t) = \delta_{Q_s} - \delta_{Q_s}^{\max} \leq 0 \\ g_3(x, t) = \delta_{dc} - \delta_{dc}^{\max} \leq 0 \\ g_4(x, t) = \delta_{i_{dg}} - \delta_{i_{dg}}^{\max} \leq 0 \\ g_5(x, t) = \delta_{i_{qg}} - \delta_{i_{qg}}^{\max} \leq 0 \\ \lambda_{n,\min} \leq \lambda_n \leq \lambda_{n,\max} \\ \alpha_{n,\min} \leq \alpha_n \leq \alpha_{n,\max}, n \in \left\{Q_s, V_{dc}, i_{dg}\right\} \end{cases} \tag{16}$$

where $J_m : \mathbb{R}_+^6 \to \mathbb{R}$ are the cost functions, $g_q : \mathbb{R}_+^6 \to \mathbb{R}$ are the problem's inequality constraints, and $\delta_{dc}, \delta_{i_{dg}}, \delta_{i_{qg}}, \delta_{P_s}$ and δ_{Q_s} are the overshoots of the DC-link voltage, grid current components, and stator power components, respectively. The terms $\delta_{dc}^{\max}, \delta_{i_{dg}}^{\max}, \delta_{i_{qg}}^{\max}, \delta_{P_s}^{\max}$ and $\delta_{Q_s}^{\max}$ indicate their maximum specified values. The terms $\lambda_{Q_s}, \alpha_{Q_s}$ represent the gains of the SOSM controller for the stator power loop, and $\lambda_{i_{dg}}$ and $\alpha_{i_{dg}}$ are the gains of the controller for the grid current loop.

In the formalism of optimal control theory, the objective functions of (16) are usually described by common performance standards, such as Integral Absolute Error (IAE), Integral Square Error (ISE), Integral Time-Weighted Absolute Error (ITAE), and Integral

Time-Weighted Square Error (ITSE) [3,17,24,25]. Therefore, the global objective function using the IAE index for all controlled dynamics is aggregated as follows:

$$J_{IAE}(\boldsymbol{x},t) = \begin{bmatrix} w_{V_{dc}} & w_{i_{dg}} & w_{i_{qg}} & w_{Q_s} & w_{P_s} \end{bmatrix} \begin{bmatrix} \int_0^T \left| e_{dc}(\boldsymbol{x},t) \right| dt \\ \int_0^T \left| e_{i_{dg}}(\boldsymbol{x},t) \right| dt \\ \int_0^T \left| e_{i_{qg}}(\boldsymbol{x},t) \right| dt \\ \int_0^T \left| e_{Q_s}(\boldsymbol{x},t) \right| dt \\ \int_0^T \left| e_{P_s}(\boldsymbol{x},t) \right| dt \end{bmatrix} \tag{17}$$

where T denotes the total simulation time, $w_r \in \{Q_s, P_s, V_{dc}, i_{dg}, i_{qg}\}$ is the weighting coefficient satisfying the convex summation $\sum_r w_r = 1$, and $e_i(.)$ is the tracking error defined as $e_{P_s}(\boldsymbol{x},t) = P_s^* - P_s(\boldsymbol{x},t)$, $e_{Q_s}(\boldsymbol{x},t) = Q_s^* - Q_s(\boldsymbol{x},t)$, $e_{dc}(\boldsymbol{x},t) = V_{dc}^* - V_{dc}(\boldsymbol{x},t)$, $e_{i_{dg}}(\boldsymbol{x},t) = i_{dg}^* - i_{dg}(\boldsymbol{x},t)$, and $e_{i_{qg}}(\boldsymbol{x},t) = i_{qg}^* - i_{qg}(\boldsymbol{x},t)$.

4. Adaptive Fuzzy ESO-Based Sliding Mode Controller

In this work, a fuzzy gain-scheduling mechanism is proposed to design an adaptive STA-SOSM controller for the DC-link voltage dynamics. An adaptive extended state observer (ESO) is designed to asymptotically estimate the disturbances and allow the STA-SOSM controller to reject them efficiently. Therefore, the adaptive fuzzy ESO is employed in the DC-link voltage loop to improve the control performance and robustness as shown in Figure 2.

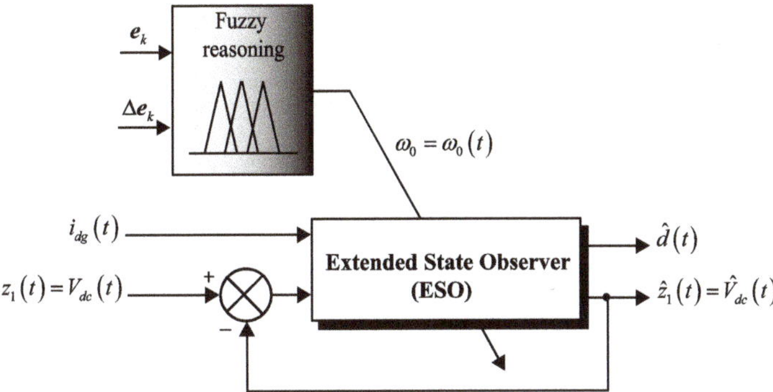

Figure 2. Adaptive fuzzy ESO-based SOSM controller for the DC-link voltage.

4.1. Concept of Extended State Observers

The ESO treats the lumped disturbances of the system as a new system state which is conceived to estimate not only the external disturbances but also the plant dynamics [30,31].

However, for any arbitrary second-order system, an ESO can be written in the following state-space form [15,30]:

$$\begin{cases} \dot{x}_1(t) = x_2(t) \\ \dot{x}_2(t) = f(x_1(t), x_2(t), u(t)) + b_0 u(t) + w(t) \\ y(t) = x_1(t) \end{cases} \tag{18}$$

where $x_1(t) \in \mathbb{R}$ and $x_2(t) \in \mathbb{R}$ are the state variables of the cascade integral form, $f(x_1(t), x_2(t), u(t)) \in \mathbb{R}$ defines the dynamics of the system which is a so-called internal disturbance, $u(t) \in \mathbb{R}$ is the control variable, $y(t) \in \mathbb{R}$ is the output variable, $w(t) \in \mathbb{R}$ is an external disturbance, and b_0 is a given constant, and $\dot{x}(t)$ denotes the time derivative of the variable $x(t)$, i.e., $\dot{x}_1(t) = \frac{dx_1(t)}{dt}$, and so on.

For a given second-order system, a three-order ESO can be used with the disturbance as an additional state variable and defined as follows:

$$\begin{cases} e(t) = z_1(t) - x_1(t) \\ \dot{z}_1(t) = z_2(t) - \beta_1 e(t) \\ \dot{z}_2(t) = z_3(t) + b_0 u(t) - \beta_2 e(t) \\ \dot{z}_3(t) = -\beta_3 e(t) \end{cases} \tag{19}$$

where β_1, β_2, and β_3 are the gains of the extended state observer.

In the state-space form (19), the ESO takes the system's output $y(t) = x_1(t)$ and control variable $u(t)$ as input and gives the state variables $(z_1(t), z_2(t), z_3(t))$ which represent the estimations of system state variables $(x_1(t), x_2(t))^T \in \mathbb{R}^2$ and total disturbances $w(t)$, respectively, i.e., $z_1(t) = \hat{x}_1(t)$, $z_2(t) = \hat{x}_2(t)$, and $z_3(t) = \hat{w}(t)$. By properly selecting the feedback coefficients β_1, β_2, and β_3, the estimation error $e(t)$ converges asymptotically to a small value. Besides, from Equation (19), it is clear that the ESO does not depend on the system parameters and thus provides strong robustness for observer dynamics [30,31].

4.2. Extended State Observer for the DC-Link Voltage Loop

To design an extended state observer (19) for the DC-link voltage loop, the first state variable of such an observer is chosen as $z_1(t) = V_{dc}(t)$. Therefore, the dynamics of the DC-link voltage (2) can be reformulated as:

$$\dot{z}_1(t) = \frac{dV_{dc}(t)}{dt} = G_{dc} u_{dc0}(t) - d(t) \tag{20}$$

where $u_{dc0}(t) = i_{dg}(t), \forall t \geq 0$ and $d(t) = -\eta_{dc}(t)$.

The ESO treats the external disturbance as an extended state. Therefore, a second-order ESO is used for the outer DC-link voltage loop, which is adopted as follows:

$$\begin{cases} e_1(t) = z_1(t) - \hat{z}_1(t) \\ \dot{\hat{z}}_1(t) = G_{dc} u_{dc0}(t) - \hat{d}(t) + \beta_1 e_1(t) \\ \dot{\hat{d}}(t) = -\beta_2 e_1(t) \end{cases} \tag{21}$$

where $\hat{z}_1(t)$ is an estimate of the output, $z_1(t)$ $\hat{d}(t)$ is an estimate of the total disturbance $d(t) = -\eta_{dc}(t)$, and β_1 and β_2 are the ESO gains chosen as follows [13,30]:

$$[\beta_1, \beta_2] = \left[2\omega_0 \; \omega_0^2\right] \in \mathbb{R}_+^2 \tag{22}$$

In Equation (22), the real $\omega_0 > 0$ denotes the observer's bandwidth that becomes the only tuning parameter of the extended state observer (21). Based on the above studies and the designed sliding-control law given in Equation (7), the proposed ESO-based control of the DC-link voltage loop is achieved by:

$$\begin{cases} i_{dg}^* = \frac{1}{G_{dc}} \left(\left(-\lambda_{dc} |s_{V_{dc}}|^{0.5} \mathrm{sgn}\left(s_{V_{dc}}\right) + y_{dc} + \frac{dV_{dc}^*}{dt} \right) + \hat{d}(t) \right) \\ \frac{dy_{dc}}{dx} = -\alpha_{dc} \mathrm{sgn}\left(s_{V_{dc}}\right) \end{cases} \tag{23}$$

The choice of the adequate bandwidth of a given ESO is a difficult task action [32,33]. The appropriate selection can improve the closed-loop system's performance, while the poor selection could degrade the time-domain performances and robustness of the controlled system. In general, the larger the ESO bandwidth, the more accurate the estimation of states will be achieved. On the other hand, the increase of such a bandwidth may lead to vulnerability against noise and loss of robustness. The design of an observer is always a trade-off between the estimation dynamics performance and the noise vulnerability [13,15]. Typically, the bandwidth of the ESO is selected to be 5 to 15 times the DC-link voltage controller's bandwidth. However, this last is limited from 1/1000 to 1/100 of the switching frequency as discussed in [34]. In this research work, the idea to design a fuzzy gains-scheduling-based observer is proposed to overcome such a complex tuning problem. Such an adaptive fuzzy supervisor is proposed to avoid the aforementioned drawbacks and achieve high convergence performance and robustness of the designed ESO under operational disturbances and high-frequency noises. Figure 2 shows such a proposed adaptive-fuzzy extended state observer for the studied wind energy converter.

The approach taken here is to exploit the Mamdani type of fuzzy rules and reasoning to generate the appropriate values of the ESO's bandwidth $\omega_0 \in \mathbb{R}_+$ of Equation (21). Let us consider as inputs of the proposed fuzzy gains scheduler the linguistic variables $e(kT_s) = e_k$ as the error between the actual and the estimated output, i.e., the sampled signals V_{dc} and $\hat{V}_{dc}$, respectively, and $\Delta e(kT_s) = \Delta e_k$ as the change of error, where T_s denotes the sampling period for the fuzzy supervisor. The designed fuzzy inference supervision mechanism produces the tuned bandwidth of the ESO denoted as ϖ_0.

For this proposed fuzzy-based tuning mechanism, a Mamdani model is applied as a type of inference mechanism. The decision-making output can be acquired using a Max-Min fuzzy inference where the crisp output is computed by the center of gravity defuzzification approach. As shown in Figure 3, all used membership functions for fuzzy sets are triangular, uniformly distributed, and symmetrical on the universe of discourse.

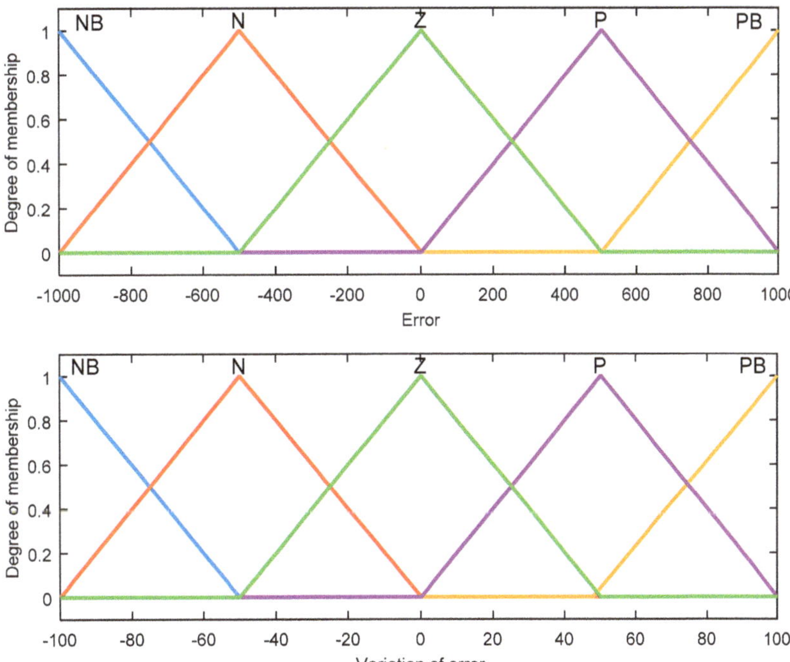

Figure 3. *Cont.*

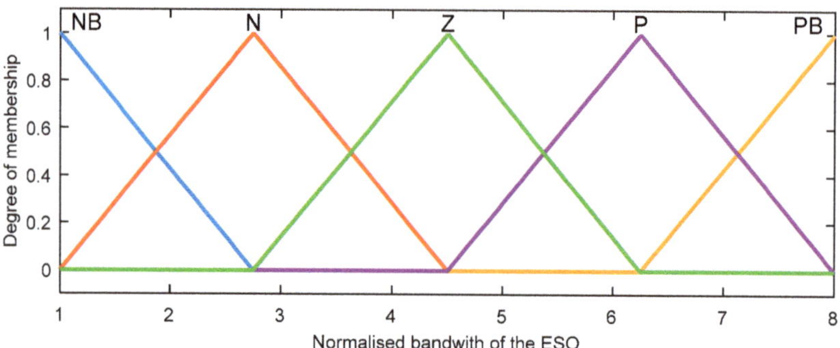

Figure 3. Membership functions of the proposed fuzzy gains-scheduling mechanism.

A set of linguistic rules in the form of Equation (24) is utilized in the fuzzy reasoning inference to define the output ω_0:

$$\text{If } e_k \text{ is } \boldsymbol{A}_i \text{ and } \Delta e_k \text{ is } \boldsymbol{B}_i, \text{ then } \omega_0 \text{ is } \boldsymbol{C}_i \tag{24}$$

where $\boldsymbol{A}_i$, $\boldsymbol{B}_i$, and $\boldsymbol{C}_i$ are the fuzzy sets of the inputs/output linguistic variables e_k, Δe_k, and ω_0, respectively. The linguistic levels assigned to the fuzzy inputs and outputs are labeled as follows: negative big (NB), negative (N), zero (ZE), positive (P), and positive big (PB). A set of 25 rules are defined for this fuzzy inference as given in Table 1.

Table 1. Fuzzy rules for the ω_0 parameter's tuning.

ω_0		Δe_k				
		NB	**N**	**ZE**	**P**	**PB**
	NB	NB	NB	NB	N	ZE
	N	NB	N	N	N	ZE
e_k	ZE	NB	N	ZE	P	PB
	P	ZE	P	P	P	PB
	PB	ZE	P	PB	PB	PB

As proposed in [32,33], and since it is assumed that the ESO's bandwidth parameter ω_0 varies in the prescribed range $[\![\omega_0^{\min}, \omega_0^{\max}]\!]$, this effective parameter is calculated according to the following linear transformation [34]:

$$\omega_0 = \left(\omega_0^{\max} - \omega_0^{\min}\right)\omega_0 + \omega_0^{\min} \tag{25}$$

where $\omega_0^{\max}$ and $\omega_0^{\min}$ are the maximum and minimum limits of ω_0, respectively.

5. Results and Discussion

The proposed STA-SOSM controllers based on an adaptive fuzzy ESO for the DC-link voltage loop are built using MATLAB/Simulink environment. The introduced algorithms GA, PSO, HSA, WCA, GOA, and TEO were switched with the equal values for the mutual factors, i.e., population size $N_{pop} = 50$ and the maximum number of iterations $N_{iter} = 100$, and run on an Intel R CoreTMi5 CPU computer at 2.5 GHz and 8 GB of RAM. The parameters of the DFIG (1.5 MW) used in this work are given in our previous works [3]. The gains of the STA-SOSM controllers for the DC-link voltage and other control loops are tuned thanks to the proposed advanced optimization algorithms for the problems (16) and (17).

Figure 4 shows the convergence curves of the cost functions for the considered time-domain performance indices. Moreover, the TEO algorithm for the ISE and ITSE criteria outperforms the other reported methods in terms of fast and non-premature convergence.

Figure 4a,c show that the TEO for the IAE and ITAE indicators display the better convergence as a second- and third-order, respectively, after the GA one.

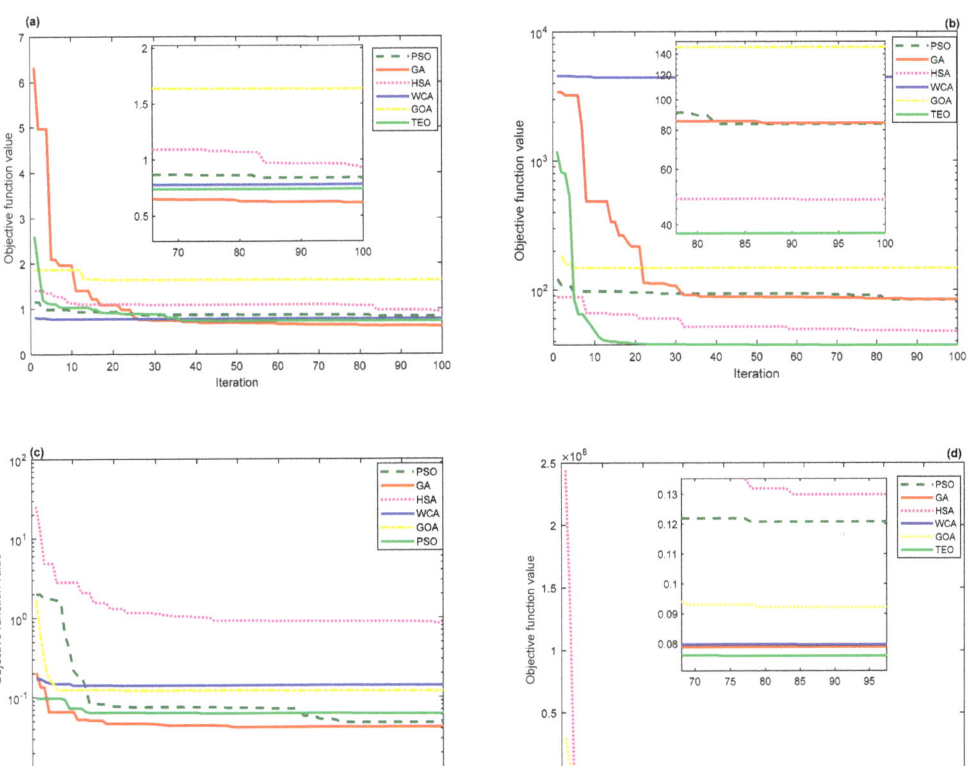

Figure 4. Convergence curves comparison: (**a**) IAE; (**b**) ISE; (**c**) ITAE; and (**d**) ITSE criterion.

Since the TEO metaheuristic outperformed all the other reported ones, the effective gains λ_{dc} and α_{dc} of the STA-SOSM controller, retained for the rest of the control strategy, are selected as the best results obtained by the TEO algorithm of the optimization problem (16) and (17). Table 2 summarizes such decision variables for each time-domain performance criterion.

Table 2. TEO-based results for the STA-SOSM controllers' gains tuning.

Performance Criteria	STA-SOSM Controllers' Gains	
	λ_{dc}	α_{dc}
IAE	26.10	14.50
ISE	17.40	93.60
ITAE	1.90	90.50
ITSE	12.30	53.90

Besides, the bandwidth of the ESO is selected to be between $1/200$ and $15/100$ of the switching frequency for the GSC circuit. The fuzzy gains-scheduling mechanism is employed to adaptively tune the bandwidth of the ESO within the predefined limits. The obtained fuzzy surface for the bandwidth gain is presented in Figure 5. On the other hand, Figure 6 shows the histories of the scheduled gain of the proposed adaptive fuzzy ESO. Figure 7 presents the performance of the DC-link voltage when different input steps are set.

It can be noted that the chattering phenomenon is reduced using the proposed TEO and fuzzy-based method in comparison with the STA-SOSM and PI controllers-based ones.

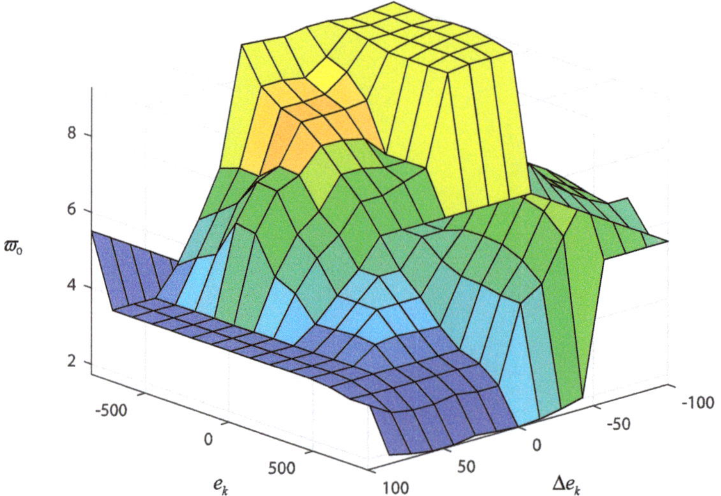

Figure 5. Fuzzy surface of the observer's bandwidth parameter ϖ_0.

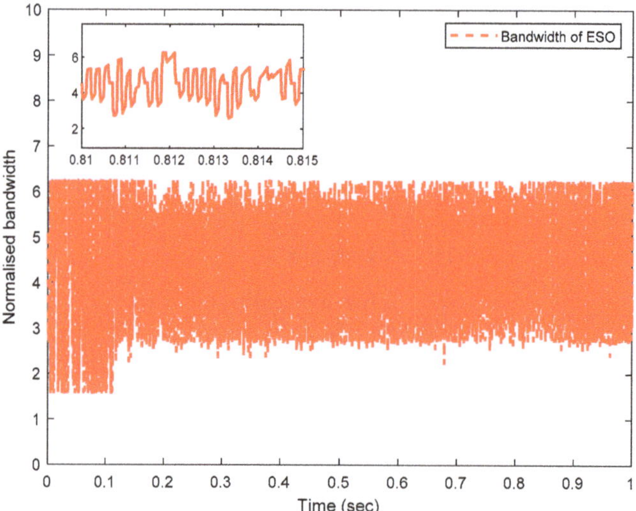

Figure 6. Time histories of the observer's bandwidth control ϖ_0.

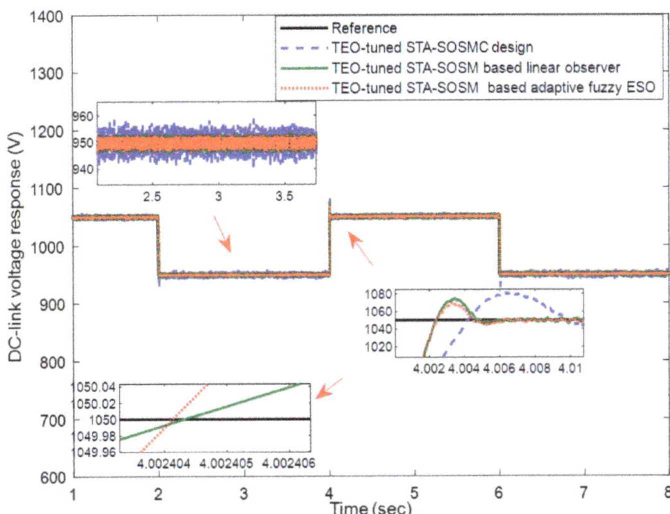

Figure 7. Time-domain performances' comparison of the DC-link voltage.

Moreover, the STA-SOSM controller based on different types of observers leads to lower DC-link voltage overshoot and faster response compared with the STA-SOSM and classical PI controllers-based cases as summarized in Table 3. Indeed, the proposed design and tuning method is compared to the STA-SOSM controller associated with a linear observer as well as without observers to show its superiority and effectiveness. Also, Figure 8 presents the time-domain performance of the DC-link voltage controller under severe voltage dips conditions. Since a 30% voltage drop is considered, the proposed adaptive fuzzy observer and TEO-based tuning method are successfully able to mitigate the voltage dip.

Table 3. Time-domain performances' comparison for the controlled DC-link voltage dynamics.

Control Strategies	Unit Step Change Response			
	$t_r(s)$	$t_s(s)$	δ	E_{ss}
TEO-tuned PI controller	0.005	4.019	2.54	0.253
TEO-tuned STA-SOSM controller	0.004	4.012	2.95	0.246
STA-SOSM controller-based linear observer	0.002	4.006	2.38	0.115
STA-SOSM controller-based adaptive fuzzy ESO	0.002	4.005	1.81	0.086

Based on these demonstrative results, one can notice that the lower amplitude of the fluctuations exists near the starting of the voltage dips in comparison with the STA-SOSM controller only. However, the proposed robust and intelligent STA-SOSM controllers based on linear and ESO observers present approximately the same performance in the case of DC-link voltage dips. Further comparison in terms of Total Harmonic Distortion (THD) variation is made in Table 4.

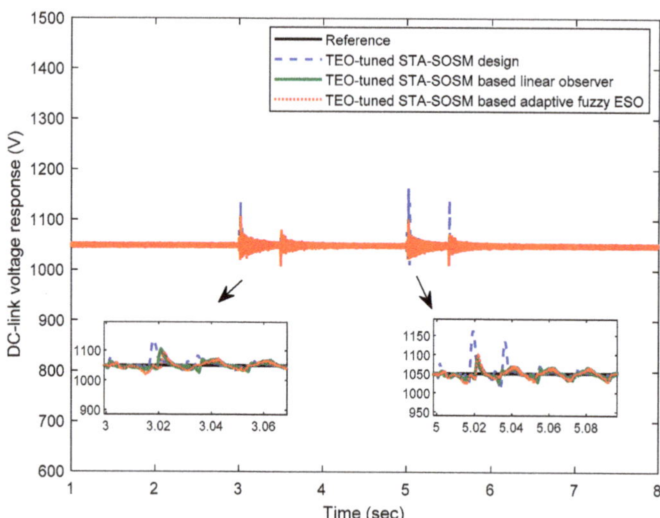

Figure 8. Time-domain responses of the controlled DC-link voltage under drop conditions.

Table 4. Comparison of stator and rotor currents' THD.

Control Strategies	THD (%)	
	Stator Current	**Rotor Current**
TEO-tuned PI controller	0.77	0.85
TEO-tuned STA-SOSM controller	0.28	0.32
STA-SOSM controller-based linear observer	0.27	0.30
STA-SOSM controller-based adaptive fuzzy ESO	0.26	0.28

The currents THD of stator and rotor for the adaptive fuzzy ESO-based SOSM controllers are about 0.26% and 0.28%, respectively. These values are better than the other reported design methods, such as TEO-tuned PI controller, TEO-tuned STA-SOSM controller, and STA-SOSM controller-based linear observer. In addition, these values agree with the limits of the IEEE519-1992 standard which stipulate that the value of the THD does not exceed 5%. The DC-link voltage states and their according estimations and observation errors are shown in Figures 9 and 10 for linear and adaptive fuzzy ESO observers, respectively. The observation errors are obtained as the difference between the actual value of the DC-link voltage and the value estimated by the observers, i.e., $e = V_{dc} - \hat{V}_{dc}$.

The estimation dynamics using the different proposed observers are perfectly ensured. However, the adaptive fuzzy ESO presents a faster convergence and reconstruction of the system's state than in the linear observer-based case, i.e., an observation error reached null value after 0.05 s with ESO compared to 0.3 sec with the linear observer. Moreover, Figure 11 gives the total disturbance estimations for the linear and adaptive fuzzy ESO observers, respectively. Such a convergence rate comparison and total disturbance estimation of observers indicate the effectiveness and superiority of the proposed adaptive fuzzy extended state observer for the DC-link voltage regulation.

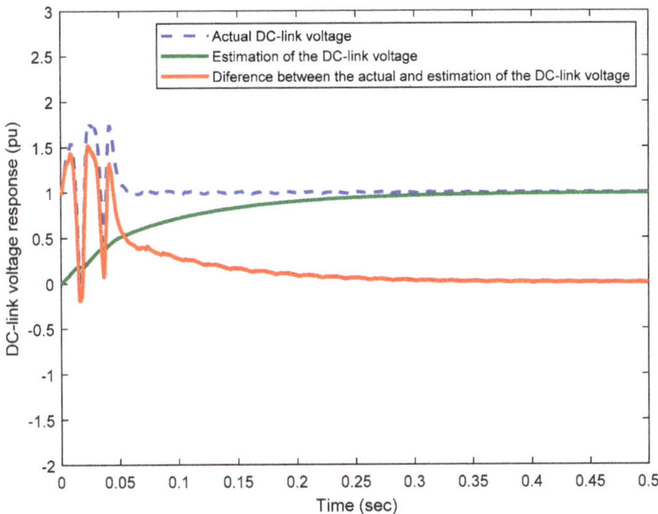

Figure 9. Convergence's dynamics of the linear observer for the DC-link voltage.

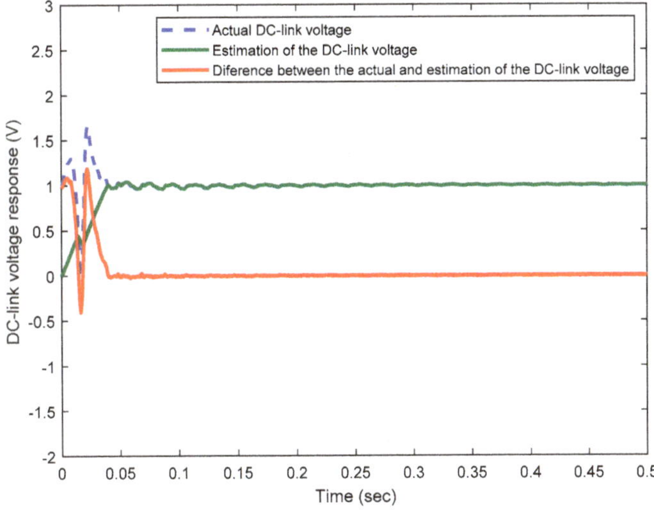

Figure 10. Dynamic performance of the adaptive fuzzy ESO for the DC-link voltage.

In a wind conversion system, the external disturbances' upper bound is unknown and is hard to be estimated in an actual doubly-fed induction generator. Therefore, the fuzzy gain-scheduled ESO is proposed in this work to compel observer parameters to vary according to the disturbances' upper bound in real-time. The uncertainty upper limit is required when developing the DC-link voltage second-order sliding mode controller, yet the upper limits are unknown in practical wind conversion systems and are difficult to be estimated. Overestimation of these upper bounds may lead to a conservative choice of controller parameters which will produce more control effectiveness, aggravate control chattering, and shorten the service cycle of wind turbines.

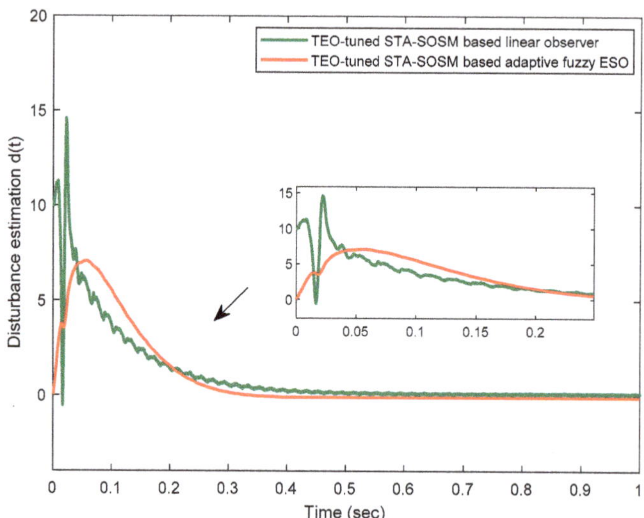

Figure 11. Total disturbance estimations' comparison of the reported observers.

6. Conclusions

This paper discussed the robust design and intelligent tuning of super twisting-based second order sliding mode (STA-SOSM) controllers for the DC-link voltage loop of a DFIG-based wind energy conversion system. An adaptive fuzzy extended state observer (ESO) is firstly proposed to estimate the external disturbance of the controlled system. Such a proposed fuzzy gains-scheduling mechanism is performed to adaptively tune the bandwidth of the ESO during the occurrence of disturbances. Such a hybrid control law based on a TEO-tuned SOSM controller and adaptive extended state observer showed high superiority and robustness under grid disturbances and model uncertainties as well as in terms of closed-loop time-domain performance and chattering issue reduction. The simulation results as well as the conducted comparisons proved that the proposed algorithm gives the best THD values for the stator and rotor current with 0.26% and 0.28%, respectively. Moreover, the time-domain performance indices were better than those with the reported methods, such as TEO-tuned PI controller, TEO-tuned STA-SOSM controller, and STA-SOSM controller-based linear observer, which had the minimal value of overshoot with 1.81% for the DC-link voltage response. This clearly highlights the effectiveness of the proposed intelligent and adaptive fuzzy-based control approach for the DC-link voltage dynamics.

The current research work is restricted by supposing that the wind speed is fixed. Moreover, the proposed approach investigated a low voltage dip condition. Stability enhancement of the DFIG system under the previous concerns will be deeply investigated in future work. This will be implemented through interactions among DFIGs with various control methods, such as adaptive gain SOSM controllers and sensor-less control methods. Indeed, these methods deserve attention because they solve the difficulty of estimating the disturbance upper bounds for WECSs.

Author Contributions: Conceptualization, M.M.A. and S.B.; methodology, M.M.A., S.B. and M.N.I.; software, M.M.A.; validation, H.R. and S.B.; formal analysis, M.N.I. and S.B.; investigation, H.R. and S.B.; resources, M.M.A., S.B. and H.R.; data curation, M.M.A.; writing—original draft preparation, M.M.A.; writing—review and editing, H.R. and S.B.; visualization, H.R. and M.N.I.; supervision, H.R. and S.B.; project administration, M.N.I. and S.B.; funding acquisition, M.N.I. and H.R. All authors have read and agreed to the published version of the manuscript.

Funding: This research received no external funding.

Institutional Review Board Statement: Not applicable.

Informed Consent Statement: Not applicable.

Data Availability Statement: No new data were created or analyzed in this study.

Conflicts of Interest: The authors declare no conflict of interest.

References

1. Khaligh, A.; Onar, O.C. *Energy Harvesting Solar, Wind, and Ocean Energy Conversion Systems*, 1st ed.; CRC Press, Taylor & Francis Group: New York, NY, USA, 2009.
2. Ahsan, H.; Mufti, M.D. Systematic development and application of a fuzzy logic equipped generic energy storage system for dynamic stability reinforcement. *Int. J. Energy Res.* **2020**, *44*, 8974–8987. [CrossRef]
3. Alhato, M.M.; Bouallègue, S. Direct power control optimization for doubly fed induction generator based wind turbine systems. *Math. Comput. Appl.* **2019**, *24*, 77. [CrossRef]
4. Naika, K.A.; Gupta, C.P.; Fernandez, E. Design and implementation of interval type-2 fuzzy logic-PI based adaptive controller for DFIG based wind energy system. *Int. J. Electr. Power Energy Syst.* **2020**, *115*, 105468. [CrossRef]
5. Pena, R.; Clare, J.; Asher, G. Doubly fed induction generator using back-to-back PWM converters and its application to variable-speed wind-energy generation. *IEE Proc. Electr. Power Appl.* **1996**, *143*, 231. [CrossRef]
6. Gagra, S.K.; Mishra, S.; Sing, M. Performance analysis of grid integrated doubly fed induction generator for a small hydro-power plant. *Int. J. Renew. Energy Res.* **2018**, *8*, 2310–2323.
7. Djilali, L.; Sanchez, E.N.; Belkheiri, M. Real-time implementation of sliding-mode field-oriented control for a DFIG-based wind turbine. *Int. Trans. Electr. Energy Syst.* **2018**, *28*, e2539. [CrossRef]
8. Merabet, A.; Eshaft, H.; Tanvir, A.A. Power-current controller based sliding mode control for DFIG-wind energy conversion system. *IET Renew. Power Gener.* **2018**, *12*, 1155–1163. [CrossRef]
9. Barambones, O.; Cortajarena, J.A.; Alkorta, P.; De Durana, J.M.G. A Real-Time Sliding Mode Control for a Wind Energy System Based on a Doubly Fed Induction Generator. *Energies* **2014**, *7*, 6412–6433. [CrossRef]
10. Xiong, L.; Li, P.; Li, H.; Wang, J. Sliding Mode Control of DFIG Wind Turbines with a Fast Exponential Reaching Law. *Energies* **2017**, *10*, 1788. [CrossRef]
11. Kim, S.-C.; Nguyen, T.H.; Lee, D.-C.; Lee, K.-B.; Kim, J.-M. Fault Tolerant Control of DC-Link Voltage Sensor for Three-Phase AC/DC/AC PWM Converters. *J. Power Electron.* **2014**, *14*, 695–703. [CrossRef]
12. Merai, M.; Naouar, M.W.; Slama-Belkhodja, I.; Monmasson, E. An improved dc-link voltage control for a three-phase PWM rectifier using an adaptive PI controller combined with load current estimation. In Proceedings of the 2017 19th European Conference on Power Electronics and Applications (EPE'17 ECCE Europe), Warsaw, Poland, 11–14 September 2017; pp. 1–10.
13. Lu, J.; Golestan, S.; Savaghebi, M.; Vasquez, J.C.; Guerrero, J.M.; Marzabal, A. An Enhanced State Observer for DC-Link Voltage Control of Three-Phase AC/DC Converters. *IEEE Trans. Power Electron.* **2018**, *33*, 936–942. [CrossRef]
14. Song, G.; Cao, B.; Chang, L.; Shao, R.; Xu, S. A Novel DC-Link Voltage Control for Small-Scale Grid-Connected Wind Energy Conversion System. In Proceedings of the 2019 IEEE Applied Power Electronics Conference and Exposition (APEC), Anaheim, CA, USA, 17–21 March 2019; pp. 2461–2466.
15. Liu, J.; Vazquez, S.; Wu, L.; Marquez, A.; Gao, H.; Franquelo, L.G. Extended State Observer-Based Sliding-Mode Control for Three-Phase Power Converters. *IEEE Trans. Ind. Electron.* **2016**, *64*, 22–31. [CrossRef]
16. Blasko, V.; Kaura, V. A new mathematical model and control of a three-phase AC-DC voltage source converter. *IEEE Trans. Power Electron.* **1997**, *12*, 116–123. [CrossRef]
17. Alhato, M.M.; Bouallègue, S. Thermal exchange optimization based control of a doubly fed induction generator in wind en-ergy conversion system. *Indones. J. Electr. Eng. Comput. Sci.* **2020**, *20*, 1–8.
18. Kaveh, A.; Dadras, A. A novel meta-heuristic optimization algorithm: Thermal exchange optimization. *Adv. Eng. Softw.* **2017**, *110*, 69–84. [CrossRef]
19. Eskandar, H.; Sadollah, A.; Bahreininejad, A.; Hamdi, M. Water cycle algorithm—A novel metaheuristic optimization method for solving constrained engineering optimization problems. *Comput. Struct.* **2012**, *110*, 151–166. [CrossRef]
20. Mohamed, M.A.; Diab, A.A.Z.; Rezk, H. Partial shading mitigation of PV systems via different meta-heuristic techniques. *Renew. Energy* **2019**, *130*, 1159–1175. [CrossRef]
21. Holland, J.H. Genetic algorithms. *Sci. Am.* **1992**, *276*, 66–72. [CrossRef]
22. Saremi, S.; Mirjalili, S.; Lewis, A. Grasshopper Optimisation Algorithm: Theory and application. *Adv. Eng. Softw.* **2017**, *105*, 30–47. [CrossRef]
23. Geem, Z.W.; Kim, J.H.; Loganathan, G. A New Heuristic Optimization Algorithm: Harmony Search. *Simulation* **2001**, *76*, 60–68. [CrossRef]
24. Alhato, M.M.; Bouallègue, S. Whale optimization algorithm for active damping of LCL-filter-based grid-connected converters. *Int. J. Renew. Energy Res.* **2019**, *9*, 986–996.
25. Alhato, M.M.; Bouallègue, S.; Rezk, H. Modeling and Performance Improvement of Direct Power Control of Doubly-Fed Induction Generator Based Wind Turbine through Second-Order Sliding Mode Control Approach. *Mathematics* **2020**, *8*, 2012. [CrossRef]
26. Matas, J.; Castilla, M.; Guerrero, J.M.; De Vicuna, L.G.; Miret, J. Feedback Linearization Of Direct-Drive Synchronous Wind-Turbines Via a Sliding Mode Approach. *IEEE Trans. Power Electron.* **2008**, *23*, 1093–1103. [CrossRef]

27. Shtessel, Y.B.; Moreno, J.A.; Plestan, F.; Fridman, L.M.; Poznyak, A.S. Super-twisting adaptive sliding mode control: A Lyapunov design. In Proceedings of the 49th IEEE Conference on Decision and Control, Atlanta, GA, USA, 15–17 December 2010; pp. 5109–5113.
28. Liu, X.; Han, Y.; Wang, C. Second-order sliding mode control for power optimization of DFIG-based variable speed wind turbine. *IET Renew. Power Gener.* **2017**, *11*, 408–418. [CrossRef]
29. Levant, A. Sliding order and sliding accuracy in sliding mode control. *Int. J. Control* **1993**, *58*, 1247–1263. [CrossRef]
30. Zheng, Q.; Gao, L.Q.; Gao, L. On stability analysis of active disturbance rejection control for nonlinear time-varying plants with unknown dynamics. In Proceedings of the 46th IEEE Conference on Decision and Control, New Orleans, LA, USA, 12–14 December 2007; pp. 3501–3506.
31. Han, J. From PID to Active Disturbance Rejection Control. *IEEE Trans. Ind. Electron.* **2009**, *56*, 900–906. [CrossRef]
32. Zhao, Z.-Y.; Tomizuka, M.; Isaka, S. Fuzzy gain scheduling of PID controllers. *IEEE Trans. Syst. Man Cybern.* **1993**, *23*, 1392–1398. [CrossRef]
33. Blanchett, T.; Kember, G.; Dubay, R. PID gain scheduling using fuzzy logic. *ISA Trans.* **2000**, *39*, 317–325. [CrossRef]
34. Zhou, D.; Blaabjerg, F. Bandwidth oriented proportional-integral controller design for back-to-back power converters in DFIG wind turbine system. *IET Renew. Power Gener.* **2017**, *11*, 941–951. [CrossRef]

Article

Multi-Objective Optimization of Switched Reluctance Machine Design Using Jaya Algorithm (MO-Jaya)

Mohamed Afifi [1,†], Hegazy Rezk [2,3], Mohamed Ibrahim [4,5,6] and Mohamed El-Nemr [1,7,*,†]

1   Electromagnetic Energy Conversion Laboratory, Tanta University, Tanta 31527, Egypt; mohamed.afifi@f-eng.tanta.edu.eg
2   College of Engineering at Wadi Addawaser, Prince Sattam Bin Abdulaziz University, Wadi Aldawaser 11991, Saudi Arabia; hegazy.hussien@mu.edu.eg
3   Electrical Engineering Department, Faculty of Engineering, Minia University, Minia 61111, Egypt
4   Department of Electromechanical, Systems and Metal Engineering, Ghent University, 9000 Ghent, Belgium; m.nabil@ugent.be
5   FlandersMake@UGent-Corelab EEDT-MP, 3001 Leuven, Belgium
6   Electrical Engineering Department, Kafrelshiekh University, Kafrelshiekh 33511, Egypt
7   Electrical Power and Machines Engineering Department, Faculty of Engineering, Tanta University, Tanta 31527, Egypt
*   Correspondence: melnemr@f-eng.tanta.edu.eg; Tel.: +20-111-3535-272
†   Current Affiliation.

Citation: Afifi, M.; Rezk, H.; Ibrahim, M.; El-Nemr, M. Multi-Objective Optimization of Switched Reluctance Machine Design Using Jaya Algorithm (MO-Jaya). *Mathematics* 2021, 9, 1107. https:// doi.org/10.3390/math9101107

Academic Editor: David Barilla

Received: 12 April 2021
Accepted: 6 May 2021
Published: 13 May 2021

Publisher's Note: MDPI stays neutral with regard to jurisdictional claims in published maps and institutional affiliations.

Abstract: The switched reluctance machine (SRM) design is different from the design of most of other machines. SRM has many design parameters that have non-linear relationships with the performance indices (i.e., average torque, efficiency, and so forth). Hence, it is difficult to design SRM using straight forward equations with iterative methods, which is common for other machines. Optimization techniques are used to overcome this challenge by searching for the best variables values within the search area. In this paper, the optimization of SRM design is achieved using multi-objective Jaya algorithm (MO-Jaya). In the Jaya algorithm, solutions are moved closer to the best solution and away from the worst solution. Hence, a good intensification of the search process is achieved. Moreover, the randomly changed parameters achieve good search diversity. In this paper, it is suggested to also randomly change best and worst solutions. Hence, better diversity is achieved, as indicated from results. The optimization with the MO-Jaya algorithm was made for 8/6 and 6/4 SRM. Objectives used are the average torque, efficiency, and iron weight. The results of MO-Jaya are compared with the results of the non-dominated sorting genetic algorithm (NSGA-II) for the same conditions and constraints. The optimization program is made in Lua programming language and executed by FEMM4.2 software. The results show the success of the approach to achieve better objective values, a broad search, and to introduce a variety of optimal solutions.

Keywords: optimal design; switched reluctance machine; MO-Jaya optimization; finite element analysis

1. Introduction

The switched reluctance machine (SRM) is the type of machines that develop output torque due to reluctance variation without using permanent magnets or rotor excitation. The switching of phases is made according to position of rotor in such a way to produce torque (induce voltage in generation mode). Reluctance variation happens with rotation as a function of certain geometric parameters. The currents of SRM are mainly in the form of pulses. The flux inside the machine is not sinusoidal. All of the previously mentioned facts of SRM give this type of machines its unique features. The SRM has shown attractive characteristics, such as simple and robust construction, low manufacturing cost, and high efficiency, over wide range of speeds [1,2]. SRM's construction simplicity and low manufacturing cost have motivated both researchers and manufacturer.

Mathematics **2021**, *9*, 1107. https://doi.org/10.3390/math9101107

However, the pulsative behaviour of currents results in torque ripples that are considered a problem of SRM and a challenge. The saliency of poles and non-sinusoidal wave-forms of flux in SRM result in noisy operation and radial vibrations. In addition, SRM structure is not simple from geometric perspective due to the wide range of probabilities for dimensions of a certain design. Moreover, the relationships between SRM's geometric parameters and performance indices are indirect and non-linear. Hence, the search for a good design is difficult despite the search for the best (optimal) design.

Optimization is the search for the best solution (the optimal solution) for a certain problem [3]. Various types of optimization algorithms exist, including traditional optimization techniques that have many limitations and advanced population-based meta-heuristic algorithms [4–6]. Optimization is needed with a switched reluctance machine (SRM) design to reach better designs. Sensitivity analysis is the study of the degree of influence of each parameter on the final objectives of the design. Subsequently, the most influencing parameters are chosen to be optimized to save computation time. Because of the large number of geometric parameters or design parameters of SRM, sensitivity analysis is usually made to make the optimization process less complicated, as in [7–9].

It is essential to have a mathematical model of SRM in order to evaluate its performance and include this evaluation in the optimization process. The performance of SRM is mainly evaluated by rated torque, torque per weight, torque per volume, rated speed, and efficiency. Other quantities can be added to give more detailed evaluation, such as torque ripples, acoustic noise, mechanical vibrations, and maximumtemperature rise, and so forth. All of these performance indices are calculated by various methods that differ from each other according to their accuracy and computational time. These method can be classified to numerical methods, such as finite elements analysis (FEA) and boundary element method (BEM), and analytical methods, such as curve-fitting methods, magnetic equivalent circuits (MECs), and Maxwell's-equation-based approaches [1]. Numerical methods provide high accuracy with the cost of its high computational time. Analytical methods can be simplified to achieve a fast calculation process. However, in most cases, this results in a reduction of accuracy. FEA is commonly used in SRM modelling to achieve accurate results, as in [10–14]. Magnetic equivalent circuit (MEC) is faster than FEA, as shown in in [15–17]. However, it is less frequent because of its reduced accuracy. Fuzzy logic and regression methods are also used, as shown in [18,19]. However, they are less reliable and have complex structures when compared with FEA and MEC. Each performance index (i.e., rated torque, torque ripples and so forth) of SRM is called "objective function" when it is used in the optimization process.

Each possible solution (candidate) of the optimization problem has a corresponding objective function value. According to this value, the candidate solution is ranked. Solution Optimization techniques may be classified according to the objective of optimization into single-objective and multi-objective. In single-objective optimization techniques, only one objective function is considered and solution candidates are ranked based on their corresponding objective function value. For example, if it is required to maximize the objective function, the optimal solution is the one that results in the maximum value of objective function (and vice versa). In multi-objective optimization techniques, more than one objective functions are considered and solution candidates are treated in a different way, as will come later. Multi-objective optimization provides a set of optimal solutions instead of one in single-objective optimization. Optimization techniques have been used together with a modelling method to obtain optimal designs. The enumeration optimization method with FEA is used in [20–22]. A genetic algorithm with FEA is used in [10,13,14]. In [18], a genetic algorithm is used with fuzzy logic. The non-dominated sorting genetic algorithm (NSGA-II) method is used in [23] with FEA. Differential evolutions are used in [24,25] with FEA. Particle swarm optimization (PSO) is used in [26–29] with different SRM modelling techniques. The increase in computation time in most approaches is due to time-consuming SRM electromagnetic modelling. Using other methods instead of FEA is even less accurate or complicated to build.

In this paper, FEA is chosen for SRM modelling to complement the mathematical formulas for average torque and core losses calculations. FEA usage is limited to inductance calculation and to obtain flux density for some points inside SRM core. This simulation has proved to be achieved in a very short time while using free software FEMM4.2. Accordingly, the computational time is reduced to an acceptable limit while achieving high accuracy. The Jaya algorithm is introduced to optimize SRM design due to its inherent characteristics, as it takes the path directly toward optimal solutions, saving computational time and achieving better objective functions values [30]. The multi-objective version of Jaya (MO-Jaya) algorithm is considered with three objective functions, and they are rated average torque, efficiency, and iron weight. The three objective functions calculation methods are presented in details. The results of optimization and the performance of the Jaya algorithm are compared with those of the non-dominated sorting genetic algorithm (NSGA-II) under the same constraints and for the same objectives [31]. The proposed method represents a general frame work for SRM multi-objective optimization. Unlimited design parameters and objective functions can be included. The objective of this study is to investigate the performance of MO-Jaya algorithm in SRM design optimization.

2. Design of SRM

Before starting the design process of switched reluctance machine, available space should be well investigated. The available space is represented as constraints of SRM dimensions values. The most important space constraints are axial length, maximum outer diameter (maximum frame size), and shaft diameter. The axial length and maximum frame size are obtained from measurements of the available space. The shaft diameter is calculated based on the maximum torque and speed. The difference between maximum outer diameter and Shaft diameter is the space that is available for SRM cores and turns. SRM cores are the stator and rotor, which are salient in producing reluctance variation. The number of poles in both stator and rotor is specified at the beginning as will come later.

In conventional design approaches, lamination dimensions, coil diameter, and number of turns are calculated analytically. Subsequently, the average torque is calculated and compared with the rated required torque. The average torque is calculated based on the flux-current ($\lambda - i$) curves for aligned and unaligned positions. Finally, the whole process is repeated with modified dimensions values and the number of turns to match the output average torque with the required torque. Other performance indices can also be considered, such as torque ripples, efficiency, and so forth. This section demonstrates calculation methods of the considered performance indices and characteristics.

2.1. Number of Poles Selection

Stator poles P_s number and rotor poles P_r number is specified mainly according to general understanding of their influence on SRM performance. A higher pole number produces higher average torque, lower torque ripples, and provides more reliable operation; however, it requires more switching devices and reduces the maximum speed. Lower poles number produces lower average torque and higher torque ripples; however, it requires less switching devices and provides higher maximum speed. For general purpose SRM design, 6/4 and 8/6 SRM configurations are commonly used. Hence, these two configurations are considered for optimization in this paper.

2.2. Poles Arcs Calculation

When considering β_s, β_r are stator pole arc and rotor pole arc, respectively. To achieve self-starting SRM design, minimum stator pole arc may be expressed—as in [32]—by:

$$min[\beta_s] = \frac{4\pi}{P_s P_r}, \qquad rad \qquad (1)$$

The following condition prevents overlap between phases:

$$\beta_s + \beta_r \leq \frac{2\pi}{P_r} \tag{2}$$

If this condition not being followed, then the SRM machine inductance will start to increase before reaching the minimum inductance value. This results in higher unaligned inductance value and the developed torque becomes lower.

2.3. Main Dimensions

Referring to Figure 1 and Table 1, the maximum value of outer diameter (D_o) and maximum value of axial length (L) are obtained from space constraints. Shaft diameter (D_{sh}) is obtained from the shaft's standard sizes that are based on output torque value. The values of outer diameter (D_o) and axial length (L) can be changed during design within their maximum values. However, keeping them at their maximum values results in the maximum output torque of SRM design. The shaft diameter (D_{sh}) is kept constant during design process. The air gap length (g) is assumed to have the commonly used value of 0.5 mm that can be obtained often in practical implementation. The rest of the dimensions are modified to reach the requirements of design.

2.4. Variables Dimensions Limits

The limits of variables are determined by the application and the available space. In this paper, outer diameter (D_o), axial length (L), shaft diameter (D_{sh}), and air gap length (g) are kept constant by making their maximum and minimum limits at the same value. The remaining limits are set by the previous experience. Table 2 shows the maximum and minimum limits values of all the variables.

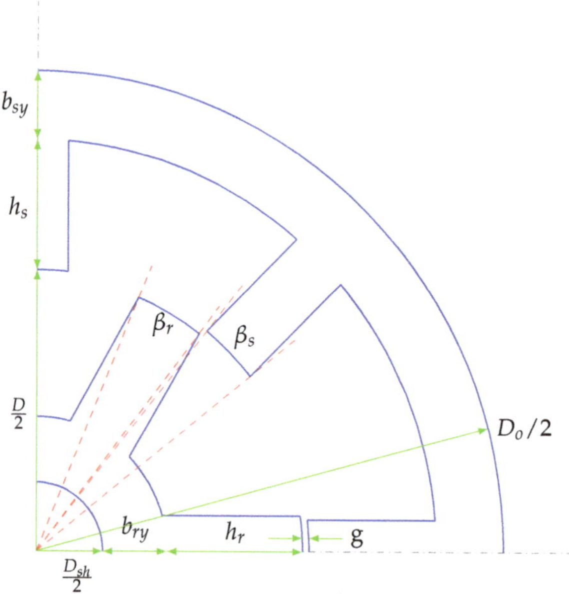

Figure 1. The lamination dimensions considered in the optimization process.

Table 1. SRM dimensions.

Dimension	Unit
outer diameter, D_o	mm
shaft diameter, D_{sh}	mm
axial length, L	mm
bore diameter, D	mm
air gap length, g	mm
stator pole length, h_s	mm
rotor pole length, h_r	mm
stator back iron length, b_s	mm
rotor back iron length, b_r	mm
stator pole arc, β_s	degree
rotor pole arc, β_r	degree
stator poles, P_s	NA
rotor poles, P_r	NA

Table 2. Limits of variables.

Variable	Min	Max	Unit
D_o	130	130	mm
L	100	100	mm
D	44	95	mm
b_{sy}	5	20	mm
b_{ry}	5	20	mm
h_s	7	52	mm
h_r	5	23	mm
D_{sh}	24	24	mm
g	0.5	0.5	mm
β_s	$0.85 \times 720/(P_s P_r)$	$0.6 \times 360/P_s$	degree
β_r	$0.85 \times 720/(P_s P_r)$	$0.6 \times 360/P_s$	degree

2.5. Winding Design

Winding design is achieved, as in [31,33]. The number of turns per phase is calculated based on SRM dimensions, such that the flux density value at the knee point of iron's B-H curve is maintained. In this paper, the value of maximum flux density for the used lamination steel is 1.65 T.

The conductor cross-sectional area is calculated from a specified maximum ampere and current density. Subsequently, the number of horizontal layers, vertical layers, and coil dimensions can be calculated. Finally, clearance between adjacent coils is calculated at the point where the two coils are the closest to each other [31].

2.6. Average Torque Calculation

The SRM average torque is calculated based on assuming that flux linkage (λ) vs. current (i) characteristics are available and the phase current is kept constant at its peak value between unaligned and aligned positions [32]. The average torque is calculated from the total work done of all strokes for one revolution, as follows:

$$T_{av} = \frac{W P_s P_r}{4\pi}, \ N.m \tag{3}$$

$$W = W_{aligned} - W_{unaligned} \tag{4}$$

where $W_{aligned}$ and $W_{unaligned}$ are the areas under $\lambda - i$ curves at the aligned position and unaligned position, respectively. W is the area of energy loop between two $\lambda - i$ curves and calculated as in [32].

2.7. Efficiency Calculation

It is essential to know the losses of switched reluctance motor in order to calculate efficiency [34]. SRM losses calculation, especially the assessment of core losses, is a very difficult task mainly due to tha fact that flux wave-forms are non-sinusoidal and different from each other based on the sector that they are located in SRM's magnetic circuit. Moreover, core losses are dependent on the type of control and operating speed (ω). In a low speed region, it is acceptable to neglect the mechanical losses. Hence, losses may be calculated as:

$$Losses(\omega) = Core\ Losses + Copper\ Losses \tag{5}$$

Once the losses are known, the efficiency may be calculated by:

$$\eta = \frac{\omega T_{av}}{\omega T_{av} + Losses(\omega)} \tag{6}$$

In this paper, efficiency is calculated at the rated speed of 1000 rpm for all SRM designs candidates.

2.8. Copper Losses

Copper losses value depends on phase current, which is determined by the control technique. When considering that N_{ph} is number of phases, R_j is phase dc resistance, and I_j is phase current, the total copper loss instantaneous value may be calculated by the equation:

$$P_{cu}(t) = \sum_{j=1}^{j=N_{ph}} I_j^2(t)R_j \tag{7}$$

The average copper losses can be calculated by equation:

$$P_{cu} = \frac{1}{T}\int_0^T P_{cu}(t)dt, \tag{8}$$

where T is the period of time for $P_s/2$ strokes. For sake of simplification, we assume no overlap between phases. Because the current of phase is not pure dc. The peak value of it (I_p) is considered for copper losses calculation as a pessimistic prediction. Copper loss is then calculated straight forwardly by the equation:

$$P_{cu} = I_p^2 R_{ph} \tag{9}$$

2.9. Eddy Currents Losses

Referring to [35], the eddy current losses in SRM can be calculated by the equation:

$$P_e = \frac{e^2}{4k_{cir}\rho_{fe}\delta}\frac{1}{T}\int\left(\frac{\partial B}{\partial t}\right)^2 dt, \qquad w/kg \tag{10}$$

where e: sheet thickness in meter, k_{cir}: constant ($1 < k_{cir} < 3$) introduced to account for the fact that paths in the interior of the lamination will have smaller emfs than those that are near the surface; ρ_{fe}: the electrical resistivity of the ferromagnetic material (in Ωm); δ: density of the ferromagnetic material (in kg/m^3).

From Equation (10), the waveform of flux density (B) for all SRM sectors must be known. Once they are available, P_e is calculated by numerical integration and differentiation. There are a lot of methods to obtain these wave-forms and many of them are time consuming. In [34], a mathematical method using matrices is introduced in order to obtain the wave-forms of all the SRM sectors in a systematic manner. The calculation of B wave-forms for all sectors is achieved by the modulation of triangular pulses. The stator poles wave-forms only consist of unipolar triangular pulses, while those of the rotor poles contain both positive and negative pulses. The stator and rotor yoke wave-forms have a more complicated relationship with the triangular pulses. This method is demonstrated in details in [34] and then used here for 8/6 and 6/4 SRMs.

2.10. Hysteresis Losses

Referring to [31], the hysteresis losses can be calculated for various sectors of SRM. The flux density wave-forms depend on the phase current waveform and the speed of the motor. The flux density wave-forms calculated in this paper rated the speed of 1000 rpm and control is by a single pulse voltag. Note that the phase current has the peak of six ampere for all SRMs design candidates.

3. Jaya Optimization Method

The population-based meta-heuristic optimization algorithms can be classified to two groups, and they are the evolutionary algorithms (EA) and swarm intelligence (SI) based algorithms. All of these algorithms have the same basic structure that is explained by Figure 2. Firstly, the initialization of solutions in the beginning is made mainly by random choice of variables. Then, the objective function values are obtained and evaluated. After that, selection of the best solutions is made to use these solutions in the production of new solutions. Finally, the termination condition of the process is checked if true, and then end or else continue.

Optimization techniques differ from each other by the different methods that are used to accomplish these steps. However, all of population-based meta-heuristic optimization algorithms have a common limitation, which is the different parameters that are required for proper working.

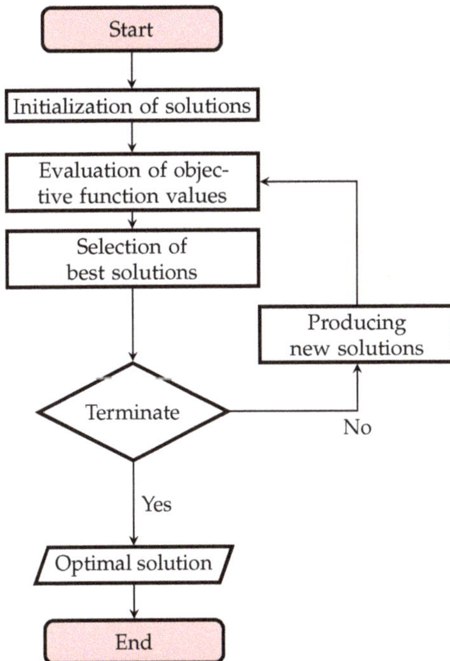

Figure 2. Flowchart of general optimization algorithm.

Referring to [30], the Jaya algorithm was introduced in 2016 by Ravipudi Venkata Rao. "Jaya" is a Sanskrit word that means victory or triumph. The algorithm is simple to implement and it does not require tuning of any parameters. In Jaya algorithm, the initial solutions are randomly generated within the search space. After that, the solutions are updated using Equation (11).

$$A(i+1,j,k) = A(i,j,k) + r(i,j,1)\Big(A(i,j,b)$$
$$- |A(i,j,k)|\Big) - r(i,j,2)\Big(A(i,j,w)$$
$$- |A(i,j,k)|\Big) \tag{11}$$

where b and w represent the index of the best and worst solutions in the population. i,j,k are the index of iteration, variable, and candidate solution, respectively. $A(i,j,k)$ is the jth variable in ith iteration of kth solution candidate. $r(i,j,1)$ *and* $r(i,j,2)$ are random generated ratios in the range of $[0,1]$ to ensure good diversification.

3.1. Single Objective Jaya Algorithm

In single objective optimization, the required is to maximize or minimize a single function. Thereby, for two solution candidates a *better* solution is whether greater or smaller in value. For objective function $f(x_1, x_2, x_3, \cdots, x_m)$, which has m variables and n population size, the solutions are represented by a data structure i.e., matrix as in Equation (12). Each solution may be represented by a column of different variables. The objective function $(f(X))$ can be represented by a matrix (F) that has one row and n columns in the case of single objective optimization, as in Equation (13). For multi objective optimization problem with number of q objective functions, matrix F is of q rows and n column as in Equation (14).

$$X = \begin{bmatrix} x_1^1 & x_1^2 & x_1^3 & \cdots & x_1^n \\ x_2^1 & x_2^2 & x_2^3 & \cdots & x_2^n \\ \vdots & \vdots & \vdots & \vdots & \vdots \\ x_m^1 & x_m^2 & x_m^3 & \cdots & x_m^n \end{bmatrix} \tag{12}$$

$$F = \begin{bmatrix} f_1^1 & f_1^2 & f_1^3 & \cdots & f_1^n \end{bmatrix} \tag{13}$$

$$F = \begin{bmatrix} f_1^1 & f_1^2 & f_1^3 & \cdots & f_1^n \\ f_2^1 & f_2^2 & f_2^3 & \cdots & f_2^n \\ \vdots & \vdots & \vdots & \vdots & \vdots \\ f_q^1 & f_q^2 & f_q^3 & \cdots & f_q^n \end{bmatrix} \tag{14}$$

Solution matrix (X) is updated using Equation (11) based on the *best* and *worst* solutions obtained by comparing all of the solution candidates with each other.

3.2. Multi Objective Jaya Algorithm

Solutions in multi objective Jaya algorithm are updated using the same Equation (11), and the results of objective functions are represented in Equation (14). However, non-dominated sorting approach and crowding distance computation mechanisms are to be used in order to handle the conflicting objectives of optimization problem. It is essential Jaya algorithm to obtain the *best* and *worst* solutions in order to produce new solutions using Equation (11).

In multi objective optimization problem, obtaining the best and worst solutions is not straightforward as in single objective optimization. After non-dominated sorting is achieved and crowding distances are computed for solutions of the same rank (front), the best solution would surely be in the first rank and the worst solution would be in the last one. However, the solutions of the same rank cannot be compared with each other and, hence, the crowding distance is used to decide the best and worst solutions among their ranks. The crowding distance is an indicator of the degree of diversity of solutions in same front; hence, solutions with higher crowding distance are preferred, which enables covering a wider search area. In [30], the solution with highest crowding distance among first rank solutions is considered to be the *best* and the solution with the lowest crowding distance among last rank solutions is considered to be the *worst*.

In this paper, the other method is used to decide the best and worst solutions. The best solution is chosen randomly from the first rank, this choice is changed four times until all population solutions are updated. In the same manner, the worst solution is chosen randomly from the last rank. This method achieves a wide variety of solutions by distributing the priority of search among different search directions. Finally, the designer has the option to choose the best design that serves the application among best ranks of solutions.

Dominance

It is required to check each solution with the rest on the dominance basis. Assuming that X_1, X_2 are two solutions, m is the number of objective functions, if $X_1 \prec X_2$ (X_1 dominates X_2) is true, the Pareto dominance conditions must all be true, and they are:

1. $f_j(X_1) \not\vartriangleright f_j(X_2) \forall j = \{1, ..., m\}$
2. $f_j(X_1) \vartriangleleft f_j(X_2) \exists j = \{1, ..., m\}$

Dominance is investigated for all of the solutions. The solution that is not dominated by any of the remaining solutions is a non-dominated solution and it is removed from the solutions matrix. This is repeated for all solutions and the resulted non-dominated solutions are considered *rank 1*. The same process is repeated for the remaining ranks.

Crowding Distance

Crowding distance is computed in the same manner, as mentioned in [30] (page 15). Crowding distance is computed for each solution using Equation (15).

$$CD_j = CD_j + \frac{f_m^{j+1} - f_m^{j-1}}{f_m^{max} - f_m^{min}} \tag{15}$$

where j is a solution in the sorted list, f_m is the objective function value of mth objective, and f_m^{max} and f_m^{min} are the population-maximum and population-minimum values of mth objective functions.

4. Multi-Objective Jaya Algorithm for SRM Design Optimization

In the optimization of SRM, the dimensions in Table 1 represent one solution. All of the solutions are stored in the matrix (X), as follows:

$$X = \begin{bmatrix} D_o^1 & D_o^2 & D_o^3 & \cdots & D_o^n \\ L^1 & L^2 & L^3 & \cdots & L^n \\ \vdots & \vdots & \vdots & \vdots & \vdots \\ \beta_s^1 & \beta_s^2 & \beta_s^3 & \cdots & \beta_s^n \end{bmatrix} \tag{16}$$

where m is the number of variables and n is the size of population in one generation.

4.1. Objective Functions

The objectives of SRM optimization depend on the application and its conditions. For general purpose SRM, average torque T_{av} and efficiency η are needed to be maximized and iron weight W_i is to be minimized. In some applications, other objectives (i.e., torque ripples, acoustic noise, vibrations $\cdots$, and so forth) are considered.

$$F = \begin{bmatrix} T_{av}^1 & T_{av}^2 & T_{av}^3 & \cdots & T_{av}^n \\ \eta^1 & \eta^2 & \eta^3 & \cdots & \eta^n \\ W_i^1 & W_i^2 & W_i^3 & \cdots & W_i^n \end{bmatrix} \tag{17}$$

4.2. Constraints

Because the SRM dimensions are significant, there must be certain constraints on them to prevent any non-logical vector of variables. The dimensions in variables matrix (X) must be modified to satisfy the following constraints:

$$D_{sh} + 2b_{ry} + 2h_r + 2g = D \tag{18}$$

$$D + 2b_{sy} + 2h_s = D_o \tag{19}$$

$$\beta_r > \beta_s \tag{20}$$

$$\frac{D}{2}\left(1 - \frac{\beta_r P_r}{2\pi}\right) \geq h_r \tag{21}$$

It is also a part of the constraints to specify certain limits to each variable(dimension). If a certain dimension needed to be fixed, this can simply achieved by setting both the minimum and maximum limits to the desired value.

Code Algorithm

The code is made using Lua programming language. The code is executed by FEMM4.2 software. The choice of Lua programming language to be used is due to its simplicity, and that it is adopted by FEMM4.2, which provides the FEA analysis in good accuracy. Figure 3 shows the code's algorithm. The code starts with inserting SRM optimization data. These data include the population size, problem variables, variables limits, and objective functions, to specify which objective function to maximize and which to minimize, maximum iterations limit and numbers of rotor and stator poles (P_r and P_s). Subsequently, solutions initialization are achieved for all solutions in the population by a random choice of variables within the search space that is stored in matrix X as in Equation (16). After that, the solutions are modified to satisfy constraints in Equations (18)–(21) by changing the values of variables that resulted from random choice. This change is to give the variable that is out of its limits one of the acceptable limits value. For example, if a variable is greater than maximum, then it is changed to the maximum value and vice versa. Subsequently, FEA Analysis is accomplished using FEMM4.2 software to calculate average torque, maximum stator and rotor poles flux densities and volume of iron. Next, remaining objective functions (η and W_i) are calculated using the results of FEA analysis. After that, non-dominated sorting is achieved and crowdingdistance is calculated for all fronts. Based on non-dominated sorting and crowding distance, the best and worst solutions are specified. Susequently, the termination condition is checked if the number of iterations exceeds the maximum limit or not. Finally, output of optimization is provided, which is the non-dominated front of all solutions. Note that the output also provides all solutions that have been produced for all iteration, which is beneficial to provide more solutions for designer to choose.

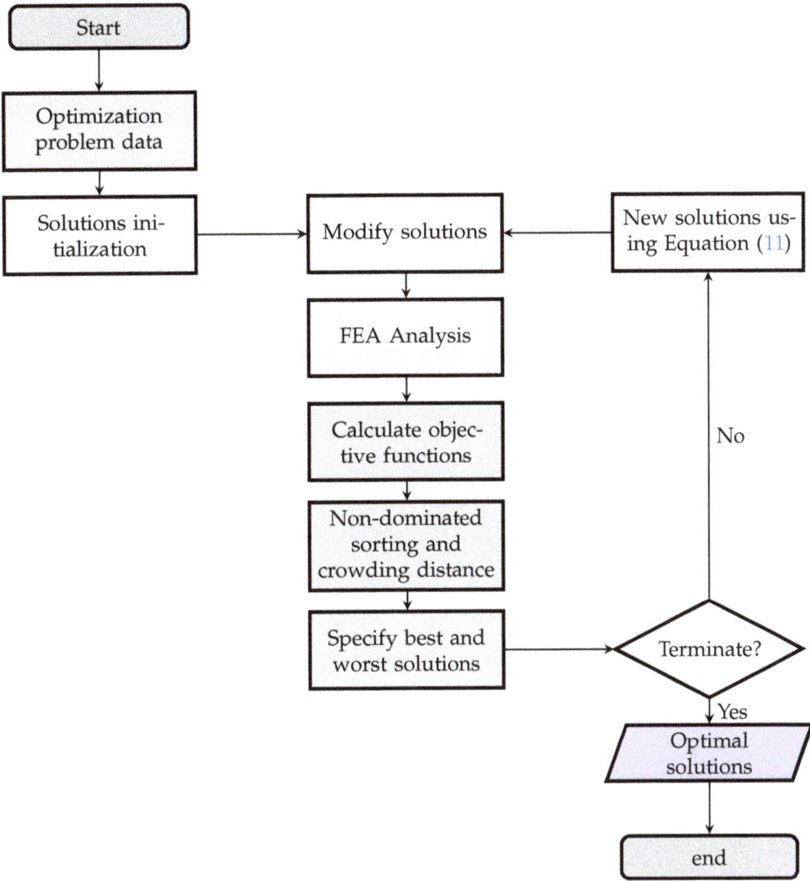

Figure 3. Flowchart of MO-Jaya algorithm.

5. Results and Discussions

The evaluation of optimization algorithm performance may be mainly achieved by two factors that are computational time and candidates. The less computational time, the better is the performance for the same results of candidates. On the other hand, the better candidates, the better performance for the same computational time. In this paper, better candidates production are of significant interest than computational time. SRM design candidates with higher average toque and efficiency and lower iron weight are considered to be better candidates. Hence, it is expected that for a successful optimization process, most of the search (crowded area) ought to be in the areas that maximize both average toque and efficiency and minimize iron weight. For example, assuming that it us desired to maximize *both* of objective functions f_1 and f_2, the results of the optimization process that are the solution candidates should be at the upper right quarter, as shown in Figure 4. The same goes for other quarters in cases of minimizing both f_1 and f_2, maximize f_1 and minimize f_2, and maximize f_2 and minimize f_1. Additionally, it is expected that the optimization program will find more solutions in the correct quarter (direction) with increased iterations until the search area is fully covered.

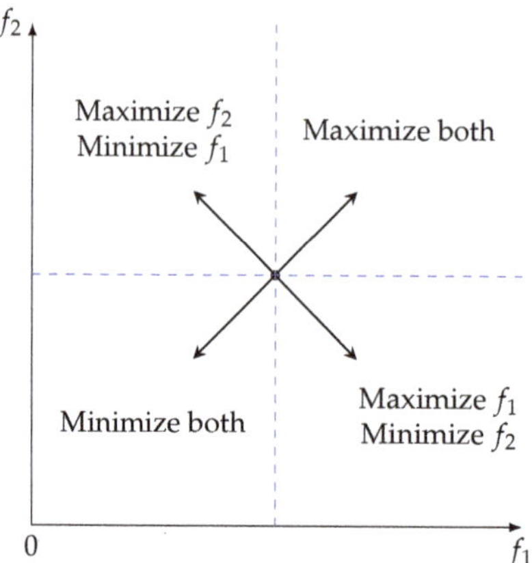

Figure 4. Search directions in multi objective optimization.

The objectives of optimization in this paper are to maximize average torque (T_{av}) and efficiency (η) and minimize iron weight (W_i). The results of optimization process are considered in pairs, as shown in Figures 5 and 6. In Figure 5c, efficiency is increased and iron weight is decreased while iterations are increasing. The search direction is correct according to Figure 4. The search direction is also correct with respect to iron weight and average torque, as shown in Figure 5b. In Figure 5a, the search direction is not clearly obvious, as in Figure 5b,c, which is due to the non-linear relationships of dimensions and objective functions (efficiency and average torque). The program changes the direction to cover all the search area to provide a various groups of solutions.

In Figure 7, objective functions of SRM are shown with respect to the number of iterations. It can be seen that the objectives are conflicting with each other. The optimization program is made, such that the iterations are continued while the algorithm is not biased for any of the objectives by changing the best and worst solutions in the way mentioned earlier in Section 3.2. Hence, better solutions are found in all of search directions (objectives). Figure 8 shows how objective functions interact with each other and confirms the wide search area.

Every point represents a solution and some solutions are invalid due to dimensions constraints, as they cause a negative clearance between windings. The penalty for these solutions is to eliminate them by making their corresponding objective functions the worst of their values. Hence, the invalid solutions take zero average torque and efficiency and iron weight of the whole volume ($2\pi DL\rho_i$), where ρ_i is the density of iron.

It can be seen that search direction is changing while iterations increases as the ups and downs in objective functions values indicate. In other words, the program is exploring new areas with more iterations. This is because of two reasons, the first reason is the random chosen ratios $r(i,j,1)$ and $r(i,j,2)$ in Equation (11). The second reason is the random choice of the best and worst solutions from the highest and lowest ranks (fronts). This is beneficial, as the algorithm does not repeat it self and provide more various solutions. However, a drawback is that optimal solutions (non-dominated front) are distributed over iterations. In other words, the best candidates do not exist in the last iteration exclusively. The designer then has to take this into consideration while choosing a design to implement.

For further evaluation, the optimization results by multi-objective Jaya algorithm (MO-Jaya) are compared with optimization results by non-dominated sorting genetic algorithm (NSGA-II) in [31]. Two optimization processes have the same constraints, objective functions, calculation methods, and common parameters (population size and maximum iterations). The results that are shown in Figures 5–7 in this paper are compared with results that are shown in Figures 4, 5k and 7 in [31]. The comparison can be summarized in the following point:

1. It can be seen that MO-Jaya provides much better coverage of search area. The results in [31] are more concentrated in small area. The reason for this is the constant parameters of cross over and mutation in NSGA-II method, unlike MO-Jaya, where the parameters are randomly changed. This is obvious by comparing Figures 5 and 6 in this paper with Figures 4 and 5 in [31].

2. Figure 7 in this paper shows that the optimal solutions may be lost with increased iterations, unlike in [31], as shown in Figure 5. This is a consequence of changing direction of search, as demonstrated earlier. This has no effect on the total design candidates that Jaya algorithm provide and the designer can choose various design candidates from any iteration's population.

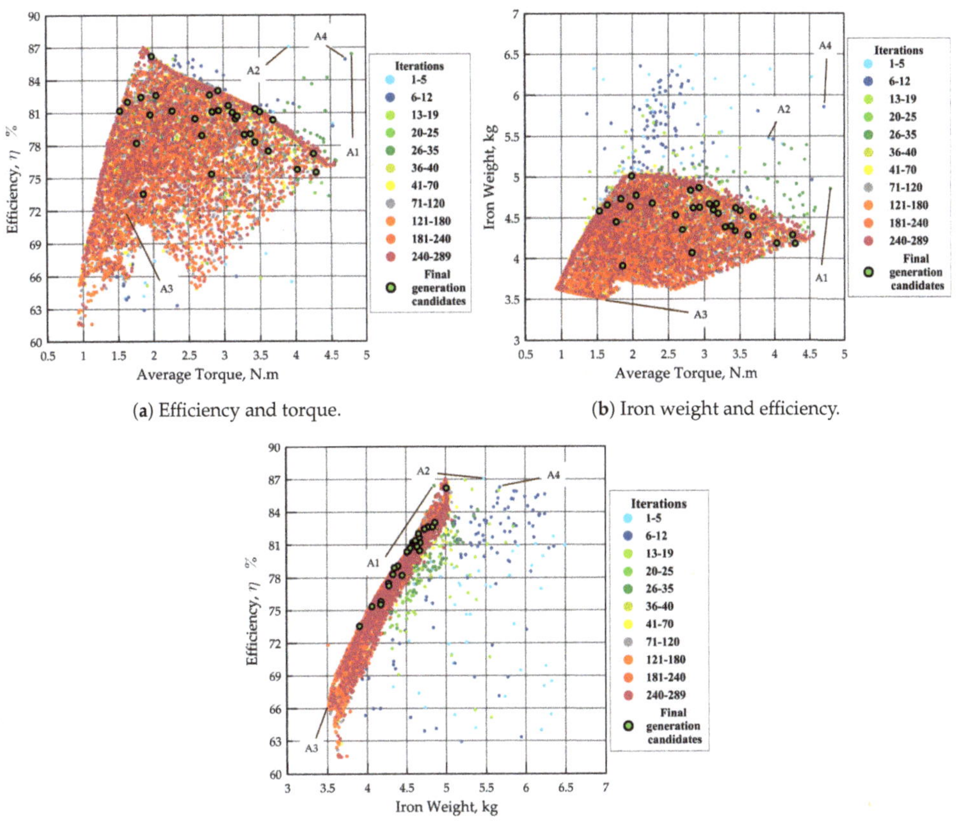

(a) Efficiency and torque.

(b) Iron weight and efficiency.

(c) Iron weight and torque.

Figure 5. Objective functions results for 8/6 SRM.

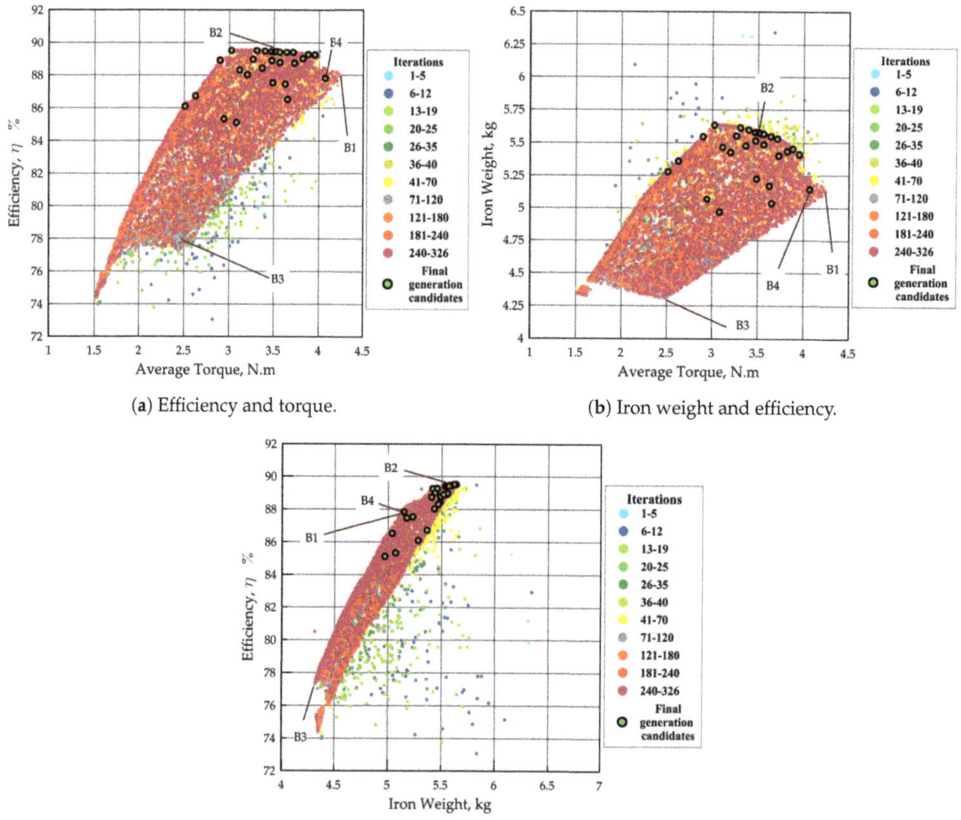

(**a**) Efficiency and torque.

(**b**) Iron weight and efficiency.

(**c**) Iron weight and torque.

Figure 6. Objective functions results for 6/4 SRM.

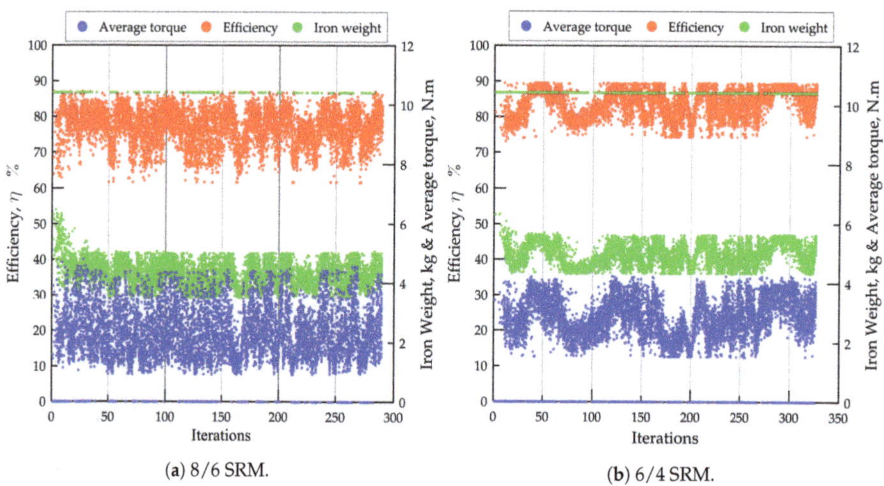

(**a**) 8/6 SRM.

(**b**) 6/4 SRM.

Figure 7. Objectives with iterations for 8/6 and 6/4 SRM using MO-Jaya.

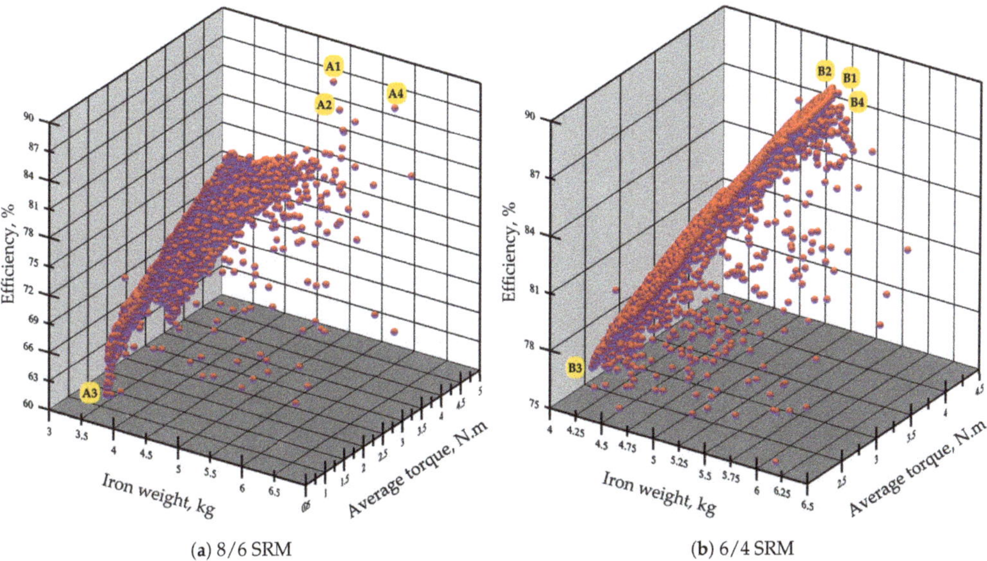

(**a**) 8/6 SRM   (**b**) 6/4 SRM

Figure 8. 3D representation of objective functions values through optimization process.

Four optimal solutions are chosen for both 8/6 and 6/4 configurations. Table 3 summarizes the selected designs and their corresponding objective functions. Most of designs do not belong to the last iteration, as they provide better characteristics. For 8/6 SRM configuration, design A1 achieves the highest average torque, A2 achieves highest efficiency at rated speed, A3 has the lowest iron weight, and A4 is a compromise design which gives a higher priority for average torque and efficiency. For the 6/4 SRM configuration, design B1 achieves both the highest average torque and efficiency at rated speed, B3 has the lowest iron weight, and B2 and B4 are the compromise designs. Figures 5 and 6 show the location of the selected designs among the remaining designs. It is worth mentioning that the resulted optimal designs by Mo-Jaya optimization method are close to these that result from NSGA-II for the same constraints. However, the Mo-Jaya method achieved better diversification than NSGA-II.

Table 4 shows the details of selected designs. It can be noticed that the parameters and dimensions of SRM are changed with iterations to produce better designs in such a way that matches with the SRM design experience. This indicates that the calculation methods of objective functions mainly succeeded to represent SRM. For example, when rotor pole length (h_r) is increased and rotor yoke thickness (b_{ry}) is reduced, the impact on SRM is to produce average torque. This result matches with design experience, as the difference between anligned and unaligned inductances is increased and, hence, energy conversion increases Equation (4).

Torque per phase is shown in Figure 9 for 8/6 and 6/4 SRM configurations before and after optimization. The peak torque is increased in optimal designs (A1 and B1) over initial designs. It can be also noticed that torque ripples also increase. This is because torque ripples minimization is not considered as an objective for optimization. However, so far, the program succeeded in achieving what is requested and specified in the objectives. More work is taking place to include torque ripple minimization in optimization objectives in the future.

Further study on selected SRM design candidates are made to evaluate efficiency over a wide range of speeds Figure 10. It can be noticed that 6/4 SRM (B group) achieves higher efficiency values. This is due to lower iron losses that are caused by flux variation in magnetic circuit of SRM. Figure 11 shows the iron losses.

Table 3. Chosen optimal candidates.

Candidate	Configuration	Average Torque (N.m)	Efficiency (%)	Iron Weight (kg)
A1	8/6	4.8	86.43	4.86
A2	-	3.79	87.1	5.48
A3	-	1.55	71.85	3.51
A4	-	4.68	85.9	5.86
B1	6/4	4.24	87.9	5.12
B2	-	3.49	88.6	5.6
B3	-	2.38	80.5	4.31
B4	-	4.1	87.8	5.144

Table 4. Parameters and objective functions values of the selected optimal designs.

Value	A1	A2	A3	A4	B1	B2	B3	B4	Unit
T_{av}	4.8	3.79	1.55	4.68	4.24	3.49	2.38	4.1	N.m
η	86.43	87.1	71.85	85.9	87.9	88.6	80.5	87.8	%
Iron weight	4.86	5.48	3.51	5.86	5.12	5.6	4.31	5.144	kg
T_{ph}	242	210	204	220	270	238	256	264	Turns
L_a	72.9	65.55	29.54	72.3	112.7	100.6	77.2	109.6	mH
L_u	11.8	15.2	10.6	12.97	13.06	17.8	18.1	12.92	mH
R_{ph}	1.6	1.4	1.4	1.46	1.43	1.26	1.35	1.39	Ω
D_o	130	130	130	130	130	130	130	130	mm
L	100	100	100	100	100	100	100	100	mm
D	75.9	68.62	44.99	80.06	78.26	62.1	47.22	76.67	mm
b_{sy}	6.01	6.42	6	6	7.2	7.31	7.2	7.2	mm
b_{ry}	5	6.17	5	9.3	5	5	5	5	mm
h_s	21.045	24.27	36.508	18.97	18.67	26.64	34.18	19.5	mm
h_r	20.46	12.78	5	17.99	21.63	13.44	6.1	20.83	mm
D_{sh}	24	24	24	24	24	24	24	24	mm
g	0.5	0.5	0.5	0.5	0.5	0.5	0.5	0.5	mm
β_s	16.46	20.24	12.75	19.04	25.5	32.12	25.5	25.5	degree
β_r	19.38	32.4	18.8	28.54	25.6	42.37	25.63	26.9	degree

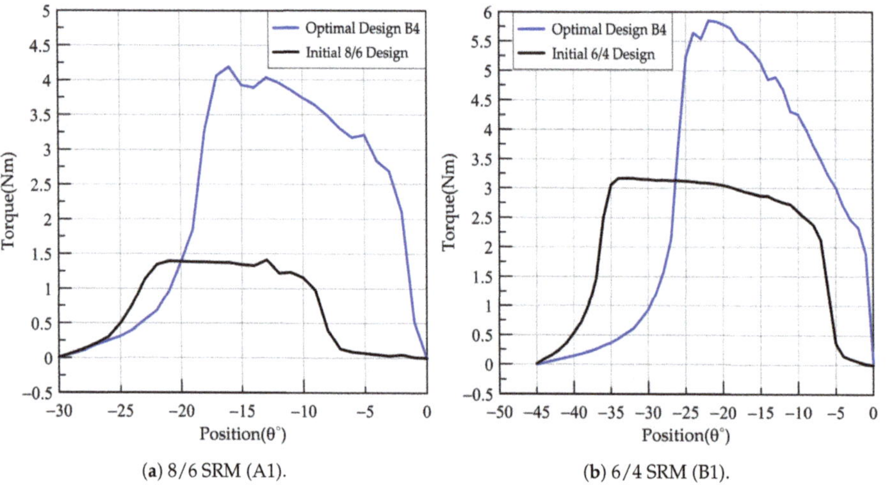

(**a**) 8/6 SRM (A1). (**b**) 6/4 SRM (B1).

Figure 9. Torque per phase of initial and optimal SRM designs.

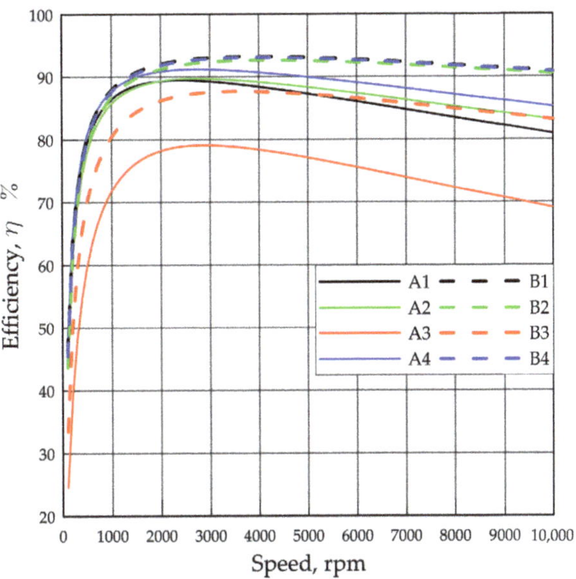

Figure 10. Efficiency over wide range of speeds.

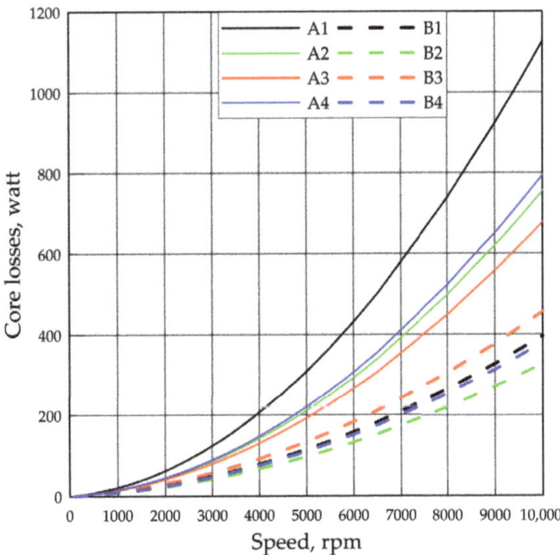

Figure 11. Core losses of selected optimal designs over wide range of speeds.

6. Conclusions

In this paper, the Jaya algorithm is introduced to solve SRM optimization problem. The proposed approach considers the multi-objectives optimization of SRM design problem. Results show the success of Jaya algorithm to achieve good diversification and good intensification.The contribution in this paper can be summarized by the comparison between MO-Jaya and NSGA-II methods for the same design requirements and constracompared-ints. MO-Jaya showed much better performance and a wider search area with results of NSGA-II. The reasons for this is the random choice of parameters, best and worst solutions in Jaya algorithm that achieve considerably more search diversity. The introduced approach represents a framework that can adopt any application requirements and calculation methods. Further work is required to develop the calculation methods and the design parameters to include other important requirements of SRM, such as torque ripple minimization, thermal modelling, and so forth. Finally, the decision is left to the designer to pick the most convenient design from a wide variety of optimal of solutions.

Author Contributions: Conceptualization, M.E.-N.; methodology, M.E.-N. and M.A.; software, M.A., M.E.-N.; validation: M.E.-N. and M.I.; formal analysis, M.E.-N. and M.A.; investigation, M.E.-N. and M.I.; resources, M.E.-N.; data curation, M.A. and M.I.; writing—original draft preparation, M.A. and M.E.-N.; writing—review and editing, M.E.-N., M.I. and H.R.; visualization, M.A. and H.R.; supervision, M.E.-N.; project administration, M.E.-N.; funding acquisition, H.R. All authors have read and agreed to the published version of the manuscript.

Funding: This research received no external funding.

Institutional Review Board Statement: Not applicable.

Conflicts of Interest: The authors declare no conflict of interest.

References

1. Li, S.; Zhang, S.; Habetler, T.G.; Harley, R.G. Modeling, Design Optimization and Applications of Switched Reluctance Machines—A Review. *IEEE Trans. Ind. Appl.* **2019**, *55*, 2660–2681. [CrossRef]
2. El-Nemr, M.K.; AI-Khazendar, M.A.; Rashad, E.M.; Hassanin, M.A. Modeling and Steady-State Analysis of Stand- Alone Switched Reluctance Generators. *IEEE Power Eng. Soc. Meet.* **2003**, *3*, 1894–1899.
3. Kreyszig, E.; Kreyszig, H.; Norminton, E.J. Unconstrained Optimization, Linear Programming. In *Advanced Engineering Mathematics*, 10th ed.; John Wiley Inc.: Chichester, UK, 2011; pp. 950–969.
4. Cao, Y.; Zhang, H.; Li, W.; Zhou, M.; Zhang, Y.; Chaovalitwongse, W.A. Comprehensive Learning Particle Swarm Optimization Algorithm With Local Search for Multimodal Functions. *IEEE Trans. Evol. Comput.* **2019**, *23*, 718–731. [CrossRef]
5. Li, W.; He, L.; Cao, Y. Many-Objective Evolutionary Algorithm With Reference Point-Based Fuzzy Correlation Entropy for Energy-Efficient Job Shop Scheduling With Limited Workers. *IEEE Trans. Cybern.* **2021**. [CrossRef] [PubMed]
6. Fu, Y.; Li, Z.; Chen, N.; Qu, C. A Discrete Multi-Objective Rider Optimization Algorithm for Hybrid Flowshop Scheduling Problem Considering Makespan, Noise and Dust Pollution. *IEEE Access* **2020**, *8*, 88527–88546. [CrossRef]
7. Besmi, M.R. Geometry Design of Switched Reluctance Motor to Reduce the Torque Ripple by Finite Element Method and Sensitive Analysis. *J. Electr. Power Energy Convers. Syst.* **2016**, *1*, 23–31.
8. Zhu, Y.; Yang, C.; Yue, Y.; Wei, W.; Zhao, C. Design and optimization of an In-wheel switched reluctance motor for electric vehicles. *IET Intell. Transp. Syst.* **2019**, *13*, 175–182.
9. Cui, X.; Sun, J.; Gan, C.; Gu, C.; Zhang, A.Z. Optimal Design of Saturated Switched Reluctance Machine for Low Speed Electric Vehicles by Subset Quasi-Orthogonal Algorithm. *IEEE Access* **2019**, *7*. [CrossRef]
10. Jiang, J.W.; Bilgin, B.; Howey, B.; Emadi, A. Design optimization of switched reluctance machine using genetic algorithm. In Proceedings of the 2015 IEEE International Electric Machines & Drives Conference (IEMDC), Coeur d'Alene, ID, USA, 10–13 May 2015; pp. 1671–1677.
11. Choi, J.-H.; Kim, S.; Shin, J.-M.; Lee, J.; Kim, S.-T. The multi-object optimization of switched reluctance motor. In Proceedings of the Sixth International Conference on Electrical Machines and Systems, Beijing, China, 9–11 November 2003; pp. 195–198.
12. Cheng, H.; Chen, H.; Yang, Z. Design indicators and structure optimisation of switched reluctance machine for electric vehicles. *IET Electr. Power Appl.* **2015**, *9*, 319–331. [CrossRef]
13. Smaka, S.; Konjicija, S.; Masic, S.; Cosovic, M. Multi-Objective design optimization of 8/14 switched reluctance motor. In Proceedings of the 2013 International Electric Machines & Drives Conference, Chicago, IL, USA, 12–15 May 2013; pp. 468–475.
14. Pisch, J.H.; Li, Y.; Kjaer, P.C.; Gribble, J.J.; Miller, T.J.E. Pareto-Optimal firing angles for switched reluctance motor control. In Proceedings of the Second International Conference on Genetic Algorithms in Engineering Systems, Glasgow, UK, 2–4 September 1997; pp. 90–96.
15. Ilea, D.; Radulescu, M.M.; Gillon, F.; Brochet, P. Multi-objective optimization of a switched reluctance motor for light electric traction applications. In Proceedings of the 2010 IEEE Vehicle Power and Propulsion Conference (VPPC 2010), Lille, France, 1–3 September 2010; pp. 1–6.
16. Li, S.; Zhang, S.; Jiang, C.; Habetler, T.G.; Harley, R.G. A fast control-integrated and multiphysics based multiobjective design optimization of switched reluctance machines. In Proceedings of the 2017 IEEE Energy Conversion Congress and Exposition (ECCE), Cincinnati, OH, USA, 1–5 October 2017; pp. 730–737.
17. Grebennikov, N.; Talakhadze, T.; Kashuba, A. Equivalent Magnetic Circuit for Switched Reluctance Motor with Strong Mutual Coupling between Phases. In Proceedings of the 26th International Workshop on Electric Drives: Improvement in Efficiency of Electric Drives (IWED), Moscow, Russia, 30 January–2 February 2019.

18. Mirzaeian, B.; Moallem, M.; Tahani, V.; Lucas, C. Multiobjective optimization method based on a genetic algorithm for switched reluctance motor design. *IEEE Trans. Magn.* **2002**, *38*, 1524–1527. [CrossRef]
19. Mousavi-Aghdam, S.R.; Feyzi, M.R.; Ebrahimi, Y. A New Switched Reluctance Motor Design to Reduce Torque Ripple using Finite Element Fuzzy Optimization. *Iran. J. Electr. Electron. Eng.* **2012**, *8*, 91–96.
20. Besbes, M.; Gasbi, M.; Hoang, E.; Lecrivian, M.; Grioni, B.; Plasse, C. SRM design for starter-alternator system. *Proc. Int. Conf. Electr. Mach.* **2000**, *6*, 1931–1935.
21. Xue, X.D.; Cheng, K.W.E.; Cheung, N.C. Multi-objective optimization design of in-wheel switched reluctance motors in electric vehicles. *IEEE Trans. Ind. Electron.* **2010**, *57*, 2980–2987. [CrossRef]
22. Omekanda, A.M. A new technique for multidimensional performance optimization of switched reluctance motors for vehicle propulsion. *IEEE Trans. Ind. Appl.* **2003**, *39*, 672–676. [CrossRef]
23. Anvari, B.; Toliyat, H.A.; Fahimi, B. Simultaneous Optimization of Geometry and Firing Angles for In-Wheel Switched Reluctance Motor Drive. *IEEE Trans. Transp. Electrif.* **2017**, *4*, 322–329. [CrossRef]
24. Öksüztepe, E. In-wheel switched reluctance motor design for electric vehicles by using pareto based multi objective differential evolution algorithm. *IEEE Trans. Veh. Technol.* **2016**, *66*, 4706–4715. [CrossRef]
25. Ma, C.; Qu, L.; Mitra, R.; Pramod, P.; Islam, R. Vibration and torque ripple reduction of switched reluctance motors through current profile optimization. In Proceedings of the 2016 IEEE Applied Power Electronics Conference and Exposition (APEC), Long Beach, CA, USA, 20–24 March 2016; pp. 3279–3285.
26. Gao, J.; Sun, H.; He, L. Optimization design of Switched Reluctance Motor based on Particle Swarm Optimization. In Proceedings of the 2011 International Conference on Electrical Machines and Systems (ICEMS), Beijing, China, 20–23 August 2011; pp. 1–5.
27. Ma, C.; Qu, L. Multiobjective optimization of switched reluctance motors based on design of experiments and particle swarm optimization. *IEEE Trans. Energy Convers.* **2015**, *30*, 1144–1153. [CrossRef]
28. Ren, Z.; Zhang, D.; Koh, C.-S. Multi-objective worst-case scenario robust optimal design of switched reluctance motor incorporated with FEM and kriging. In Proceedings of the 2013 International Conference on Electrical Machines and Systems (ICEMS), Busan, Korea, 26–29 October 2013; pp. 716–719.
29. Zhang, S.; Li, S.; Dang, J.; Habetler, T.G.; Harley, R.G. Multi-objective design and optimization of generalized switched reluctance machines with particle swarm intelligence. In Proceedings of the 2016 IEEE Energy Conversion Congress and Exposition (ECCE), Milwaukee, WI, USA, 18–22 September 2016; pp. 1–7.
30. Rao, R.V. *Jaya: An Advanced Optimization Algorithm and Its Engineering Applications*; Springer: Berlin/Heidelberg, Germany, 2018.
31. El-Nemr, M.; Afifi, M.; Rezk, H.; Ibrahim, M. Finite Element Based Overall Optimization of Switched Reluctance Motor Using Multi-Objective Genetic Algorithm (NSGA-II). *Mathematics* **2021**, *9*, 576. [CrossRef]
32. Krishnan, R. *Switched Reluctance Motor Drives, Modeling, Simulation, Analysis, Design, and Applications*; CRC Press: Boca Raton, FL, USA, 2001.
33. Vijayraghavan, P. *Design of Switched Reluctance Motors and Development of a Universal Controller for Switched Reluctance and Permanent Magnet Brush-Less DC Motor Drives*; Virginia Tech: Blacksburg, VA, USA, 2001.
34. Hayashi, Y.; Miller, T.J.E. A New Approach to Calculating Core Losses in the SRM. *IEEE Trans. Ind. Appl.* **1995**, *31*, 5. [CrossRef]
35. Torrent, M.; Andrada, P.; Blanque, B.; Martinez, E.; Perat, J.I.; Sanchez, J.A. Method for estimating core losses in switched reluctance motors. *Eur. Trans. Electr. Power* **2011**, *21*, 757–771. [CrossRef]

mathematics

Article

Improving Reliability and Energy Efficiency of Three Parallel Pumps by Selecting Trade-Off Operating Points

Safarbek Oshurbekov, Vadim Kazakbaev, Vladimir Prakht * and Vladimir Dmitrievskii

Department of Electrical Engineering, Ural Federal University, 620002 Yekaterinburg, Russia; safarbek.oshurbekov@urfu.ru (S.O.); vadim.kazakbaev@urfu.ru (V.K.); vladimir.dmitrievsky@urfu.ru (V.D.)
* Correspondence: va.prakht@urfu.ru; Tel.: +7-343-375-45-07

Abstract: Reliability, along with energy efficiency, is an important characteristic of pump units in various applications. In practical pump applications, it is important to strike a balance between reliability and energy efficiency. These indicators strongly depend on the applied control method of the pump unit. This study analyzes a trade-off method for regulating a system with three parallel pumps equipped with only one frequency converter (multi-pump single-drive system). A typical operating cycle of a pumping system with variable flow rate requirements is considered. The proposed trade-off method is compared with the traditional regulation, when a change in the operating point of the pump is achieved only by changing the rotation speed, and with the method for maximum reliability. It is shown that the proposed trade-off method makes it possible to ensure sufficient reliability of the multi-pump system operation without a significant increase in energy consumption.

Keywords: centrifugal pump; energy efficiency; induction motor; parallel pumps; throttling; variable speed pump

Citation: Oshurbekov, S.; Kazakbaev, V.; Prakht, V.; Dmitrievskii, V. Improving Reliability and Energy Efficiency of Three Parallel Pumps by Selecting Trade-Off Operating Points. *Mathematics* **2021**, *9*, 1297. https://doi.org/10.3390/math9111297

Academic Editor: Nicu Bizon

Received: 5 April 2021
Accepted: 3 June 2021
Published: 5 June 2021

Publisher's Note: MDPI stays neutral with regard to jurisdictional claims in published maps and institutional affiliations.

1. Introduction

Pumps consume about 20% of the electricity generated worldwide [1]. Most of the life cycle cost of a pump is the cost of the electricity it consumes. Therefore, pumps are one of the most promising applications for the implementation of energy-saving technologies. At the same time, maintenance and repairs account for a significant part of the life cycle costs of a pump unit (Figure 1a). Therefore, when optimizing the total life cycle costs, in addition to the initial cost of equipment and electricity costs, it is necessary to take into account also the costs of maintenance and repairs, which are affected by the reliability of the pump [2]. It was shown in [3,4] that mean time between failure (MTBF) can be used to quantify pump reliability. In turn, MTBF depends on the deviation of the pump flow rate Q from the Best Efficiency Point (BEP, Figure 1b) [5,6].

Parallel pumps are widely used in many applications, such as when high flow rates or wide flow control ranges are required. When using parallel pumps, it is possible to significantly reduce all components of the life cycle cost of a pumping station, in comparison with a single-pump unit of the same rated power [3,7]. At the same time, due to a large number of variable parameters and the nonlinearity of such systems, the problems of optimizing the energy consumption of parallel pumps, taking into account the reliability and cost of the life cycle, are complex and still not considered very often in the literature.

Many studies analyze the energy consumption of parallel pumps, without taking into account their reliability. Thus, in [8], parallel operation of two pumps with the same rated head and different rated flow is investigated. It is shown that the overall efficiency of the system decreases with an increase in the ratio of the rated flow rates of individual pumps. This study compared the performance of 28 different combinations of the pumping system parameters. In [9], parallel operation of a multi-pump system consisting of four pumps is considered, in which the flow is regulated by three different methods (throttling, bypass

151

and speed control). A genetic algorithm and various optimization criteria were used to find the most economical control strategy. It was shown that the lowest energy consumption is achieved when the rotational speed of all pumps is simultaneously controlled at the same value. In [10], an optimization of the total cost of systems of three parallel and three serial pumps without rotational speed control is considered, taking into account the cost of the pump, pipeline and energy consumption. In [11], an optimization of energy consumption of a system of seven parallel pumps with various ratings, some of which are equipped with variable speed drive (VSD), is considered. In [12], energy consumption of three parallel pumps when changing the number of VSDs is compared. In [13], energy consumption of pumping systems with 2–4 parallel pumps without speed control is compared. In [14], a predictive control algorithm is proposed for a system of 3 parallel pumps, each of which is equipped with a VSD, which allows, based on look-up tables, to increase the system efficiency.

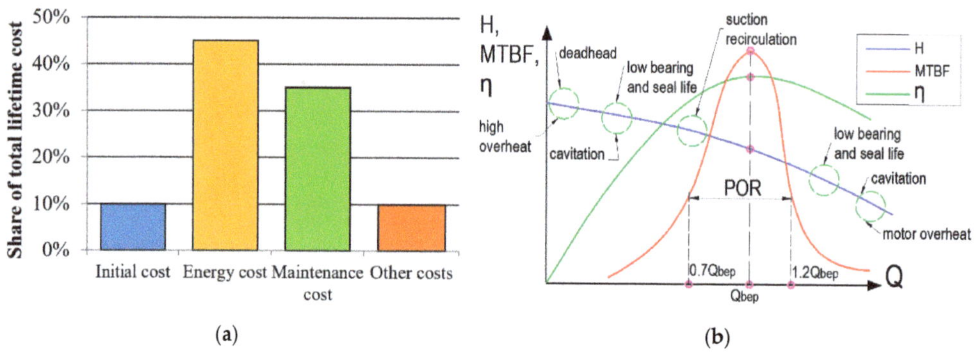

Figure 1. Pump statistics: (**a**) Components of the pump lifetime cost [2]; (**b**) Pump head (*H*), reliability (MTBF) and efficiency (η) curves versus flow rate *Q* [3].

Several works on parallel pumps consider not only energy consumption, but also the deviation of the pump operating points from the BEP. In [15], an optimization of the energy consumption of a system of two parallel pumps using a genetic algorithm is considered. The results show that the lowest energy consumption of the system is obtained by equalizing the flows of the two pumps. Furthermore, [15] compares the energy consumption of the two pumps equipped with a different number of VSD: one or two. It is shown that in the latter case, it is possible to achieve a lower energy consumption. The deviation of the operating point of the pumps from the BEP in various considered cases is compared, but not optimized.

In [6], an optimization of a single-pump unit is considered to increase its reliability using a genetic algorithm. However, some aspects were not taken into account: the optimization criterion is only the maximum reliability of the pump unit. Achieving a trade-off between reliability and power consumption is not considered. The study also does not consider the use of bypass regulation to obtain a better operating point. The static head in the hydraulic system is assumed to be zero, which also reduces the range of cases to which the results of this study are applicable.

In [16], a control strategy is proposed for a variable speed multi-pump system to reduce energy consumption. It was shown that the system of parallel pumps has the highest efficiency when the pumps operate at the same speed and flow rate than when one of the pumps operates at the rated speed and the speed of the other one is adjusted to obtain the required flow rate. It is also shown that in the former case, the deviation of the pump operating point from the BEP is less.

In [17], an analysis of a single-pump unit with a power rating of 11 kW is carried out and the issues of reducing energy consumption and increasing reliability are investigated.

A trade-off regulation is proposed that provides good pump reliability with the energy consumption close to the minimum. The energy consumption is calculated for three cases: the conventional rotational speed regulation, regulation with maximum reliability and trade-off regulation. It is shown that when using the trade-off regulation, it is possible to significantly reduce the energy consumption of the pump unit in comparison with the case of maximum reliability, while maintaining all pump operating points in the preferred operating range with sufficient reliability.

In [3], an optimal trade-off control method using particle swarm optimization is proposed for a system of two identical parallel pumps, each equipped with a frequency converter (FC). The optimization parameters are the number of simultaneously operating pumps and the rotational speed of the pumps. The optimization criterion is the minimization of energy consumption. The deviation of the pump operating point from the BEP is used as an optimization constraint. It is assumed that both pumps rotate at the same speed at each operating point, and the throttle is used to regulate the flow in the common pipeline section of the parallel pumps. However, the results of this study are only applicable when the speed of both parallel pumps is controlled by VSD. Meanwhile, in practice, parallel pumping systems are often used in which some of the pumps do not have a VSD and are powered directly from the mains [18]. The literature overview carried out shows that not all configurations of parallel pump systems have been analyzed from the point of view of reliability. In particular, multi-motor pumping stations with a single frequency converter (multi-pump single-drive systems), which are also actively used in practice to reduce the total cost of a multi-pump system, were not considered.

This paper analyzes the effectiveness of applying the trade-off regulation method proposed in [17] to a system of 3 parallel pumps. This trade-off method is compared with the traditional regulation and regulation with maximum reliability proposed in [3]. In contrast to [3], a pumping system of the "multi-pump single-drive" type is considered, in which only one frequency converter is used to alternately drive several pump units (in this case, 3 units). Such systems are widely used in parallel pumping stations equipped with low-power electric motors. However, analysis of such systems is not very common in the literature [7].

In such pumping systems, one frequency converter controls two or more pumps. At the same time, in contrast to systems without a frequency converter, a smooth start-up of each pump unit and a smooth flow/pressure adjustment are ensured. In contrast to the case where each pump unit is equipped with an individual frequency converter, the capital cost of the system is significantly reduced. This advantage is especially important if the system uses low-power pump units, for which the cost of the frequency converter is the largest part of the total cost, as well as in systems containing a large number of pumping units [18,19].

Since in multi-pump single-drive systems it is possible to control the rotational speed of only one pump at a time, the load range of the pumps in such a system is significantly different from the case when each of the pumps is equipped with a VSD. Therefore, the conclusions carried out in [3] cannot be directly applied to the multi-pump single-drive systems.

Therefore, the contribution of this article is as follows: the effectiveness of the control principle, which provides a trade-off between reliability and energy efficiency, was theoretically verified in the case of a multi-pump single-drive system, which was not covered in previous studies on the topic.

It should be noted that many studies discuss the selection and setting of dynamic controllers for pumping systems [6,20–25]. Several studies show the effectiveness of fractional-order proportional-integral-derivative control (FOPID) [20] for improving the dynamics of complex non-linear pumping systems. Thus, [21] provides a theoretical and experimental comparative study of a simple single-pump system with different types of regulators, including PI, PID and FOPID. Several papers [22,23] show the advantages of the FOPID over conventional PIDs in controlling pumping systems that provide liquid level in a connected tank system. Further improvement of pumping system dynamics can be achieved by adding fuzzy logic to the FOPID [24,25].

Our paper does not discuss the selection and setting of dynamic controllers; however, it evaluates the benefits of applying the proposed approach to the selection of operating points for pumps in a multi-pump system. The novelty of the proposed approach is that it considers both the energy efficiency of the system and its reliability. At the same time, it is assumed that a controller is used that provides a satisfactory quality of transients.

Using the traditional mathematical model of parallel pumps and polynomial interpolation of data from the manufacturer's catalogs, the energy consumption and reliability of the considered multi-pump single-drive system using different regulation methods are compared. The conventional rotational speed regulation and the previously proposed method providing maximum reliability [6] are considered. The benefits of using the proposed trade-off regulation method for the considered pumping system are also evaluated.

The aim of this study is to develop a regulation method that increases the reliability of multi-pump single-drive systems, in which, as will be shown below, when using the traditional control method, individual pumps can operate for a long time in conditions that shorten their service life. Reliability issues have already been discussed in detail in the literature for both single pump systems and systems of parallel pumps, each of which is equipped with a frequency converter [3,6,15]. The objective of this study is a comparative analysis of a particular case of a pumping system consisting of three parallel low-power pump units, one of which is powered by a frequency converter and the other two are powered directly from the grid, using different control methods.

2. Mathematical Equations of the System and Methodology of the Study

The article uses the traditional analytical method of mathematical analysis of pumping systems, based on the calculation of Q-H (head versus flow) and Q-P (mechanical shaft power versus flow) characteristics, to analyze the performance of the pumping system with various regulation methods. The characteristics corresponding to the rated rotational speed are interpolated using polynomials according to the pump datasheet. The characteristics at the arbitrary rotational speed are calculated using the characteristics at the rated speed and the affinity laws [26]. This study examines the change in energy consumption of the considered multi-pump single drive system when applying to it various restrictions on the deviation of its operating points from the BEP curve which is also the maximum reliability curve.

The parallel pump system serves to provide the required flow rate Q_{req} with a constant static head H_{st} in an open hydraulic system from point A to point B (Figure 2) [26].

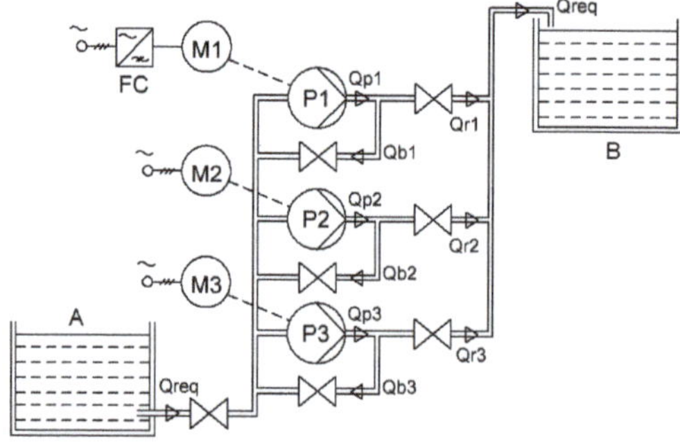

Figure 2. Hydraulic diagram of the pumping system.

Depending on the geometric dimensions of the pipelines (length, section, shape, etc.), the physical properties of the pumped liquid (density, viscosity, etc.) and the difference in heights of the basins *A* and *B*, a curve of the hydraulic system is constructed, which is described by the Equation (1) [26]:

$$H_{REQ} = H_{ST} + k \cdot Q_{REQ}{}^2, \tag{1}$$

where H_{REQ} is the required hydraulic head; H_{ST} is the static head; k is the hydraulic friction coefficient.

Figure 3 shows the various characteristics of the hydraulic system. Curves labeled with 1 and 2 are the curves of the system at a non-zero value of the static head H_{ST} at different values of k. Numbers 3 and 4 mark the curves of the hydraulic system, at different values of k and $H_{ST} = 0$. Number 5 marks the dashed curve on which the BEP points lie at different pump speed (BEP curve). The maximum reliability and maximum efficiency of the pump are reached when the system curve matches the BEP curve (dashed curve 5 in Figure 3). The BEP curve is defined according to the affinity laws by the following equation [3]:

$$H = k_{BEP} \cdot Q^2, \tag{2}$$

where Q_{BEP} and H_{BEP} are the flow and the head of the pump at the BEP with $n = n_{rate}$; $k_{BEP} = H_{BEP}/Q_{BEP}$. Q_{BEP} and H_{BEP} are defined according to the pump datasheet.

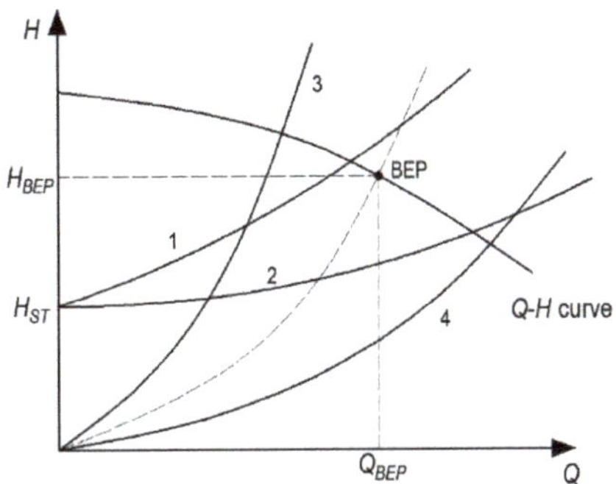

Figure 3. *Q-H* curve of the pump, hydraulic load curves (marked with numbers 1–4) and BEP curve (marked with number 5).

The selection of the pump unit is carried out based on the maximum values of the required flow rate and head.

In practice, however, they do not match, and the pump operating point may be far from the BEP curve. Therefore, depending on the parameters of the hydraulic load curve (1), it is necessary to shift the pump operating point closer to the BEP curve to increase the pump reliability.

If the hydraulic system curve is located to the right of the BEP curve (as curve 4 in Figure 3), then to increase the pump reliability, it is necessary to regulate the water flow by adjusting the rotational speed and throttling [6]. If the hydraulic system curve is located to the left of the BEP curve (as curve 3 in Figure 3), then bypass also must be applied. However, the use of throttling and bypass along with the regulation of the pump rotational speed leads to additional energy consumption.

This study considers a hydraulic system with a curve like curve 2 in Figure 3. With this relationship between the system curve and the BEP curve, for maximum reliability, it is necessary to regulate the water flow by bypassing and throttling along with speed variation. It is assumed that in the considered parallel system centrifugal monoblock pumps Calpeda-B-50/12A with a power rating of 4 kW and a rated rotational speed of 2900 rpm are used. Table 1 and Figure 4 show the catalogue characteristics of this pump [27].

Table 1. Catalogue characteristics of centrifugal pump Calpeda-B-50/12A (4 kW, 2900 rpm).

Flow Q, m³/h	30	33	37.8	42	48	54	60	66
Head H, m	24	24	23	22.5	21	19.5	17.5	15
Efficiency η, %	63.4	66.9	69.4	72.4	73.8	74.2	73.6	69.6

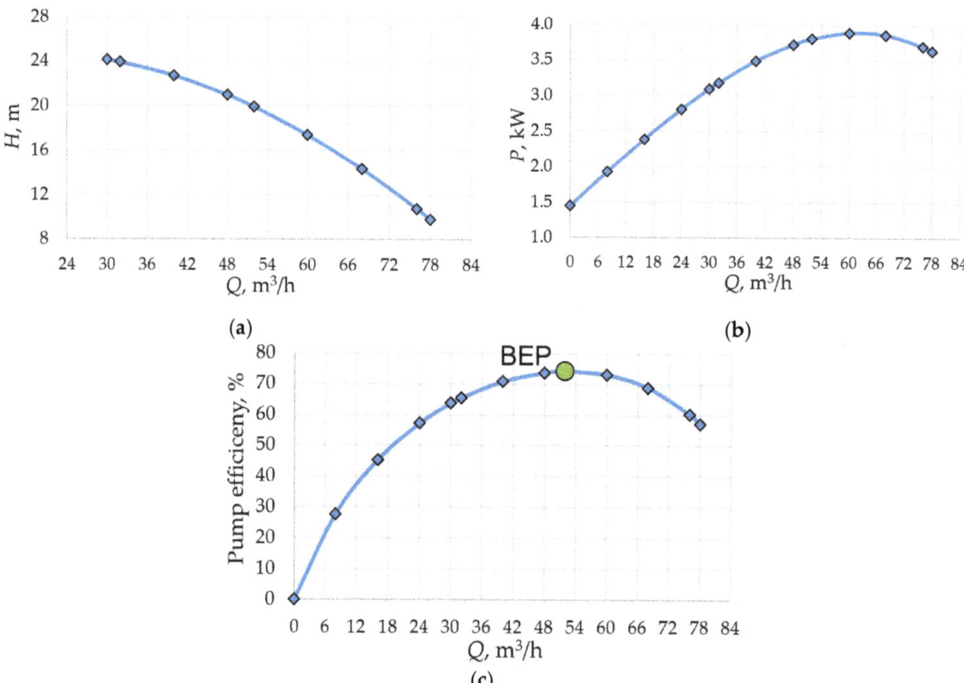

Figure 4. Data sheet characteristics for the pump Calpeda-B-50/12A versus flow rate (**a**) head; (**b**) mechanical power; (**c**) efficiency [27].

The maximum required flow rate of the pumping system is Q_{max} = 120 m³/h. To simplify calculations, the Q-H pump characteristic and characteristic of the pump required mechanical power $P = f(Q, s)$ are interpolated as polynomials with two variables of the 2nd and 3rd order, respectively [10]:

$$H = a{\cdot}Q^2 + b{\cdot}Q{\cdot}s + c{\cdot}s^2;$$

(3)

$$P = c_0{\cdot}Q^3 + c_1{\cdot}Q^2{\cdot}s + c_2{\cdot}Q{\cdot}s^2 + c_3{\cdot}s^3,$$

(4)

where $s = n/n_{rate}$ is the relative rotational speed; $a = -0.0012$, $b = 0.166$, $c = 21.505$, $c_0 = -0.0047$, $c_1 = -0.053$, $c_2 = 60.76$, $c_3 = 1449$ are the coefficients of Equations (3) and (4).

Figure 5 shows the pump H-Q curve at the rated speed (n_{rate}), the hydraulic system curve and the BEP curve. The BEP curve is defined according to Equation (2).

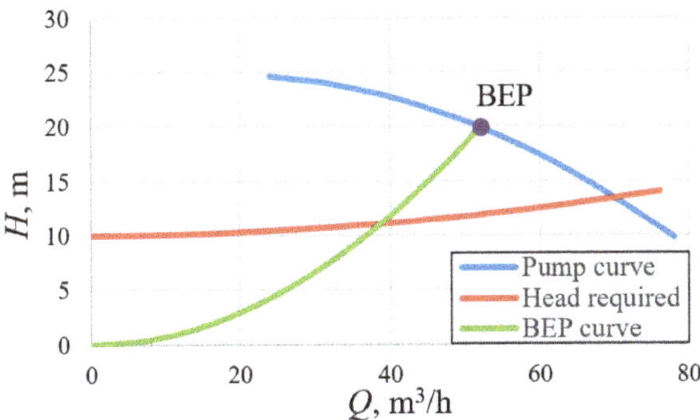

Figure 5. *Q-H* pump characteristic, hydraulic load characteristic and BEP curve.

3. Calculation of Characteristics of the Pumping System at Different Methods of Regulation

This section describes the results of calculating the characteristics of the pumping system under consideration with three considered regulation methods: maximum efficiency, maximum reliability and trade-off.

Due to the wide range of variation in water flow and head in a real pumping system, the pump operating point is very rarely near the BEP. However, to ensure an acceptable service life, the pump operating point must be in the so-called preferred operating region (POR). According to [5], the POR is located between the points $0.7 \cdot Q_{BEP}$ and $1.2 \cdot Q_{BEP}$ on the Q-H characteristic. This condition can be considered as a constraint when regulating the pumping system. Therefore, this paper compares the energy consumption of the pumping system in three cases:

(1) maximum efficiency (minimum power consumption) regulation, without applying any reliability constraints. In this case, the flow rate is regulated by changing only the rotational speed of the regulated (first) pump and throttling of the unregulated (second and third) pumps. When two and three pumps operate together, an equal distribution of the flow rate between them is achieved. In this case, energy consumption is minimized, but at the same time not all pump operating points are in the POR [6].

(2) maximum reliability regulation. In this case, the flow rate is regulated by changing the rotation speed of the first pump, throttling and bypass regulation of all three pumps. With this regulation, the operation of all turned-on pumps is achieved at points with maximum reliability. Due to the use of throttling and bypass, the energy consumption of the pumping system increases in comparison with the first case.

(3) trade-off regulation taking into account the POR reliability constraint. Flow rate is regulated by changing the rotation speed of the first pump, throttling and bypass the regulation of all three pumps. With this regulation, the operation of all turned-on pumps is achieved at operating points in the range from $0.7 \cdot Q_{BEP}$ to $1.2 \cdot Q_{BEP}$ in the entire required control range.

In all three considered cases, when the flow rate changes from 0% to 40%, only the regulated pump runs, then the second (unregulated) pump turns on. When the flow rate reaches 80%, the third (unregulated) pump also turns on.

The rest of this section describes the results of calculating the pump operating points for each of the considered regulation methods.

3.1. Characteristics of the Pumping System with the Maximum Efficiency Regulation (Conventional Speed Regulation)

The rotational speed of the first pump in the parallel system can be adjusted. The speeds of the second and third pumps running in parallel are not adjusted. With this control, the operating points of the variable speed pump are along the hydraulic load curve ("head required" curve) and the operating points of the fixed speed pumps are along the catalog *Q-H* curve (Figure 6).

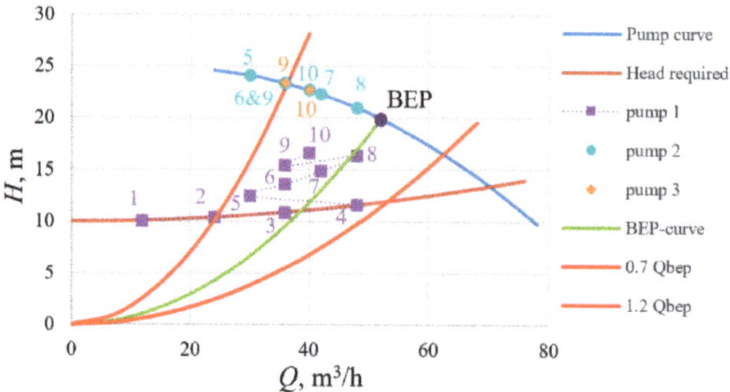

Figure 6. Operating points of the pumps when using the conventional rotational speed regulation.

The characteristics of the pumps when using this regulation method are determined as follows. The head of the variable speed pump $H_1 = H_{REQ}$ is determined by Equation (1), the head of the unregulated pumps is determined by Equation (3), the pump mechanical power P is determined by Equation (4). The rotational speed of the variable speed pump n is determined by Equation (5) [7], the pump efficiency is determined by Equation (6) and the deviation of the operating point from the BEP is determined by the Equation (7):

$$n_1 \;=\; n_{rate} \cdot \frac{-b \cdot Q_1 + \sqrt{(b \cdot Q_1)^2 - 4 \cdot c(a \cdot Q_1{}^2 - H_{req})}}{2 \cdot c}; \tag{5}$$

$$\eta_j = g \cdot \rho \cdot Q_j \cdot H_j / P; \tag{6}$$

$$\theta_j = (Q_j - Q_{BEP})/Q_{BEP}, \tag{7}$$

where $g = 9.81$ m/s^2 is the gravitational acceleration; $\rho = 1000$ kg/m^3 is the water density; Q_i and H_i are the flow and the head of *j*-th pump; η_j and θ_j are the efficiency and the operating poit deviation of *j*-th pump; $j = 1 \dots 3$; $Q_{BEP} = f(n)$ is the flow of the pump at BEP according the BEP curve Equation (2); *a*, *b* and *c* are coefficients from Equation (3).

Table 2 shows the calculation results for flow rate in the range from 10% to 100% of the maximum flow rate $Q_{max} = 120$ m^3/h. In Table 2 and below the following characteristics of pumps 1–3 are indicated: $Q_1 \dots Q_3$ are their flow rates; $Q_{req} = Q_1 + Q_2 + Q_3$; $H_1 \dots H_3$ are their hydraulic heads; $P_1 \dots P_3$ are their mechanical powers; $n_1 \dots n_3$ are their rotational speeds; $P\Sigma = P_1 + P_2 + P_3$; $\eta_1 - \eta_3$ are their efficiencies; $\theta_1 \dots \theta_3$ are the deviations of the operating point of the pumps. From Figure 6 and Table 2, it can be seen that the estimated deviations θ exceed 30.8% and reach 63.9%, which are outside the POR in Table 2.

Such points with the large deviations occur at all stages of the flow regulation: when the required water flow is less than 20% of the Q_{max} (when only the first pump is running), at the flow rate in the range of 50–80% (the second pump turns on), and also when the flow rate is more than 80% (the third pump turns on). If the flow rate is below $0.7 \cdot Q_{BEP}$ pump operation may become unstable due to the flatness of the *Q-H* curve: a small deviation in the rotational speed leads to a large deviation in the flowrate. In addition, due to the unstable shape of the *Q-H* characteristic, the latter intersects the hydraulic system curve in two points which leads to the occurrence of a surge. In this case, the pump operates alternately in different operating points, the whole system is unstable, the loading on the pump changes and hydraulic shocks occur [28,29].

Table 2. The characteristics of the parallel pumping system using the conventional speed control.

Point Number i	Q_{req}, %	Q_{req}, m³/h	Q_1, m³/h	Q_2, m³/h	Q_3, m³/h	$H_1 = H_{req}$, m	H_2, m	H_3, m	n_1, rpm	n_2, rpm	n_3, rpm	P_1, kW	P_2, kW	P_3, kW	$P\Sigma$, kW
1	10	12	12	-	-	10.1	-	-	1851	-	-	0.66	-	-	0.66
2	20	24	24	-	-	10.4	-	-	1934	-	-	0.99	-	-	0.99
3	30	36	36	-	-	10.9	-	-	2110	-	-	1.45	-	-	1.45
4	40	48	48	-	-	11.6	-	-	2351	-	-	2.07	-	-	2.07
5	50	60	30	30	-	12.5	24.1	-	2155	2900	-	1.44	3.09	-	4.53
6	60	72	36	36	-	13.6	23.4	-	2301	2900	-	1.83	3.35	-	5.17
7	70	84	42	42	-	14.9	22.3	-	2466	2900	-	2.31	3.56	-	5.87
8	80	96	48	48	-	16.4	21.0	-	2644	2900	-	2.89	3.72	-	6.61
9	90	108	36	36	36	15.4	23.4	23.4	2421	2900	2900	2.09	3.35	3.35	8.79
10	100	120	40	40	40	16.7	22.7	22.7	2549	2900	2900	2.48	3.49	3.49	9.47

Point number i	η_1, %	η_2, %	η_3, %	θ_1, %	θ_2, %	θ_3, %
1	50.0	-	-	−63.9	-	-
2	68.5	-	-	−30.8	-	-
3	73.9	-	-	−4.8	-	-
4	73.3	-	-	13.9	-	-
5	71.1	63.7	-	−22.4	−42.3	-
6	73.0	68.5	-	−12.8	−30.8	-
7	73.9	71.8	-	−5.0	−19.2	-
8	74.2	73.7	-	1.2	−7.7	-
9	72.3	68.5	68.5	−17.1	−30.8	−30.8
10	73.1	70.9	70.9	−12.5	−23.1	−23.1

3.2. Characteristics of the Pumping System with the Maximum Reliability Regulation

For maximum pump reliability, the operating points of the variable speed pump must be on the BEP curve (dashed line in Figure 3), and fixed-speed pumps must always run at the BEP point on the catalog pump Q-H curve, regardless of the required flow rate. This can only be achieved by using all three of the above-mentioned water flow regulation methods together.

As shown in Figure 6, variable speed pump operating points 1–3, 5–7 and 9–10 are located to the left of the BEP curve. To implement the condition of maximum reliability, these points are to move horizontally until they coincide with the BEP curve. All operating points of the fixed-speed pumps are also located to the left of the BEP curve. They need to be moved along the catalog Q-H curve up to the BEP. This means that the pump flow rate Q_i will increase from $Q_i{'}$ to Q_{BEP} at the same head. Excess flow $Q_i - Q_i{'}$ flows back through the bypass to the suction pipe, where. In this case $Q_{req} = Q_{1'} + Q_{2'} + Q_{3'}$.

Points 4 and 5 are located to the right of the BEP curve. To move these points up to the BEP curve, at a constant flow rate Q_{req}, the head at the pump outlet must be increased by throttling (Figure 7).

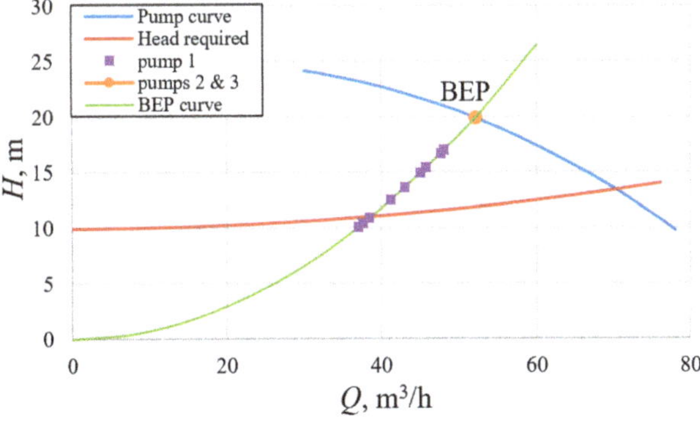

Figure 7. Operating points of the pump when using the maximum reliability regulation.

The required head H_{req} is determined by the Equation (1). H_{req} is the same for all switched on pumps as they operate in parallel. The flow Q_j and the head H_j are determined by Equation (8) when regulating the bypass together with speed regulation or by Equation (9) when regulating only the pump rotational speed. The rest of the parameters are determined in the same way as in the previous section:

$$Q_j = \sqrt{(H_{req}/k_{BEP})}; \ H_j = H_{req}, \tag{8}$$

$$Q_j = Q_{req}; \ H_j = k_{BEP} \cdot Q_{req}{}^2, \tag{9}$$

where Q_{req} and H_{req} are the required output flow and head; Q_j and H_j are the flow and head of j-th pump.

Table 3 shows the calculated characteristics of the pump system with the maximum reliability regulation. As the results in Figure 7 in Table 3 show, when the throttling and bypass are used in combination with the speed control, the maximum pump reliability can be achieved. Operating point deviation θ is zero.

Table 3. The characteristics of the parallel pumping system using the maximum reliability regulation.

Point Number i	Q_{req}, m³/h	H_{req}, m	Pump 1 (j = 1)				Pump 2 (j = 2)				Pump 3 (j = 3)			
			$Q_{1'}$, m³/h	Q_1, m³/h	H_1, m	n_1, rpm	$Q_{2'}$, m³/h	Q_2, m³/h	H_2, m	n_2, rpm	$Q_{3'}$, m³/h	Q_3, m³/h	H_3, m	n_3, rpm
1	12	10.1	12	37.0	10.1	2066	-	-	-	-	-	-	-	-
2	24	10.4	24	37.6	10.4	2096	-	-	-	-	-	-	-	-
3	36	10.9	36	38.5	10.9	2146	-	-	-	-	-	-	-	-
4	48	11.6	48	48.0	16.9	2677	-	-	-	-	-	-	-	-
5	60	12.5	30	41.2	12.5	2298	30	52.0	19.9	2900	-	-	-	-
6	72	13.6	36	43.0	13.6	2397	36	52.0	19.9	2900	-	-	-	-
7	84	14.9	42	45.0	14.9	2509	42	52.0	19.9	2900	-	-	-	-
8	96	16.4	48	48.0	16.9	2677	48	52.0	19.9	2900	-	-	-	-
9	108	15.4	36	45.7	15.4	2551	36	52.0	19.9	2900	36	52.0	19.9	2900
10	120	16.7	40	47.6	16.7	2654	40	52.0	19.9	2900	40	52.0	19.9	2900

Point number i	P_1, kW	P_2, kW	P_3, kW	$P\Sigma$, W	η_1, %	η_2, %	η_3, %	θ_1, %	θ_2, %	θ_3, %
1	1.37	-	-	1.37	74.2	-	-	0	-	-
2	1.44	-	-	1.44	74.2	-	-	0	-	-
3	1.54	-	-	1.54	74.2	-	-	0	-	-
4	2.99	-	-	2.99	74.2	-	-	0	-	-
5	1.89	3.80	-	5.70	74.2	74.2	-	0	0	-
6	2.15	3.80	-	5.95	74.2	74.2	-	0	0	-
7	2.46	3.80	-	6.27	74.2	74.2	-	0	0	-
8	2.99	3.80	-	6.79	74.2	74.2	-	0	0	-
9	2.59	3.80	3.80	10.2	74.2	74.2	74.2	0	0	0
10	2.91	3.80	3.80	10.5	74.2	74.2	74.2	0	0	0

The pump efficiency throughout the entire flow regulation range remains at its maximum (74.2%). However, power consumption increases compared to the conventional speed regulation due to the increase in Q_{PUMP} when using the bypass and the increase in H_{PUMP} when using the throttling.

3.3. Characteristics of the Pumping System with the Trade-Off Regulation

It is also possible to apply a trade-off regulation method to reduce energy consumption and, at the same time, reach sufficiently high reliability. To achieve this, it is necessary to ensure that all pump operating points are located within the POR. According to Figure 6, when using only the speed control, points 1, 2 of the first regulated pump, points 5, 6 of the second pump and point 9 of the third pump are outside of the POR.

In this case, as in the case of the maximum reliability regulation, points 1, 2 of the first pump are moved horizontally but only to the right to the border of the POR ($1.2 \cdot Q_{BEP}$ line) using a bypass (Figure 8). Furthermore, points 5, 6 of the second pump and point 9 of the third pump move along the Q-H-curve to the left border of the POR ($0.7 \cdot Q_{BEP}$ line). Thus, all operating points are restricted by the POR. As in the previous case, H_{req} is the same for all switched on pumps as they operate in parallel and $Q_{req} = Q_{1'} + Q_{2'} + Q_{3'}$. This ensures high reliability and low power consumption. Throttling is not applied in this case. The characteristics of pumps with the trade-off regulation are determined as follows. The required head of the variable speed pump H_{req} is determined using Equation (1).

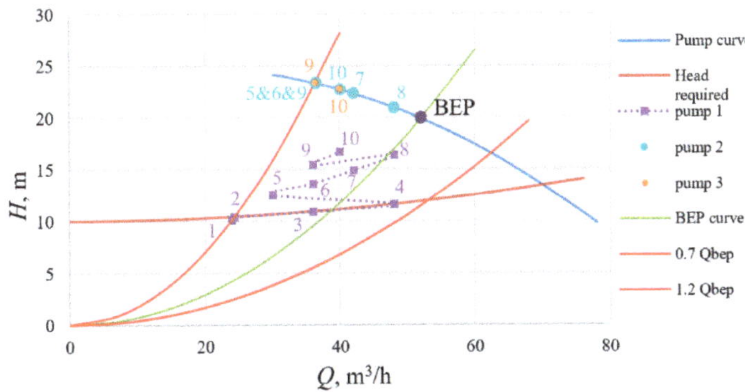

Figure 8. Operating points of the pumps when using the trade-off regulation.

The flow through the pump Q_j and the head H_j are determined by Equation (10) with bypass regulation. The rest of the parameters are determined in the same way as in the case of the conventional speed regulation:

$$Q_j = \sqrt{(H_{REQ}/k_{0.7BEP})}; \; H_j = H_{REQ}, \tag{10}$$

where $k_{0.7BEP} = H(0.7 \cdot Q_{BEP})/(0.7 \cdot Q_{BEP})$.

Table 4 and Figure 8 show the calculated characteristics of the pump system applying the proposed trade-off regulation. As the results in Table 4 show, with the proposed trade-off regulation method, the efficiency of the pumps in the entire range of the flow control is not less than 68.8%. The flow deviation θ of the operating points from the BEP is no more than 30%, and the energy consumption is reduced compared to the maximum reliability regulation method.

Table 4. The characteristics of the parallel pumping system using the trade-off regulation.

Point Number i	Q_{req}, m³/h	H_{req}, m	Pump 1 (j = 1)				Pump 2 (j = 2)				Pump 3 (j = 3)			
			$Q_{1'}$, m³/h	Q_1, m³/h	H_1, m	n_1, rpm	$Q_{2'}$, m³/h	Q_2, m³/h	H_2, m	n_2, rpm	$Q_{3'}$, m³/h	Q_3, m³/h	H_3, m	n_3, rpm
1	12	10.1	12	24.0	10.1	1909	-	-	-	-	-	-	-	-
2	24	10.4	24	24.3	10.4	1937	-	-	-	-	-	-	-	-
3	36	10.9	36	36.0	10.9	2110	-	-	-	-	-	-	-	-
4	48	11.6	48	48.0	11.6	2351	-	-	-	-	-	-	-	-
5	60	12.5	30	30.0	12.5	2155	30	36.4	23.3	2900	-	-	-	-
6	72	13.6	36	36.0	13.6	2301	36	36.4	23.3	2900	-	-	-	-
7	84	14.9	42	42.0	14.9	2466	42	42.0	22.3	2900	-	-	-	-
8	96	16.4	48	48.0	16.4	2644	48	48.0	21.0	2900	-	-	-	-
9	108	15.4	36	36.0	15.4	2421	36	36.4	23.3	2900	36	36.4	23.3	2900
10	120	16.7	40	40.0	16.7	2549	40	40.0	22.7	2900	40	40.0	22.7	2900

Point number i	P_1, kW	P_2, kW	P_3, kW	$P\Sigma$, W	η_1, %	η_2, %	η_3, %	θ_1, %	θ_2, %	θ_3, %
1	0.96	-	-	0.96	68.8	-	-	−30.0	-	-
2	1.00	-	-	1.00	68.8	-	-	−30.0	-	-
3	1.45	-	-	1.45	73.9	-	-	−4.8	-	-
4	2.07	-	-	2.07	73.3	-	-	13.9	-	-
5	1.44	3.36	-	4.80	71.1	68.8	-	−22.4	−30.0	-
6	1.83	3.36	-	5.19	73.0	68.8	-	−12.8	−30.0	-
7	2.31	3.56	-	5.87	73.9	71.8	-	−5.0	−19.2	-
8	2.89	3.72	-	6.61	74.2	73.7	-	1.2	−7.7	-
9	2.09	3.36	3.36	8.82	72.3	68.8	68.8	−17.1	−30.0	−30.0
10	2.49	3.49	3.49	9.47	73.1	70.9	70.9	−12.5	−23.1	−23.1

4. Comparison of Energy Consumption and Reliability of the Parallel System with Different Regulation Methods

Figures 9–12 summarize the results of Tables 2–4. Figure 9 compares the flow rates of individual pumps with different control methods. With the maximum efficiency method, the flow of each pump is minimized, but the deviations of θ at some points are below the POR limit. When applying maximum reliability regulation, the pump flow is always increased by applying a bypass to keep all pumps running at the BEP. When applying the trade-off regulation, if $\theta < -30\%$, then the pump flow increases so that $\theta = -30\%$.

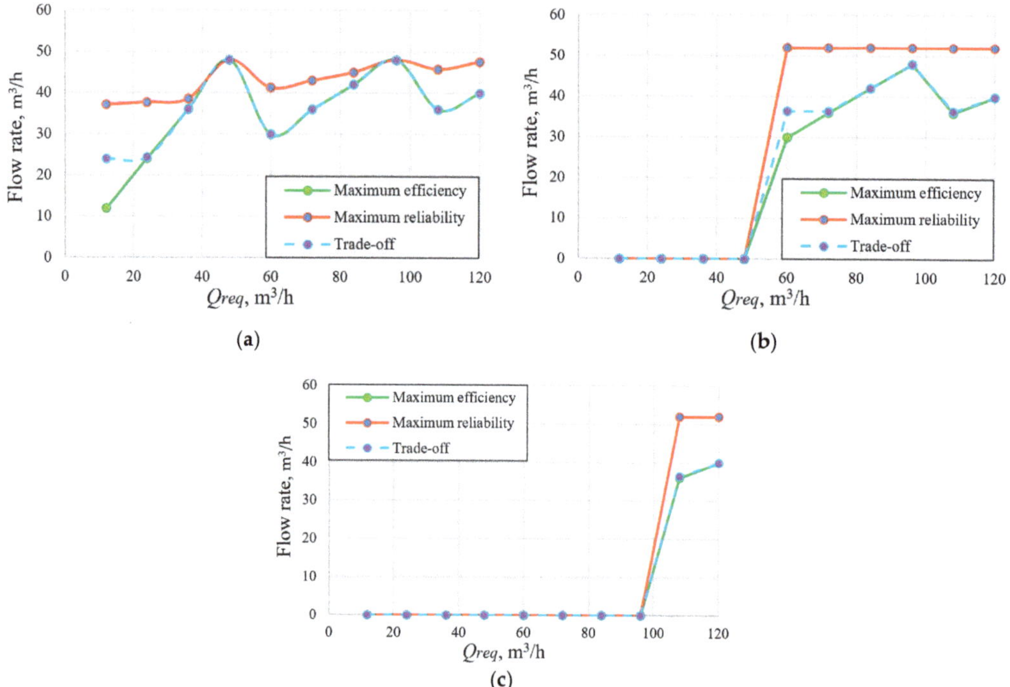

Figure 9. Comparison of flow rates of individual pumps with different regulation methods (**a**) pump 1 ($j = 1$); (**b**) pump 2 ($j = 2$); (**c**) pump 3 ($j = 3$).

Figure 10 compares the efficiency of individual pumps with different control methods. With the maximum reliability method, the efficiency of all pumps is always maximized since they always run at the BEP. However, the total energy consumption of the pumping system in this case is also maximum (Figure 11) since the additional energy consumption from throttling and bypass exceeds the benefit from improving the efficiency of the pumps.

Figures 11 and 12 compare the required mechanical power and the pump operating point deviation from BEP θ at the various regulation methods considered. Figure 11 shows that when using the maximum reliability regulation, the required mechanical power significantly increases compared to the conventional speed regulation due to the use of bypass and throttling. At the same time, when using the trade-off regulation, the required mechanical power only slightly increases at two operating points.

According to the proposed trade-off regulation principle, it is necessary to correct only the operating points which deviations θ are outside the POR boundaries (marked in red in Table 2). As Figure 6 and Table 2 show, most of the operating points of the three pumps are already within the POR when the "maximum efficiency" control is applied. Therefore, when the trade-of regulation is applied, most of the operating points remain unchanged, in comparison with the maximum efficiency regulation method. In addition, 4 of 6 points with an unacceptable deviation θ have a deviation value of 30.8%, which is very close to the permissible limit value of 30%. Therefore, the correction of these points when using the trade-off regulation does not require significant additional energy consumption. Significant deviations θ only need to be corrected for the first pump ($j = 1$) at point 1 ($i = 1$) and for the second pump ($j = 1$) at point 5 ($i = 5$). This results in a very similar shape of the graphs for PΣ and θ in Figures 11 and 12. Only for the correction of operating points 1 and 5 significant additional power consumption is required, which is reflected in Figure 11d. Figure 12 shows that the proposed trade-off regulation maintains pump duty points within POR over the entire required flow range.

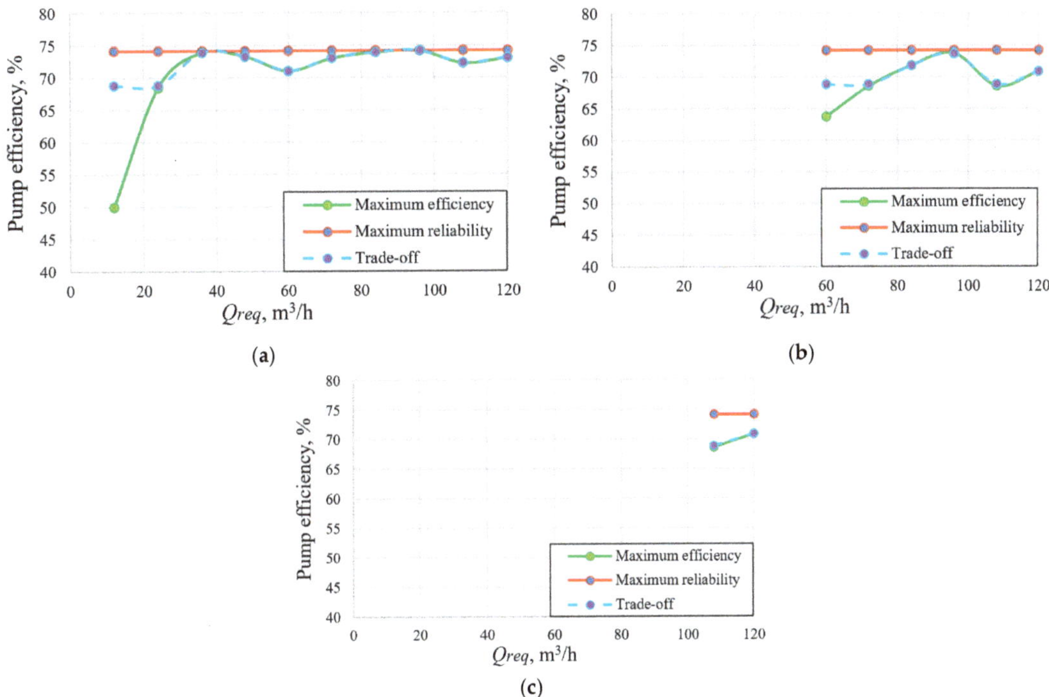

Figure 10. Comparison of efficiencies of individual pumps with different regulation methods (**a**) pump 1 ($j = 1$); (**b**) pump 2 ($j = 2$); (**c**) pump 3 ($j = 3$).

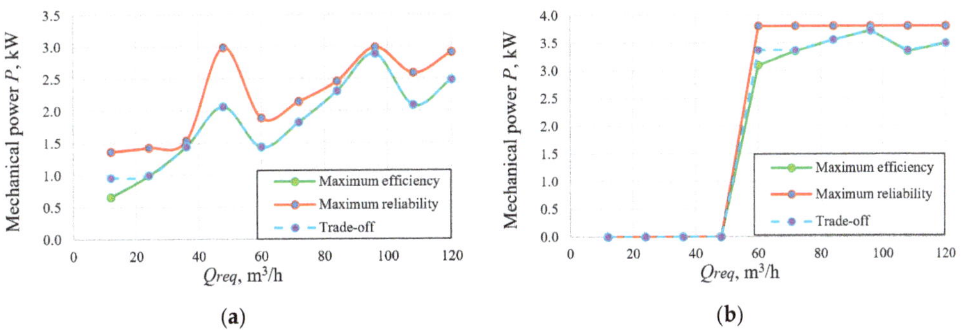

Figure 11. *Cont.*

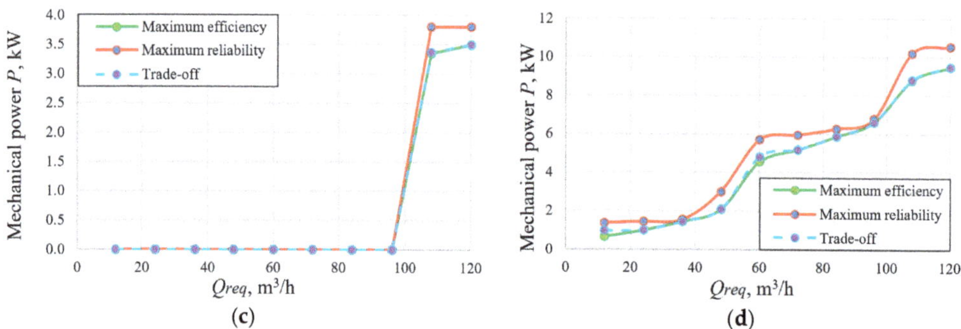

Figure 11. Comparison of the mechanical power of the pumps with different regulation methods (**a**) pump 1 (P_1); (**b**) pump 2 (P_2); (**c**) pump 3 (P_3); (**d**) sum of mechanical power ($P\Sigma$).

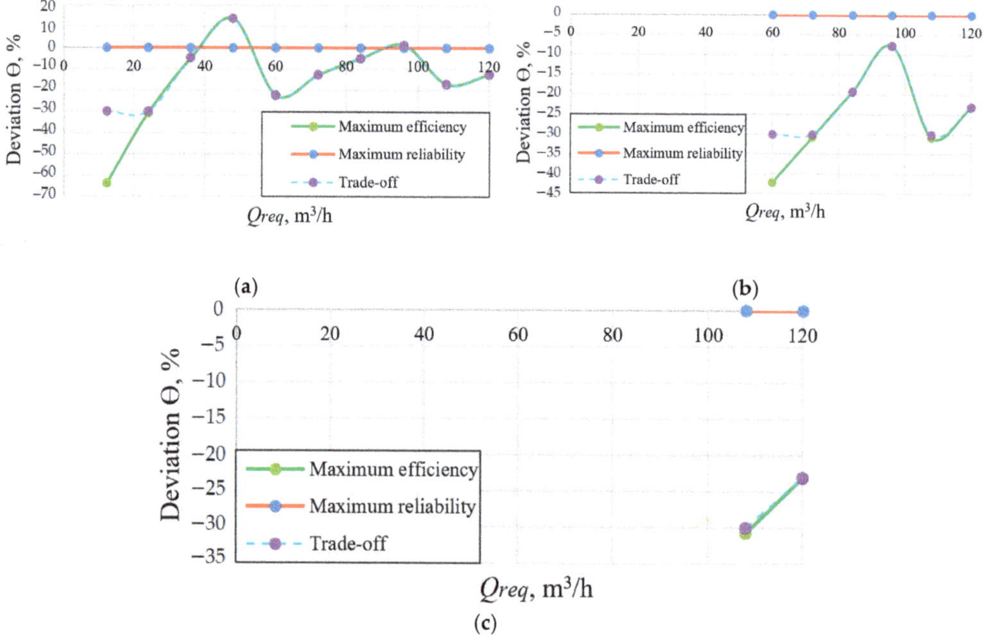

Figure 12. Comparison of deviations θ of individual pumps with different regulation methods (**a**) pump 1 ($j = 1$); (**b**) pump 2 ($j = 2$); (**c**) pump 3 ($j = 3$).

Based on the obtained results (Figure 11), the energy consumption of the parallel pumping system is compared when operating with a duty cycle corresponding to a typical open-type pumping system [26]. Figure 13 shows the diagram of the duty cycle. The duty cycle period is 24 h. Electricity consumption is compared at three regulation methods under consideration.

When calculating the electrical power consumption of a pump unit, it is also necessary to consider the characteristics of the components of its electric drive. It was assumed that Sinamics G120C frequency converter with the rated power of 4 kW [30] and Simotics 1LE1001-1CA6 induction motors with the power of 4 kW and rated speed of 2955 rpm [31] are used in the drive of the pump system. Data from SinaSave software [32] in 8 standard loading points are used to determine the power losses and efficiency of the drive (motor plus frequency converter) in the operating points with given values of the shaft torque

T and speed n that were calculated in the previous section (Tables 2–4). The standard loading points are determined according to IEC 61800-9-2, "Adjustable Speed Electrical Power Drive Systems—Part 9-2: Ecodesign for Power Drive Systems, Motor Starters, Power Electronics and Their Driven Applications—Energy Efficiency Indicators for Power Drive Systems and Motor Starters" [33]. These data are used because the standard [33] requires manufacturers to declare the loss values for variable frequency drives at these 8 operating points. Table 5 shows the results of calculating the losses in the electric drive using the SinaSave program at the standard points.

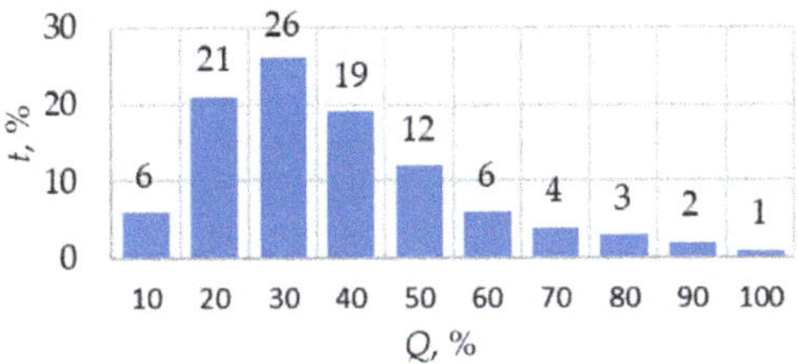

Figure 13. Flow-time diagram.

Table 5. Loss data for the 4 kW electric drive at the standard loading points.

Standard Operating Point	1	2	3	4	5	6	7	8
$n_{rate\ drive}$, %	100	100	50	50	50	0	0	0
$T_{rate\ drive}$, %	100	50	100	50	25	100	50	25
ΔP, kW	0.81	0.59	0.5	0.42	0.28	0.21	0.21	0.15

In Table 5, $T_{rate\ drive}$ = 100% = 12.93 N·m is the rated torque of the electric drive; $n_{rate\ drive}$ = 100% = 2955 rpm is the rated rotational speed of the electric drive.

Using the data from Table 5, the sum of losses in the three pump drives $\Delta P\Sigma$ at the considered operating points (Tables 2–4) using polynomial interpolation [34,35] were found. The results of this calculation are shown in Table 6. Table 6 also shows the required total mechanical power $P\Sigma$ from Tables 2–4.

Table 6. Results of the loss calculation in the electric drive of the pump system applying the different regulation methods.

	Q_{req}, %	10	20	30	40	50	60	70	80	90	100
	Q_{req}, m³/h	12	24	36	48	60	72	84	96	108	120
$\Delta P\Sigma$, kW	Maximum efficiency	0.24	0.27	0.33	0.43	0.93	1.05	1.18	1.33	1.75	1.88
	Maximum reliability	0.32	0.33	0.35	0.60	1.17	1.21	1.27	1.37	2.06	2.12
	Trade-off	0.27	0.27	0.33	0.43	0.99	1.05	1.18	1.33	1.75	1.88
$P\Sigma$, kW	Maximum efficiency	0.66	0.99	1.45	2.07	4.53	5.17	5.87	6.61	8.79	9.47
	Maximum reliability	1.37	1.44	1.54	2.99	5.70	5.95	6.27	6.79	10.19	10.52
	Trade-off	0.96	1.00	1.45	2.07	4.80	5.19	5.87	6.61	8.82	9.47

Due to the results of calculating the mechanical power and interpolation of losses in the electric drive from Table 6, it can be concluded that the losses in the electric drive also make up a significant part of the energy consumption, about 20–30%, and strongly depend on the operating point of the pump. Therefore, the losses in the electric drive must also be taken into account when calculating the total energy consumption of the pumping system.

Using the results obtained (Tables 2–6), it is possible to calculate the electrical power consumed from the grid (P_{elec}), the daily consumed electrical energy (E_{day}), the annual consumed electrical energy (E_{year}), the annual energy cost (C_{year}) and the cost of energy over the entire life cycle of the pumping system (C_{LLC}) [36]:

$$P_{elec} = P\Sigma + \Delta P\Sigma; \tag{11}$$

$$E_{day} = \frac{t_\Sigma}{1000} \cdot \sum_{i=1}^{10} \left(P_{elec}(i) \cdot \frac{t_i}{t_\Sigma} \right); \tag{12}$$

$$E_{year} = E_{day} \cdot 365; \tag{13}$$

$$C_{year} = E_{year} GT; \tag{14}$$

$$C_{LCC} = \sum_{k=1}^{w} \left(\frac{C_{year\,k}}{(1+[y-p])^k} \right), \tag{15}$$

where $P\Sigma$ is the mechanical power required by the pumps; ΔP is the loss in the electric drive; $t_\Sigma = 24$ h is the whole operating period; t_i is the operation time of i-th loading point; $GT = 0.2036$ €/kWh is the applied grid tariffs for non-household consumers for Germany in the second half of 2019 [37]; C_{year} is the annual electricity cost; $C_{year\,k}$ is the annual electricity cost for k-th year; $y = 0.06$ is the interest rate; $p = 0.04$ is the expected annual inflation; $w = 20$ years is the lifetime of the pump system.

Annual and life cycle cost savings S_{year} and S_{LCC} for a given regulation method compared to the maximum reliability method is calculated as:

$$S_{year} = C_{year\,max.\,reliab.} - C_{year}; \tag{16}$$

$$S_{LCC} = C_{LCC\,max.\,reliab.} - C_{LCC}, \tag{17}$$

where $C_{year\,max.\,reliab.}$ is the annual energy cost with the maximum reliability control; $C_{LCC\,max.\,reliab}$ is the lifetime energy cost with the maximum reliability control.

S_{year} year S_{LCC} percentage are calculated according to:

$$S_{year} = 100\% \cdot (C_{year\,max.\,reliab.} - C_{year})/C_{year\,max.\,reliab}; \tag{18}$$

$$S_{LCC} = 100\% \cdot (C_{LCC\,max.\,reliab.} - C_{LCC})/C_{LCC\,max.\,reliab}, \tag{19}$$

Tables 7 and 8 show the calculation results based on Equations (10)–(17).

Table 7. Results of the electric power calculation in the pump system at the different regulation methods.

	Q_{req}, %	10	20	30	40	50	60	70	80	90	100
	Q_{req}, m³/h	12	24	36	48	60	72	84	96	108	120
	Maximum efficiency	0.90	1.27	1.78	2.50	5.47	6.22	7.04	7.94	10.53	11.35
P_{elec}, kW	Maximum reliability	1.70	1.77	1.89	3.59	6.87	7.16	7.53	8.16	12.25	12.64
	Trade-off	1.23	1.28	1.78	2.50	5.79	6.24	7.04	7.94	10.57	11.35

Table 8. Results of the lifetime electricity costs in the pump system at the different regulation methods.

Control Method	Maximum Efficiency	Maximum Reliability	Trade-Off
E_{day}, kWh	75.16	91.63	76.70
E_{year}, kWh	27,433	33,445	27,980
C_{year}, €	5585	6809	5697
S_{year}, €	1224	-	1112
$S_{year.}$, %	18.0	-	16.3
C_{LLC}, k€	91.3	111.3	93.1
S_{LLC}, k€	20.0	-	18.2
S_{LLC}, %	18.0	-	16.3

5. Discussion and Results

From the results obtained, as shown by the diagram in Figure 14, it can be concluded that the highest electrical power consumption is required for maximum reliability control under all considered load conditions. When applying trade-off regulation, a significant increase in electrical power consumption is required only at a flow rate of 10% and 50%.

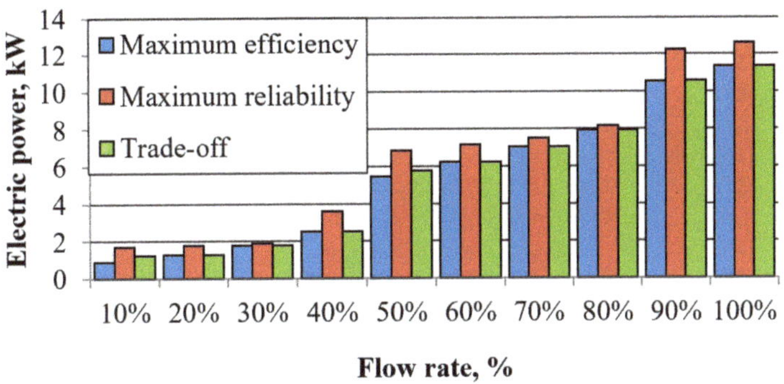

Figure 14. Electric power depending on flow rate and regulation method.

Calculation of long-term energy indicators also confirms these conclusions. Figure 15 shows that the reduction in lifetime energy consumption compared to the maximum reliability method differs only slightly in cases of maximum efficiency regulation and trade-off regulation (18% and 16.3%, respectively). At the same time, the trade-off method allows all pump operating points to be kept within the POR (Figure 12), which will significantly increase the reliability of the pumping system.

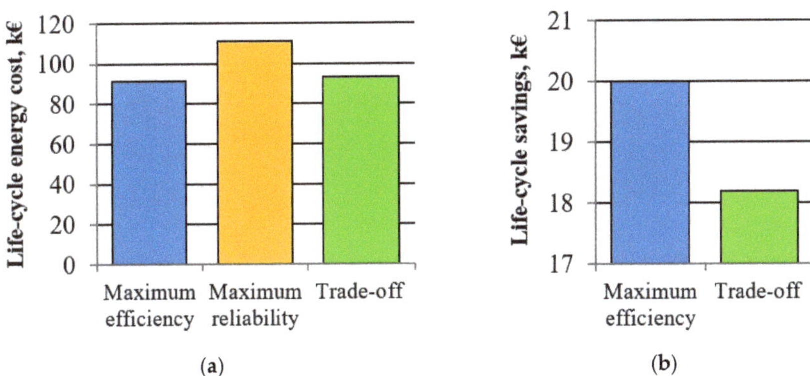

Figure 15. Energy savings: (**a**) Life-cycle energy cost depending on the regulation method; (**b**) life-cycle savings depending on the regulation.

6. Conclusions

In this paper, an analysis of the energy consumption of a pumping system consisting of three 4 kW pumps operating in parallel is carried out, considering the reliability constraints. One 4 kW variable frequency induction motor and two 4 kW fixed frequency induction motors are used in the drive of the pumping system.

Energy consumption of the system when three methods of flowrate control regulation are applied is compared: with minimum power consumption, with maximum reliability and trade-off regulation. Comparison of energy consumption in the considered cases shows that energy consumption is 18% higher with maximum reliability regulation compared to minimum energy consumption regulation without any reliability constraints. At the same time, when trade-off regulation is applied, the power consumption is only 2.3% higher. It can be concluded that the proposed trade-off regulation is promising in terms of minimizing the overall lifetime costs of the considered multi-pump single-drive system when both energy costs and maintenance and repair costs are considered.

In future works, it is planned to simulate the dynamics and select a controller for the multi-pump system under consideration when applying the proposed trade-off regulation principle.

Author Contributions: Conceptual approach, S.O. and V.P.; data curation V.D. and V.K.; software S.O. and V.K.; calculations and modeling, S.O., V.K. and V.P.; writing of original draft S.O., V.D., V.K. and V.P.; visualization, S.O. and V.K.; review and editing, S.O., V.D., V.K. and V.P. All authors have read and agreed to the published version of the manuscript.

Funding: The work was partially supported by the Ministry of Science and Higher Education of the Russian Federation (through the basic part of the government mandate, Project No. FEUZ-2020-0060).

Institutional Review Board Statement: Not applicable.

Informed Consent Statement: Not applicable.

Data Availability Statement: Data is contained within the article.

Acknowledgments: The authors thank the editors and reviewers for careful reading and constructive comments.

Conflicts of Interest: The authors declare no conflict of interest.

References

1. Shankar, V.K.A.; Umashankar, S.; Paramasivam, S.; Hanigovszki, N. A comprehensive review on energy efficiency enhancement initiatives in centrifugal pumping system. *Appl. Energy* **2016**, *181*, 495–513. [CrossRef]
2. Frenning, L. (Ed.) *Pump Life Cycle Costs: A Guide to LCC Analysis for Pumping Systems, Executive Summary*; Hydraulic Institute: Parsippany, NJ, USA; Europump: Brussels, Belgium; Office of Industrial Technologies Energy Efficiency and Renewable Energy, U.S. Department of Energy: Washington, DC, USA, 2001; pp. 1–19. Available online: https://www.energy.gov/sites/prod/files/2014/05/f16/pumplcc_1001.pdf (accessed on 5 April 2021).
3. Lai, Z.; Li, Q.; Zhao, A.; Zhou, W.; Xu, H.; Wu, D. Improving Reliability of Pumps in Parallel Pump Systems Using Particle Swam Optimization Approach. *IEEE Access* **2020**, *8*, 58427–58434. [CrossRef]
4. Barringer, H.P. A life cycle cost summary. In Proceedings of the International Conference of Maintenance Societies, Perth, Australia, 20–23 May 2003; pp. 20–23.
5. ANSI HI 9.6.3-2012. Rotodynamic Centrifugal and Vertical Pump. Guideline for Allowable Operating Region. Hydraulic Institute Standards. Available online: http://www.reliability-centred-maintenance.com/articles/LifeCycleCostSummary.pdf (accessed on 5 April 2021).
6. Lai, Z.; Wu, P.; Yang, S.; Wu, D. A Control Method to Balance the Efficiency and Reliability of a Time-Delayed Pump-Valve System. *Math. Probl. Eng.* **2016**, *2016*, 5898209. [CrossRef]
7. Oshurbekov, S.; Kazakbaev, V.; Prakht, V.; Dmitrievskii, V.; Gevorkov, L. Energy Consumption Comparison of a Single Variable-Speed Pump and a System of Two Pumps: Variable-Speed and Fixed-Speed. *Appl. Sci.* **2020**, *10*, 8820. [CrossRef]
8. Wen, Y.; Zhang, X.; Wang, P. The Relationship between the Maximum Efficiency and the Flow of Centrifugal Pumps in Parallel Operation. *J. Press. Vessel Technol.* **2010**, *132*, 034501. [CrossRef]
9. Olszewski, P. Genetic optimization and experimental verification of complex parallel pumping station with centrifugal pumps. *Appl. Energy* **2016**, *178*, 527–539. [CrossRef]
10. Pandey, S.; Singh, R.P.; Mahar, P.S. Optimal Pipe Sizing and Operation of Multistage Centrifugal Pumps for Water Supply. *J. Pipeline Syst. Eng. Pract.* **2020**, *11*, 04020007. [CrossRef]
11. Jia, M.; Zhang, J.; Xu, Y. Optimization Design of Industrial Water Supply Pump Station Considering the Influence of Atmospheric Temperature on Operation Cost. *IEEE Access* **2020**, *8*, 161702–161712. [CrossRef]
12. Luo, Y.; Xiong, Z.; Sun, H.; Guo, Y. Research on energy-saving operation control model of the multi-type configuration centrifugal pump system with single invert. *Adv. Mech. Eng.* **2017**, *9*, 1–10. [CrossRef]
13. Sike, H.; Xuejing, J.; Huifen, G. Optimization of the Number of Multiple Pumps Running Simultaneously in Open Cycle Cooling Water System in Power Plant. *Energy Procedia* **2012**, *17*, 1161–1168. [CrossRef]
14. Bakman, I.; Gevorkov, L.; Vodovozov, V. Predictive control of a variable-speed multi-pump motor drive. In Proceedings of the 2014 IEEE 23rd International Symposium on Industrial Electronics (ISIE), Istanbul, Turkey, 1–4 June 2014; pp. 1409–1414. [CrossRef]
15. Wu, P.; Lai, Z.; Wu, D. Optimization Research of Parallel Pump System for Improving Energy Efficiency. *J. Water Resour. Plan. Manag.* **2014**, *141*. [CrossRef]
16. Viholainen, J.; Tamminen, J.; Ahonen, T.; Ahola, J.; Vakkilainen, E.; Soukka, R. Energy-efficient control strategy for variable speed-driven parallel pumping systems. *Energy Effic.* **2012**, *6*, 495–509. [CrossRef]

17. Oshurbekov, S.; Kazakbaev, V.; Prakht, V.; Dmitrievskii, V.; Gevorkov, L. Extending Pump Unit Service Life Using Combined Pump Control. In Proceedings of the 2021 28th International Workshop on Electric Drives: Improving Reliability of Electric Drives (IWED), Moscow, Russia, 27–29 January 2021; pp. 1–6. [CrossRef]

18. Saggewiss, G.; Kotwitz, R.; McIntosh, D. AFD synchronizing applications: Identifying potential methods and benefits. In Proceedings of the 2001 Petroleum and Chemical Industry Technical Conference (Cat. No.01CH37265), Toronto, ON, Canada, 26 September 2001; pp. 83–89. [CrossRef]

19. Automation Pump, Pump Genius. Product Description, WEG. Document Code: 50059602, Revision 04. January 2019. Available online: https://static.weg.net/medias/downloadcenter/hb3/hda/WEG-pump-genius-50059602-brochure-en.pdf (accessed on 5 April 2021).

20. Amirahmadi, A.; Rafiei, M.; Tehrani, K.; Griva, G.; Batarseh, I. Optimum Design of Integer and Fractional-Order PID Controllers for Boost Converter Using SPEA Look-up Tables. *J. Power Electron.* **2015**, *15*, 160–176. [CrossRef]

21. Rajesh, R. Optimal tuning of FOPID controller based on PSO algorithm with reference model for a single conical tank system. *SN Appl. Sci.* **2019**, *1*, 758. [CrossRef]

22. Bhamre, P.; Kadu, C. Design of a smith predictor based fractional order PID controller for a coupled tank system. In Proceedings of the 2016 International Conference on Automatic Control and Dynamic Optimization Techniques (ICACDOT), Pune, India, 9–10 September 2016; pp. 705–708. [CrossRef]

23. Tepljakov, A.; Petlenkov, E.; Belikov, J. Gain and order scheduled fractional-order PID control of fluid level in a multi-tank system. In Proceedings of the ICFDA'14 International Conference on Fractional Differentiation and Its Applications 2014, Catania, Italy, 23–25 June 2014; pp. 1–6. [CrossRef]

24. Ismail, M.; Bendary, A. FOPID Controller Based AC Pump Supplied from PV Standalone Source Tuned using Fuzzy Logic Type 2. *Indones. J. Electr. Eng. Comput. Sci.* **2016**, *4*, 10–19. [CrossRef]

25. Wu, X.; Xu, Y.; Liu, J.; Lv, C.; Zhou, J.; Zhang, Q. Characteristics Analysis and Fuzzy Fractional-Order PID Parameter Optimization for Primary Frequency Modulation of a Pumped Storage Unit Based on a Multi-Objective Gravitational Search Algorithm. *Energies* **2020**, *13*, 137. [CrossRef]

26. Stoffel, B. *Assessing the Energy Efficiency of Pumps and Pump Units. Background and Methodology*, 1st ed.; Elsevier: Amsterdam, The Netherlands, 2015; pp. 8, 9, 109, 113. [CrossRef]

27. *NM, NMS. Close Coupled Centrifugal Pumps with Flanged Connections*; Catalogue; Calpeda: Montorso Vicentino, Italy, 2018; Available online: https://www.calpeda.com/system/pdf/catalogue_en_50hz.pdf (accessed on 5 April 2021).

28. Dutta, N.; Palanisamy, K.; Subramaniam, U.; Padmanaban, S.; Holm-Nielsen, J.B.; Blaabjerg, F.; Almakhles, D.J. Identification of Water Hammering for Centrifugal Pump Drive Systems. *Appl. Sci.* **2020**, *10*, 2683. [CrossRef]

29. Tripathy, A.K.; Nambiar, P.; Pereira, A.; D'souza, S.; Rodrigues, L.; D'souza, A.; D'souza, B.; D'mello, B. Pressure surge analysis in pump systems. In Proceedings of the 2015 International Conference on Technologies for Sustainable Development (ICTSD), Mumbai, India, 4–6 February 2015; pp. 1–5. [CrossRef]

30. SINAMICS G120C. The Compact and Versatile Drive with Optimum Functionality. Available online: https://assets.new.siemens.com/siemens/assets/api/uuid:2c65250d-a8fc-4e40-bbb3-550543d54a1f/version:1573939918/sinamics-g120c-brochure.pdf (accessed on 5 April 2021).

31. SIMOTICS GP, SD, XP, DP Low-Voltage Motors Catalog. Available online: https://cache.industry.siemens.com/dl/files/197/109749197/att_955119/v1/simotics-gp-sd-xp-dp-catalogue-d-81-1-en-2018.pdf (accessed on 5 April 2021).

32. Siemens. SinaSave. Available online: https://www.sinasave.siemens.com/#/ru/pump (accessed on 5 April 2021).

33. IEC. *Adjustable Speed Electrical Power Drive Systems—Part 9-2: Ecodesign for Power Drive Systems, Motor Starters, Power Electronics and Their Driven Applications—Energy Efficiency Indicators for Power Drive Systems and Motor Starters*; IEC 61800-9-2/Ed1; IEC: Geneva, Switzerland, 2017.

34. Kazakbaev, V.; Prakht, V.; Dmitrievskii, V.; Ibrahim, M.N.; Oshurbekov, S.; Sarapulov, S. Efficiency Analysis of Low Electric Power Drives Employing Induction and Synchronous Reluctance Motors in Pump Applications. *Energies* **2019**, *12*, 1144. [CrossRef]

35. Safin, N.; Kazakbaev, V.; Prakht, V.; Dmitrievskii, V.; Sarapulov, S. Interpolation and analysis of the efficiency of a synchronous reluctance electric drive at various load points of a fan profile. In Proceedings of the 25th International Workshop on Electric Drives: Optimization in Control of Electric Drives (IWED 2018), Moscow, Russia, 31 January–2 February 2018; pp. 1–5. [CrossRef]

36. Goman, V.; Oshurbekov, S.; Kazakbaev, V.; Prakht, V.; Dmitrievskii, V. Energy Efficiency Analysis of Fixed-Speed Pump Drives with Various Types of Motors. *Appl. Sci.* **2019**, *9*, 5295. [CrossRef]

37. Eurostat Data for the Industrial Consumers in Germany. Available online: http://ec.europa.eu/eurostat/statistics-explained/index.php/Electricity_price_statistics#Electricity_prices_for_industrial_consumers (accessed on 5 April 2021).

 *mathematics*

MDPI

Article

Design Optimization of a Traction Synchronous Homopolar Motor

Vladimir Dmitrievskii [1], Vladimir Prakht [1], Alecksey Anuchin [2],* and Vadim Kazakbaev [1]

[1] Department of Electrical Engineering, Ural Federal University, 620002 Yekaterinburg, Russia; vladimir.dmitrievsky@urfu.ru (V.D.); va.prakht@urfu.ru (V.P.); vadim.kazakbaev@urfu.ru (V.K.)

[2] Department of Electric Drives, Moscow Power Engineering Institute, 111250 Moscow, Russia

* Correspondence: anuchinas@mpei.ru; Tel.: +7-905-538-19-10

Abstract: Synchronous homopolar motors (SHMs) have been attracting the attention of researchers for many decades. They are used in a variety of equipment such as aircraft and train generators, welding inverters, and as traction motors. Various mathematical models of SHMs have been proposed to deal with their complicated magnetic circuit. However, mathematical techniques for optimizing SHMs have not yet been proposed. This paper discusses various aspects of the optimal design of traction SHMs, applying the one-criterion unconstrained Nelder–Mead method. The considered motor is intended for use in a mining dump truck with a carrying capacity of 90 tons. The objective function for the SHM optimization was designed to reduce/improve the following main characteristics: total motor power loss, maximum winding current, and torque ripple. One of the difficulties in optimizing SHMs is the three-dimensional structure of their magnetic core, which usually requires the use of a three-dimensional finite element model. However, in this study, an original two-dimensional finite element model of a SHM was used; it allowed the drastic reduction in the computational burden, enabling objective optimization. As a result of optimization, the total losses in the motor decreased by up to 1.16 times and the torque ripple decreased by up to 1.34 times; the maximum armature winding current in the motor mode decreased by 8%.

Keywords: optimal design; Nelder–Mead method; synchronous homopolar machine; synchronous homopolar motor; traction drives; traction motor

Citation: Dmitrievskii, V.; Prakht, V.; Anuchin, A.; Kazakbaev, V. Design Optimization of a Traction Synchronous Homopolar Motor. *Mathematics* **2021**, *9*, 1352. https://doi.org/10.3390/math9121352

Academic Editor: Nicu Bizon

Received: 2 May 2021
Accepted: 9 June 2021
Published: 11 June 2021

Publisher's Note: MDPI stays neutral with regard to jurisdictional claims in published maps and institutional affiliations.

1. Introduction

Hybrid electric powertrains are widely used in mining trucks; a combustion engine rotates a wound rotor synchronous generator, producing AC voltage, which is then rectified. This electric energy supplies traction to the electric motors mounted in the wheels of a truck. Nowadays, both DC and AC motors are utilized as traction drives of these mining trucks. The significant drawback of DC motors is well known: an unreliable brush-collector unit. The rapid development of power semiconducting devices has led to the creation of reliable frequency converters and the feasibility of using brushless AC traction motors in truck powertrains. Currently, the induction AC motor is the most widespread solution for mining trucks [1], and DC motor powertrains have been discontinued. The usage of traction induction motors significantly increases the reliability of traction electric drives in comparison with the brushed DC motor and reduces the operating costs associated with the maintenance and replacement of brushes. However, induction traction motors in mining trucks have the following main disadvantages: reduced reliability due to the high risk of failure of the welded rotor winding [2], increased overheating due to high losses in the rotor [3], reduced speed control range in comparison with synchronous machines [4], impossibility of reliable sensorless control over the entire speed range due to the inapplicability of the self-sensing position estimation methods [3,5], and limitations in the use of pure electric brakes during a standstill due to the thermocycling of semiconducting devices.

To eliminate the above-described disadvantages of powertrains with induction electric motors, a traction synchronous homopolar motor (SHM) with the rated power of 320 kW was developed for the BELAZ 75570 mining truck (manufacturer is Belarusian Automobile Plant) with a carrying capacity of 90 tons. Two traction SHMs are installed in the two rear wheels of the mining truck. In [4,5], the inverter was described and the development of sensorless control algorithms for the traction SHM was highlighted.

The SHM has a complex magnetic core layout, which requires the calculation of a three-dimensional magnetic field, which makes its magnetic analysis challenging. The magnetic flux flows axially in the rotor sleeve and in the stator yoke; however, it changes its direction to transverse in the stator and rotor laminated cores.

Three kinds of models were proposed for the evaluation of the characteristics of the SHMs: the first is the 3D finite element method (FEM) [6,7]; the second is the 2D FEM [8,9], where the axial and radial fluxes are evaluated using a magnetic circuit; and the third is the 1D [6,10] lumped parameters-equivalent circuit. The 3D FEM provides the most accurate field calculation; however, the use of any method of mathematical optimization together with it is barely possible due to the very long time required by one calculation. Two-dimensional FEM models of SHMs described in [8,9] have a much shorter computation time, but they are not as accurate, due to the introduction of virtual windings into the computational area, imitating the axial excitation flux due to the substitution of the SHM with a salient-pole synchronous machine, which causes an additional error. One-dimensional-equivalent circuits provide the shortest calculation time, but they do not take into account the details of the machine geometry, and they give the largest error.

The article [11] described a new method of the mathematical modeling of SHMs, which is based on the 2D FEM. In contrast to the mathematical models of SHMs based on the 3D FEM [7,8], this method requires less computation time and is less demanding on the available computing resources. The calculation results obtained using this model were in good agreement with the experimental results. However, the traction SHM was developed without using any optimal design methods such as the genetic algorithm or the Nelder–Mead method. Therefore, the characteristics of the traction SHM described in [11] can be improved.

Synchronous homopolar machines have been known for a long time and are used in various equipment such as generators in aircrafts and trains [12], welding inverters [13], and flywheel energy storage systems [14]. Moreover, in [4,5,11,15,16], SHMs were presented as traction motors. In [17], the design of an SHM for a flywheel energy storage was considered with the use of 'manual' optimization of the SHM parameters based on a lumped model to obtain a higher efficiency of the machine. However, no mathematical methods of optimal design have been adopted for traction SHMs yet.

This paper discusses various aspects of the optimal design of traction SHM, applying the Nelder–Mead method. The objective function for the SHM optimization was designed to reduce/improve the following main characteristics: total motor power loss, maximum winding current, and torque ripple.

2. SHM Design Features

Figure 1a shows a sketch of the SHM with the number of stator and rotor stack combinations (SRSCs) equal to 3. The rotor stacks are mounted on the sleeve pressed onto the motor shaft. The stator stacks are pressed into the housing (back iron). The excitation coils are located in the gaps between SRSCs. A single stator winding is placed in the slots of all stator stacks. Each rotor core has 6 teeth, which corresponds to the number of pole pairs $p = 6$ of the stator armature winding. The motor electric frequency can be expressed through rotational speed n given in revolutions per minute by the formula: $f = p \times n/60$. The mechanical and electrical angular frequencies are defined as $\Omega = 2 \times \pi \times n/60$ and $\omega = 2 \times \pi \times f$, respectively. The stator has $Z_s = 54$ slots. The electromagnetic analysis was carried out for 2 poles and $Z_s/p = 9$ stator slots.

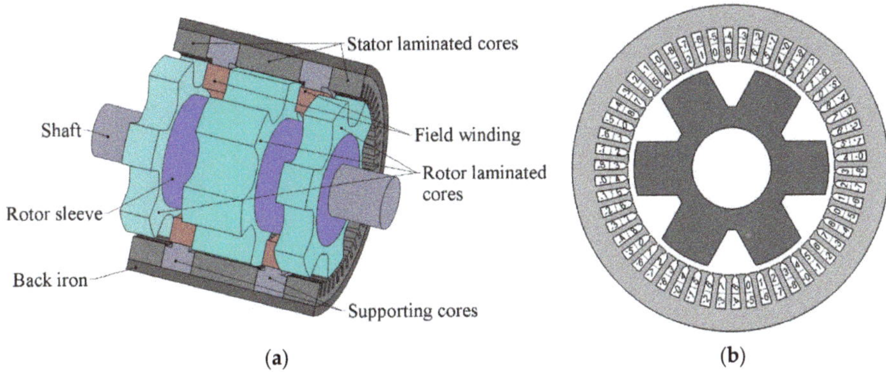

Figure 1. (**a**) Sketch of the SHM design. Armature winding is not shown to avoid cluttering; (**b**) sketch of the SHM cross-section.

To reduce the maximum current in the semiconductor switches and reduce the cost of the traction inverter, a nine-phase armature winding whose phases are indicated by the numbers 0–8 in Figure 1b was used, which consists of three separate three-phase windings, each of which has its own neutral point. The currents in the adjacent phases are shifted by $360°/9 = 40$ electrical degrees. The distributed double-layer winding has a coil pitch of four stator slots. The analysis assumes that the phase currents are sinusoidal.

The SHM has two sets of SRSCs with the same mutual orientation of the stator and rotor cores. In the considered case, the first set consists of only the central SRSC, and the second set consists of two lateral SRSCs. The angular position of the lateral rotor stacks is displaced relative to the position of the central rotor stack by 30 mechanical degrees (which is $30 * p = 180$ electrical degrees) so that SRSCs produce unidirectional electromotive forces (EMF) in the armature winding. The total stack length of one set must be approximately equal to the total stack length of the other set. For this reason, the length of the central SRSC must be twice those of the lateral SRSCs, as the flux of both field coils flows through it, and the flux of only one of the field coils flows through the lateral SRSCs.

In the case of sinusoidal armature currents, the two sets of SRSC under consideration make the same contribution to the active and reactive power, as well as to the torque. However, their instantaneous values of EMF and torque are not exactly the same. For this reason, the calculation method of SHM performances includes two steps. In the first step, it is assumed that the SHM has only one SRSC, the length of which is equal to the sum of the lengths of all SRSCs. The dependences of torque, voltage, etc. on the rotor position are calculated using a set of two-dimensional problems of magnetostatics. In the second step, symmetrization is applied to take into account that the torque ripple and the total harmonic distortion of the voltage wave produced by single SRSCs partially extinguish each other, and these parameters of the total machine are much less than those obtained in the first step. These magnetostatic problems are similar to those usually used in modeling radial motors, except that the excitation field is modeled by a magnetic monopole. Then, the symmetrization procedure is applied to spread the results to the real SHM. A detailed description of the mathematical model of SHM is given in [11].

Figure 2 shows the inverter circuit diagram for the traction SHM. The considered nine-phase inverter consists of 3 separate three-phase inverters, and it also has a one-phase chopper for powering the field winding.

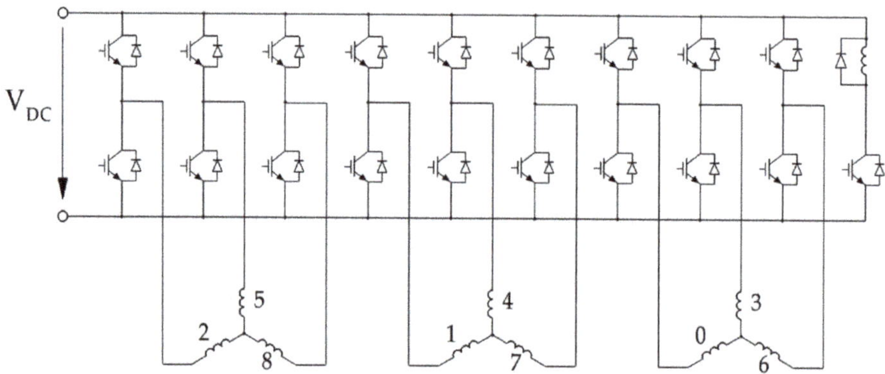

Figure 2. Inverter schematic.

3. Construction of an Objective Function for Nine-Phase Traction Synchronous Homopolar Motors

Figure 3 shows the traction characteristic of the electric drive of the BELAZ 75570 mining truck [3], limited by the maximum rotational speed and the maximum torque. The constant power speed ranges from 400 to 4000 rpm (10:1). The maximum mechanical power of the machine in the motor mode is 370 kW.

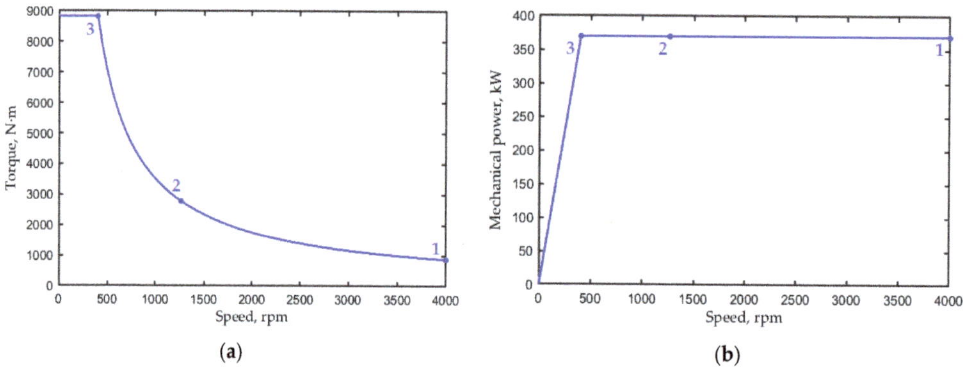

(a)   **(b)**

Figure 3. Traction characteristics of the motor. (**a**) Torque vs. rotational speed; (**b**) output mechanical power vs. rotational speed.

When optimizing the motor, three operating points were considered: the points with speeds of 400 rpm (maximum torque) and 4000 rpm (maximum speed), as well as the point with the geometric averages of speed and torque. These operating points are shown in Table 1.

Table 1. Operating points of the traction SHM considered during the optimization.

Mode Number	Torque, N·m	Rotational Speed, rpm	Mechanical Power, W
1	883	4000	370
2	2793	1265	370
3	8833	400	370

It is assumed that the SHM can operate with equal probability in the subranges 1–2 and 2–3. It is assumed that the average losses in the subranges are equal to the arithmetic mean of the losses at their boundaries (points 1 and 2 and points 2 and 3, respectively). Therefore, as the first optimization objective, the weighted average loss is chosen:

$$<P_{losses}> = (P_{losses1} + 2 \times P_{losses2} + P_{losses3})/4. \tag{1}$$

The other two optimization objectives are maximum symmetrized and nonsymmetrized torque ripples: $\max(TR)$ and $\max(TRsym)$, respectively. A nonsymmetrized torque ripple is produced by single SRSCs. A symmetrized torque ripple is produced by a SHM as a whole. Details of TR and $TRsym$ are given in [11].

The remaining optimization objective is the maximum stator armature winding current I_3 (it is achieved in operating point 3). Therefore, the objective function for the traction SHM optimization is:

$$F_0 = \ln(<P_{losses}>) + 0.7 \ln(I_3) + 0.05 \times \ln[\max(TR_{sym})] + 0.025 \times \ln[\max(TR)]. \tag{2}$$

This expression indicates that $<P_{losses}>$ is considered as the most valuable objective. I_3 is also a valuable objective. Decreasing I_3 by 1% is as valuable as decreasing $<P_{losses}>$ by 0.7%. $\max(TR_{sym})$ and $\max(TR)$ are much less valuable objectives. Decreasing $\max(TR_{sym})$ and $\max(TR)$ by 1% is as valuable as decreasing $<P_{losses}>$ by 0.05% and 0.025%, respectively.

In addition, the following constraints were adopted during the optimization:

$$U_{DC1} < 1000 \text{ V}; B_3 < 1.65 \text{ T}, \tag{3}$$

where U_{DC1} is the maximum voltage that is reached at operating point 1 and B_3 is the maximum flux density in the nonlaminated sections of the magnetic core (the rotor sleeve or the housing).

The one-criterion Nelder–Mead method is applied in this study to optimize the SHM design. The Nelder–Mead method belongs to unconstrained optimization methods. The optimization constraints could be specified simply by assigning infinite values to the objective function when the constraint conditions are not met. However, this would lead to a rapid decrease in the volume of the simplex. For this reason, objective function (2) is modified by using the 'soft constraints' with the penalty growing rapidly in the forbidden area:

$$F = F_0 + k_1 \times f_1(U_{DC1}/1000[\text{V}] - 1) + k_2 \times f_1(B_3/1.65[\text{T}] - 1), \tag{4}$$

where $f_1(x) = \begin{cases} x, x > 0 \\ 0 \end{cases}$.

Thus, during the optimization process, the constraint conditions can be violated, which prevents a rapid decrease in the simplex. At the same time, if k_1 and k_2 are large enough—they exceed the corresponding Lagrange multipliers—the optimized design will satisfy the constraints. In this study, it was assumed that $k_1 = k_2 = 1.5$, and it turned out that the optimized design satisfied constraints (3).

4. Initial Design and Parameters Varied during Optimization

Figure 4 demonstrates the main geometric parameters of the traction SHM. In the initial design, the lengths of the stator stacks were $L_{stat1} = 101$ mm, $L_{stat2} = 197$ mm, and $L_{stat3} = 101$ mm. The total length of the stator lamination was $L_{stat} = L_{stat1} + L_{stat2} + L_{stat3} = 399$ mm. The lengths of the rotor stacks were less than the lengths of the corresponding stator stacks and were equal to $L_{rot1} = 92$ mm, $L_{rot2} = 184$ mm, and $L_{rot3} = 92$ mm. The total length of the rotor lamination was $L_{rot} = L_{rot1} + L_{rot2} + L_{rot3} = 368$ mm. The parameters changed during optimization, and some fixed parameters are shown in Tables 2 and 3, and Figure 4. The main dimensions of the machine (outer radius of the stator housing $R_{housing} = 367$ mm and machine length without end winding parts $L = 545$ mm) remained unchanged during the optimization.

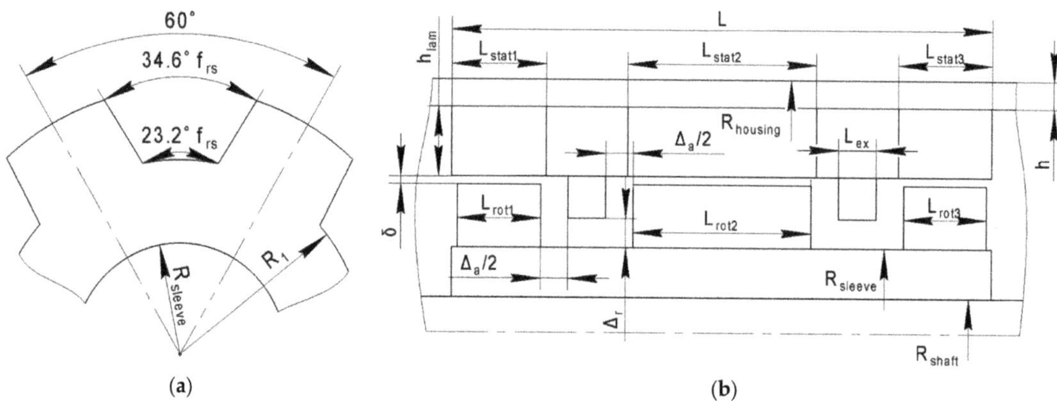

Figure 4. SHM geometric parameters. (**a**) Rotor parameters; (**b**) stator and rotor parameters.

Table 2. Some geometric parameters fixed during the optimization.

Parameter	Value
Machine length without end winding parts L, mm	545
$L_{stat} - L_{rot}$, m	31
Axial clearance between excitation winding and rotor, Δ_a, mm	25
Radial clearance between field winding and rotor Δ_r, mm	20
Number of turns per stator armature coil	5
Number of parallel strands per turn of the stator armature coil	2
Rotor sleeve outer radius R_{sleeve}, mm	161
Rotor slot bottom radius R_1, mm	183.8
Shaft radius R_{shaft}, mm	70
Stator lamination height h_{lam}, mm	65
External radius of the stator housing $R_{housing}$, mm	367

Table 3. Geometric parameters varied during the optimization.

Parameter	Initial Design	Optimized Design
Housing thickness h, mm	36	32
Total stator stacks length L_{stator}, mm	399	445
Airgap width δ, mm	2.3	3.2
Rotor slot factor f_{rs}	1	1.084
Angles of field weakening at operating points 1,2,3, electrical radians	0.61; 0.3; 0.25	0.68; 0.32; 0.22
Magnetic monopole densities at operating points 1,2,3, Wb/m	0.48; 0.63; 1.2	0.38; 0.67; 1.07

The shape of the stator yoke, stator yoke thickness, and rotor yoke thickness did not change. As the external radius of the stator housing $R_{housing}$ was fixed, the inner radius of the stator changed as the housing thickness h changed. The outer rotor radius was also influenced by the air gap width δ. As the outer radius of the rotor sleeve R_{sleeve} and the thickness of the rotor yoke $R_1 - R_{sleeve}$ were fixed, the depth of the rotor slot changed with the change in the outer diameter of the rotor.

The angular distances between the rotor teeth on the outer radius of the rotor and on the inner radius of the rotor slot R_1 were changed consistently by multiplying with the coefficient f_{rs}. As the outer radius of the rotor sleeve R_{sleeve} and the radial clearance Δ_r were fixed, the inner radius of the field winding was also fixed. The axial length of the field winding is $L_{ex} = (L - L_{stat})/2 - \Delta_a$ and changes with the variation of L_{stat}. $L_{ex} = 48$ mm at the initial design. The resistance of the excitation winding was assumed to be 10.2 Ohms at the initial design. During the optimization, this resistance changed, depending on the length of the field winding as 10.2 Ohm·48 mm/L_{ex}. The number of turns of the field winding equal to 340 did not change.

The lengths of the stator stacks were slightly longer than the lengths of the rotor stacks. As a result, the excitation field, which is constant in the rotor reference frame, coming out of the ends of the rotor stacks, was closed on the inner surface of the laminated stator. Therefore, additional eddy current losses due to the penetration of the magnetic field into the end surfaces of the stator laminations did not arise. However, the use of the 2D FEM model required the calculation of the equivalent total length of the stator lamination L_{equ} [18]. In this study, it was assumed that the equivalent length increased over the total length of the rotor stacks by $1.26 \times \delta$ at each edge of the rotor stack; therefore, $L_{equ} = L_{rot} + 6 \times 1.2 \times \delta$. The steel filling factor of the stator and rotor laminations was assumed to be $k_{steel} = 0.95$. The following magnetization curves were adopted for the rotor and stator laminations:

$$H_{stator} = H_0\left(\frac{B\,L_{equ}}{k_{steel}L_{stator}}\right),$$

$$H_{rot} = H_0\left(\frac{B\,L_{equ}}{k_{steel}L_{rot}}\right),$$

(5)

where $H_0(B)$ is the catalog steel magnetization curve.

5. Traction SHM Optimization Results and Discussion

The Nelder–Mead algorithm described in [19] was used in designing the traction SHM. The number of optimization parameters was ten (listed in Table 3). The mathematical model described in [11] was used to evaluate the objectives included in the optimization function F (3).

Figure 5 shows plots of flux density magnitude up to 2 T at the initial and optimized designs for the most saturated operating point 3. The contours of the regions of extreme saturation with flux density greater than 2 T were also outlined. As a result of optimization, the overall saturation and the areas of extreme saturation decreased.

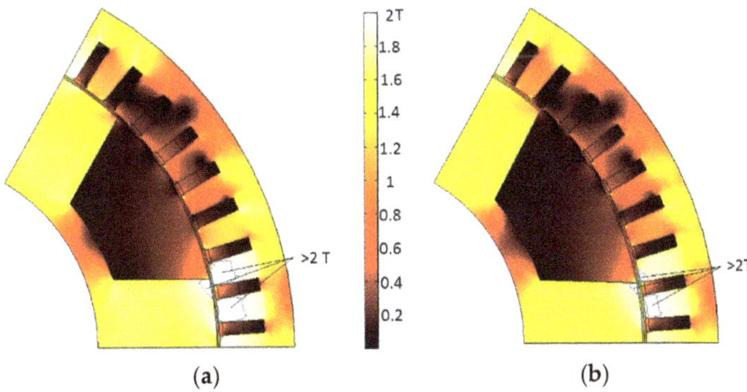

(a)   (b)

Figure 5. SHM geometry and flux density magnitude; black contours mark the regions in which the magnitude of the flux density vector exceeds 2 T (extreme saturation): (**a**) before optimization; (**b**) after optimization; it can be seen that after optimization, the regions in which the flux density exceeds 2 T have noticeably decreased.

Table 3 (see above) shows the varied design parameters of the traction SHM before and after the optimization. Among the varied parameters, the air gap width changed most significantly: it increased 1.4 times. There is a trade-off when choosing the air gap width of the SHM. On the one hand, reducing the gap makes it easier to produce the useful excitation flux. This flux interaction with the current in the armature winding creates the torque. On the other hand, with a small air gap, the leakage flux of the armature winding increases; it does not create the useful torque but only saturates the magnetic core and increases the reactive power. The synchronous homopolar machine has an effective excitation system, in which one excitation winding with ring-shaped coils magnetizes all the poles of the machine. Therefore, the trade-off in choosing the air gap width shifts from increasing the excitation flux to reducing the reactive power and saturation. For this reason, the air gap in the SHM must be increased compared to other types of electrical machines such as induction and reluctance machines. In addition, increasing the air gap width of the SHM increases its robustness, simplifies assembly, improves reliability, and decreases torque ripple.

Figure 6 shows the change in the total losses $<P_{losses}>$ and the maximum current of the armature winding I_3 in the SHM during optimization. Figure 7a shows the change in the maximum symmetrized TR_{sym} and nonsymmetrized torque ripples TR during optimization. Figure 7b shows the change in the objective function F during optimization. Table 4 shows the main results of the optimization of the traction SHM.

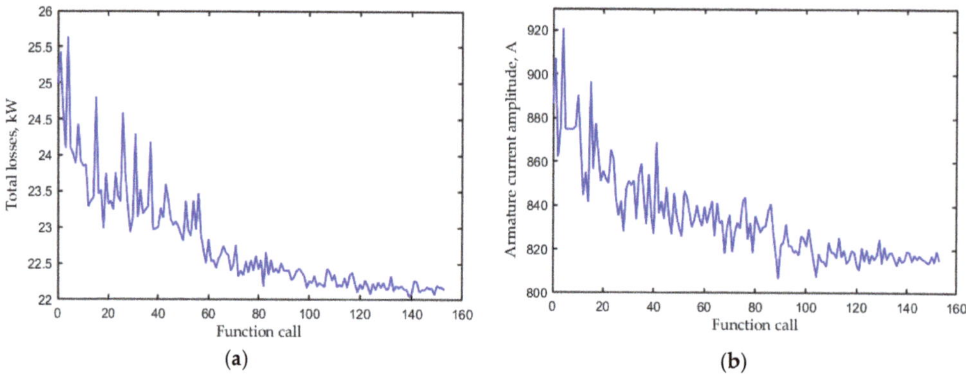

Figure 6. (**a**) Variation in the total losses $<P_{losses}>$ during the optimization; (**b**) variation in the maximum current amplitude in the armature winding during the optimization.

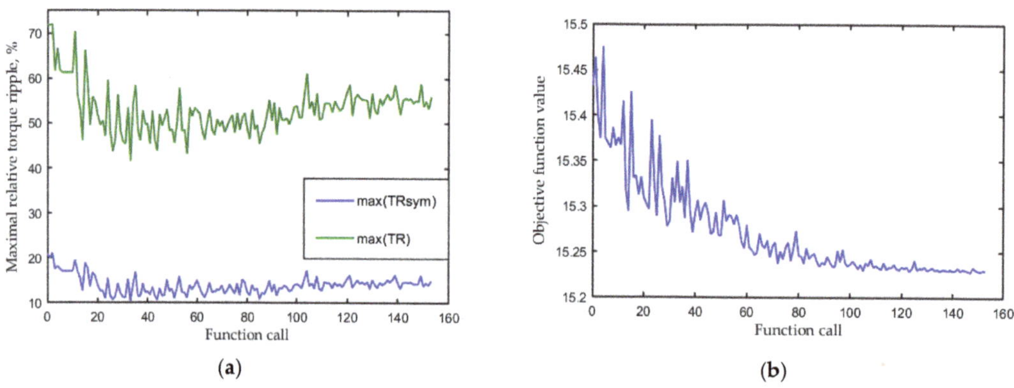

Figure 7. (**a**) Variation in the average efficiency during the optimization; (**b**) variation in the objective function during the optimization.

Table 4. Optimization results.

Value	Before Optimization				After Optimization			
Operating point	1	2	3	Brake mode	1	2	3	Brake mode
Speed, rpm	4000	1265	400	1100	4000	1265	400	1100
Current, A ampl	197	408	886	643	223	356	816	589
Mechanical power, kW	370	370	370	−540	370	370	370	−540
Mechanical losses, kW	17.57	0.65	0.05	0.45	17.57	0.65	0.05	0.45
Conductive winding losses, kW	1.49	6.40	30.25	15.96	1.92	4.89	25.68	13.40
Eddy-current winding losses, kW	6.58	3.15	1.39	5.59	6.89	2.42	1.24	4.47
Stator core losses, kW	13.44	6.45	2.19	8.41	8.29	4.91	2.44	6.50
Rotor core losses, kW	2.53	0.63	0.09	0.59	0.98	0.37	0.06	0.33
Excitation losses, kW	0.32	0.67	7.06	1.16	0.74	1.92	11.39	2.43
Active power, kW	412	387	404	−508	406	383	399	−512
Efficiency, %	89.8	95.4	90.0	93.8	91.0	96.1	90.1	94.5
Total losses, kW	41.93	17.95	41.04	32.15	36.40	15.15	40.86	27.57
Line voltage, V ampl	940	472	196	462	772	462	211	446
Not symmetrized torque ripple, N·m	71.94	61.53	24.13	42.09	55.18	51.91	24.61	40.43
Symmetrized torque ripple, N·m	20.97	12.44	2.81	8.43	14.42	9.38	2.01	7.44
Excitation current, A	5.6	8.1	26.3	10.7	6.3	10.1	24.6	11.4
Flux density in nonlaminated parts of the magnetic core, T	0.59	0.77	1.46	0.77	0.59	1.04	1.65	0.98

In the motor mode, at operating points 1 and 2, the total losses reduced by 1.13 and 1.16 times, respectively. At operating point 3, only a slight reduction in the total losses was achieved. This loss reduction was probably due to an increase in the air gap width, leading to a weakening in the leakage flux of the armature winding and a decrease in saturation and reactive power. An increase in the air gap width also leads to a weakening in the high harmonics of the flux density in the air gap, which makes it possible to reduce the torque ripple. At operating points 1 and 2 of the motor mode, the symmetrized torque ripple reduced by 1.45 and 1.32 times, respectively. The maximum armature winding current occurring at operating point 3 of the motor mode decreased by 8%. The flux density in the nonlaminated parts of the magnetic core reached a maximum value of 1.65 T at operating point 3.

In [20], it was reported that in in the braking (generator) mode of the considered application, the highest torque was reached at a speed of 1100 rpm (5200 N·m, 540 kW). For this reason, the calculation results for the braking mode are also presented in Table 4. In the braking mode, the total losses were also significantly reduced after optimization, although this operating point was not optimized and was not included in the objective function (2), as a mining truck, when driving from a mountain, dissipates all the energy into the braking resistors. As seen in Table 4, in the braking mode, the voltage and currents were within an acceptable range.

6. Conclusions

This paper discusses various aspects of the optimal design of a traction SHM, applying the one-criterion unconstrained Nelder–Mead method. This SHM is intended for use in a mining dump truck with a carrying capacity of 90 tons.

The objective function for the SHM optimization was designed to reduce/improve the following main characteristics: total motor power loss, maximum winding current, and torque ripple. Optimization was carried out by taking into account the characteristics of the SHM at three loading modes. The constraints of the supplied voltage and of the maximum magnetic flux density in the nonlaminated parts of the magnetic core were imposed.

Among the varied parameters, after the optimization, the air gap width changed most significantly; it increased 1.4 times, which makes it possible to reduce the saturation of the magnetic circuit, reduce the reactive power of the motor, increase the reliability of the motor, simplify assembly, and also to reduce the torque ripple.

As a result of optimization, in the motor mode, at operating points 1 and 2, the total losses reduced by 1.13 and 1.16 times, respectively. At operating point 3, only a slight reduction in the total losses was achieved. At operating points 1 and 2 of the motor mode, the symmetrized torque ripple reduced by 1.45 and 1.32 times, respectively. The flux density in the nonlaminated parts of the magnetic core reached a maximum value of 1.65 T at operating point 3. The maximum armature winding current in the motor mode decreased by 8%. In the braking mode, the total losses of the SHM were also significantly reduced after optimization, although this operating point was not optimized and was not included in the objective function.

In addition, after the optimization, the regions of the motor magnetic core with extreme saturation noticeably decreased, which is one of the reasons for the decrease in losses and an increase in efficiency.

In future works, the SHM will theoretically be considered in other applications, for example, as a traction motor for a light electric vehicle and electric bus. In addition, a theoretical comparison of the SHM with a traction induction motor will be carried out.

Author Contributions: Conceptual approach, A.A., V.D. and V.P.; data duration, V.D. and V.K.; software, V.D. and V.P.; calculations and modeling, A.A., V.D., V.K. and V.P.; writing—original draft, A.A., V.D., V.K. and V.P.; visualization, V.D. and V.K.; review and editing, A.A., V.D., V.K. and V.P. All authors have read and agreed to the published version of the manuscript.

Funding: The research was performed with the support of the Russian Science Foundation grant (Project No. 21-19-00696).

Institutional Review Board Statement: Not applicable.

Informed Consent Statement: Not applicable.

Data Availability Statement: All data are contained within the article.

Acknowledgments: The authors thank the editors and reviewers for their careful reading and constructive comments.

Conflicts of Interest: The authors declare no conflict of interest.

Glossary

List of Abbreviations

AC	Alternating current
DC	Direct current
EMF	Electromotive force
FEM	Finite element method
SHM	Synchronous homopolar motor
SRSC	Stator and rotor stack combination

List of Mathematical Symbols

B	Flux density, T
B_3	Maximum flux density in the nonlaminated parts of the magnetic core, T
f	Electric frequency, Hz

F	Second objective function
F_0	First objective function
f_1	Auxiliary function
f_{rs}	Rotor slot factor
h	Housing thickness, mm
H_0	Magnetic field strength according to the catalog magnetization curve, A/m
H_{rotor}	Rotor magnetic field strength, A/m
H_{stator}	Stator magnetic field strength, A/m
h_{lam}	Stator lamination height, mm
I_3	Amplitude of the maximum stator armature winding current, A
k_1, k_2	Multipliers of terms of an objective function
k_{steel}	Steel fill factor
L	Total machine length without end winding parts, mm
L_{equ}	Equivalent total length of the stator lamination, mm
L_{ex}	Axial length of the field winding, mm
L_{rot}	Total length of the rotor lamination, mm
$L_{rot1}, L_{rot2}, L_{rot3}$	Lengths of individual rotor stacks, mm
L_{stat}	Total length of the stator lamination, mm
$L_{stat1}, L_{stat2}, L_{stat3}$	Lengths of individual stator stacks, mm
n	Rotational frequency, revolution per minute
p	Number of pole pairs
P_{losses}	Total power losses, kW
R_1	Thickness of the rotor yoke, mm
$R_{housing}$	Outer radius of the stator housing, mm
R_{sleeve}	Outer radius of the rotor sleeve, mm
R_{shaft}	Shaft radius, mm
TR	Torque ripple, %
TR_{sym}	Symmetrized torque ripple, %
U_{DC1}	Maximum voltage, V
Z_s	Number of stator slots
δ	Air gap width, mm
Δ_a	Axial clearance between excitation winding and rotor, mm
Δ_r	Radial clearance between field winding and rotor, mm
ω	Electrical angular speed, radian per second
Ω	Mechanical angular speed, radian per second

References

1. Koellner, W.; Brown, G.; Rodriguez, J.; Pontt, J.; Cortes, P.; Miranda, H. Recent advances in mining haul trucks. *IEEE Trans. Ind. Electron.* **2004**, *51*, 321–329. [CrossRef]
2. Bernatt, J.; Gawron, S.; Glinka, T.; Polak, A. Traction induction motor. In Proceedings of the 13th International Conference Modern Electrified Transport (MET), Warsaw, Poland, 5–7 October 2017; pp. 1–5. [CrossRef]
3. Anuchin, A. Development pf Digital Systems for Efficient Control of Traction Electric Equipment for Hybrid Electric Vehicles. Ph.D. Thesis, Moscow Power Engineering Institute, Moscow, Russia, 2018; pp. 1–445. Available online: https://mpei.ru/diss/Lists/FilesDissertations/365-%D0%94%D0%B8%D1%81%D1%81%D0%B5%D1%80%D1%82%D0%B0%D1%86%D0%B8%D1%8F.pdf (accessed on 2 June 2021). (In Russian).
4. Lashkevich, M.; Anuchin, A.; Aliamkin, D.; Briz, F. Control strategy for synchronous homopolar motor in traction applications. In Proceedings of the 43rd Annual Conference of the IEEE Industrial Electronics Society (IECON), Beijing, China, 29 Octobe–1 November 2017; pp. 6607–6611. [CrossRef]
5. Lashkevich, M.; Anuchin, A.; Aliamkin, D.; Briz, F. Self-sensing control capability of synchronous homopolar motor in traction applications. In Proceedings of the 2017 IEEE 58th International Scientific Conference on Power and Electrical Engineering of Riga Technical University (RTUCON), Riga, Latvia, 12–13 October 2017; pp. 1–5. [CrossRef]
6. Belalahy, C.; Rasoanarivo, I.; Sargos, F. Using 3D reluctance network for design a three phase synchronous homopolar machine. In Proceedings of the 2008 34th Annual Conference of IEEE Industrial Electronics, Orlando, FL, USA, 10–13 November 2008; pp. 2067–2072. [CrossRef]
7. Cheshmehbeigi, H.; Afjei, E. Design Optimization of a Homopolar Salient-Pole Brushless DC Machine: Analysis, Simulation, and Experimental Tests. *IEEE Trans. Energy Convers.* **2013**, *28*, 289–297. [CrossRef]

8. Ye, C.; Yang, J.; Xiong, F.; Zhu, Z.Q. Relationship between homopolar inductor machine and wound-field synchronous machine. *IEEE Trans. Ind. Electron.* **2020**, *67*, 919–930. [CrossRef]
9. Yang, J.; Ye, C.; Liang, X.; Xu, W.; Xiong, F.; Xiang, Y.; Li, W. Investigation of a Two-Dimensional Analytical Model of the Homopolar Inductor Alternator. *IEEE Trans. Appl. Supercond.* **2018**, *28*, 5205205. [CrossRef]
10. Severson, E.; Nilssen, R.; Undeland, T.; Mohan, N. Magnetic Equivalent Circuit Modeling of the AC Homopolar Machine for Flywheel Energy Storage. *IEEE Trans. Energy Convers.* **2015**, *30*, 1670–1678. [CrossRef]
11. Dmitrievskii, V.; Prakht, V.; Anuchin, A.; Kazakbaev, V. Traction Synchronous Homopolar Motor: Simplified Computation Technique and Experimental Validation. *IEEE Access* **2020**, *2020*. *8*, 185112–185120. [CrossRef]
12. Bindu, G.; Basheer, J.; Venugopal, A. Analysis and control of rotor eccentricity in a train-lighting alternator. In Proceedings of the 2017 IEEE International Conference on Power, Control, Signals and Instrumentation Engineering (ICPCSI), Chennai, India, 21–22 September 2017; pp. 2021–2025. [CrossRef]
13. Bianchini, C.; Immovilli, F.; Bellini, A.; Lorenzani, E.; Concari, C.; Scolari, M. Homopolar generators: An overview. In Proceedings of the 2011 IEEE Energy Conversion Congress and Exposition, Phoenix, AZ, USA, 17–22 September 2011; pp. 1523–1527. [CrossRef]
14. Kalsi, S.; Hamilton, K.; Buckley, R.G.; Badcock, R.A. Superconducting AC Homopolar Machines for High-Speed Applications. *Energies* **2019**, *12*, 86. [CrossRef]
15. Sugitani, N.; Chiba, A.; Fukao, T. Characteristics of a doubly salient-pole homopolar machine in a constant-power speed range. In Proceedings of the 1998 IEEE Industry Applications Conference. Thirty-Third IAS Annual Meeting (Cat. No.98CH36242), St. Louis, MO, USA, 12–15 October 1998; pp. 663–670. [CrossRef]
16. Lee, S.-H.; Hong, J.-P.; Kwon, Y.-K.; Jo, Y.-S.; Baik, S.-K. Study on homopolar superconductivity synchronous motors for ship propulsion applications. *IEEE Trans. Appl. Supercond.* **2008**, *18*, 717–720. [CrossRef]
17. Tsao, P.; Senesky, M.; Sanders, S. A synchronous homopolar machine for high-speed applications. In Proceedings of the 2002 IEEE Industry Applications Conference. 37th IAS Annual Meeting (Cat. No.02CH37344), Pittsburgh, PA, USA, 13–18 October 2002; pp. 406–416. [CrossRef]
18. Ruuskanen, V.; Nerg, J.; Niemelä, M.; Pyrhönen, J.; Polinder, H. Effect of Radial Cooling Ducts on the Electromagnetic Performance of the Permanent Magnet Synchronous Generators With Double Radial Forced Air Cooling for Direct-Driven Wind Turbines. *IEEE Trans. Magn.* **2013**, *49*, 2974–2981. [CrossRef]
19. Nelder, J.A.; Mead, R. A Simplex Method for Function Minimization. *Comput. J.* **1965**, *7*, 308–313. [CrossRef]
20. Vinogradov, A.; Gnezdov, N.; Korotkov, A.; Chistoserdov, V. Features of traction electrical equipment of a mining truck with a carrying capacity of 90 tons. In Proceedings of the X International Conference on the Automated Electric Drive (AEP), Novocherkassk, Russia, 3–6 October 2018; pp. 194–197. Available online: https://elibrary.ru/item.asp?id=36585977 (accessed on 2 June 2021). (In Russian)

mathematics

Communication

High-Harmonic Injection-Based Brushless Wound Field Synchronous Machine Topology

Syed Sabir Hussain Bukhari [1,2], Fareed Hussain Mangi [1], Irfan Sami [2], Qasim Ali [1] and Jong-Suk Ro [2,3,*]

[1] Department of Electrical Engineering, Sukkur IBA University, Sukkur 65200, Sindh, Pakistan; sabir@iba-suk.edu.pk (S.S.H.B.); fareed.mangi@iba-suk.edu.pk (F.H.M.); qasim-ali@iba-suk.edu.pk (Q.A.)
[2] School of Electrical and Electronics Engineering, Chung-Ang University, Seoul 06910, Korea; irfansamimwt@gmail.com
[3] Department of Intelligent Energy and Industry, Chung-Ang University, Seoul 06910, Korea
* Correspondence: jsro@cau.ac.kr

Abstract: This paper discusses the design and analysis of a high-harmonic injection-based field excitation scheme for the brushless operation of wound field synchronous machines (WFSMs) in order to achieve a higher efficiency. The proposed scheme involves two inverters. One of these inverters provides the three-phase fundamental-harmonic current to the armature winding, whereas the second inverter injects the single-phase high-harmonic i.e., 6th harmonic current in this case, to the neutral-point of the Y-connected armature winding. The injection of the high-harmonic current in the armature winding develops the high-harmonic magnetomotive force (MMF) in the air gap of the machine beside the fundamental. The high-harmonic MMF induces the harmonic current in the excitation winding of the rotor, whereas the fundamental MMF develops the main armature field. The harmonic current is rectified to inject the direct current (DC) into the main rotor field winding. The main armature and rotor fields, when interacting with each other, produce torque. Finite element analysis (FEA) is carried out in order to develop a 4-pole 24-slot machine and investigate it using a 6th harmonic current injection for the rotor field excitation to both attain a brushless operation and analyze its electromagnetic performance. Later on, the performance of the proposed topology is compared with the typical brushless WFSM topology employing the 3rd harmonic current injection-based field excitation scheme.

Keywords: high-harmonic injection; brushless field excitation; wound field synchronous machines

Citation: Bukhari, S.S.H.; Mangi, F.H.; Sami, I.; Ali, Q.; Ro, J.-S. High-Harmonic Injection-Based Brushless Wound Field Synchronous Machine Topology. *Mathematics* **2021**, *9*, 1721. https://doi.org/10.3390/math9151721

Academic Editors: Vladimir Prakht, Mohamed N. Ibrahim and Aleksey S. Anuchin

Received: 26 June 2021
Accepted: 20 July 2021
Published: 22 July 2021

Publisher's Note: MDPI stays neutral with regard to jurisdictional claims in published maps and institutional affiliations.

1. Introduction

In contrast to the widespread range of applications and the speedy production increase of the permanent magnet (PM) machines due to their high efficiency, power density, and power factor, the price of rare-earth permanent magnets with high magnetic properties have ridiculously increased. On the other hand, the rare-earth material industry is facing serious environmental issues in keeping the prolonged and stable supply of rare-earth materials to the PM machines manufacturing industry [1].

In order to deal with these problems and to attain high flux-weakening-based performances of electrical machines in electrical vehicle (EV) and hybrid vehicle (HV) applications, other possibilities, such as PM-assisted synchronous reluctance machines and wound field synchronous machines, have sparked interest among the researchers [2–10].

The rotor field is excited in a typical WFSM, utilizing an excitation approach that includes brushes, slip rings, and a separate exciter. On the rotor side, a brushes and slip rings assembly connects the machine field winding to the excitation system [11–14]. The usage of brushes in a typical WFSM increases its maintenance cost due to their periodic replacement and continuous sparking. On the other hand, an additional exciter rises the overall size and cost of the system [15–18].

Therefore, to get rid of the obligation of brushes, slip rings, and additional exciters, a number of brushless topologies have been presented by various researchers in the available literature. In general, for the brushless operation of WFSM, a distinct excitation winding is housed in the rotor in addition to the field winding. However, another winding, named the auxiliary winding, is housed in the stator along with the main armature winding. This auxiliary winding produces an additional MMF in the air gap that induces a harmonic current in the excitation winding of the rotor. The induced harmonic current is then rectified and supplied to the field winding of the rotor. This field current generates a rotor field that becomes coupled with the armature field to produce torque. A basic diagram of such a system is presented in Figure 1.

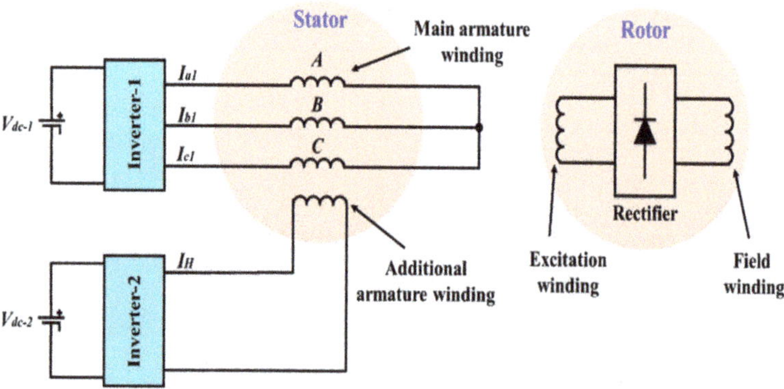

Figure 1. Typical 3rd harmonic injection-based brushless WFSM topology employing two dual-stator winding configurations.

Many researchers have proposed various schemes to attain the brushless operation of WFSMs by modifying the typical brushless topology in order to attain the maximum advantages of brushless excitation. These schemes are either based on space–harmonics generation or on time–harmonics generation.

A brushless WFSM topology based on a sub-harmonic field excitation scheme is proposed in [6]. In the proposed brushless topology, a dual-inverter configuration is utilized to provide two different magnitudes of current to the armature winding that is divided into two halves with a distinct star connection distributed in a symmetric arrangement. The difference in the magnitudes of the current supplied by the inverters produces sub-harmonics, besides the fundamental-harmonic, which can be utilized to induce currents in the excitation winding of the rotor. This topology is further investigated using a single inverter with a different number of turns for each half of the armature winding in [7]. In order to generate unbalanced radial forces associated with the sub-harmonically excited BL–WFSM topology and to attain its variable speed operation, an eight-pole machine topology is proposed in [8].

In [19], a brushless WFSM topology based on the 3rd harmonic field excitation scheme is presented. In this topology, two inverters are utilized to develop an armature current shape which comprises of a fundamental and 3rd harmonic. The required armature current shape is attained using thyristor switches operating at $-180°$ phase-shifted after the completion of each cycle of the phase. The 3rd harmonic component of the armature current is utilized to induce a harmonic current in the excitation winding of the rotor in order to attain a brushless operation. The proposed topology is further investigated using an open-winding pattern in [13]. In this topology, the required armature current shape is attained using a dual-inverter configuration without employing the thyristor switches. The proposed topology is also realized by using a modified inverter that has a different number of switches when compared to the typical three-phase inverter topology in [20].

In [14], a 3rd harmonic-based brushless WFSM topology using zero-sequence currents is proposed. In this topology, the armature winding is supplied current from an inverter by the thyristor switches that operate a zero-crossing of each phase, generating a zero-sequence current which predominantly contains a 3rd harmonic current component for the armature winding. The generated 3rd harmonic current is induced in the exciter winding of the rotor to provide DC to the main rotor field winding after rectification in order to attain brushless operation.

A brushless WFSM topology that employs an additional armature winding beside the main winding is proposed in [15]. Both windings are electrically connected by a rectifier and are supplied current from a single inverter. The main armature winding current produces the main armature field, whereas the rectified current flowing in the additional armature winding produces the harmonic field that is not coupled with the main armature field and is utilized in order to induce a harmonic current in the rotor exciter winding.

In general, the typical brushless WFSM topologies employ either sub-harmonic or the 3rd harmonic field excitation schemes in order to achieve a brushless operation.

In this paper, a high-harmonic injection-based field excitation scheme for the brushless operation of WFSMs is proposed. In the proposed topology, two inverters are utilized. One of the inverters gives the regular three-phase current i.e., fundamental-harmonic to the armature winding, whereas the second inverter injects a single-phase 6th harmonic current to the neutral-point of the Y-connected armature winding. The inverters are of customary design, without holding any modification in their structure. The fundamental-harmonic current produces the main armature field, whereas the high-harmonic i.e., 6th harmonic current, in this case, develops the high-harmonic field in the air gap of the machine. The high-harmonic field induces the harmonic current in the exciter winding of the rotor that is connected to the field winding by means of the rectifier. The induced harmonic current is rectified and utilized to excite the main rotor field winding. The interaction of the main armature and rotor fields produces electromagnetic torque. Unlike the conventional brushless WRSM topologies, which require dual three-phase inverter configurations with a parallel or open-winding pattern-based operation, or a modification in the stator armature winding to hold a dual-winding configuration, the proposed topology just utilizes customary three-phase and single-phase inverters. The operation of the single-phase inverter does not require any sophisticated control strategy, which eliminates the complications related to the control of the inverters associated with conventional brushless topologies based on an open-winding pattern. In addition, the operation of the proposed topology does not require any additional power electronics components, such as thyristor switches, as in the case of the conventional brushless topologies presented in [14] and [18]. The proposed topology along with its operating principle and electromagnetic analysis are discussed in subsequent sections. A comparative performance analysis of the proposed topology with the typical brushless WFSM employing the 3rd harmonic injection-based field excitation scheme is carried out and is presented in these sections.

2. Methodology

2.1. Proposed Topology

The proposed high-harmonic injection-based WFSM topology is presented in Figure 2. This topology employs two inverters, namely inverter-1 and inverter-2. Inverter-1 provides the regular three-phase input armature currents (I_{abc1}) to the armature winding (ABC) of the machine, whereas inverter-2 injects the single-phase high-harmonic current (I_H) i.e., 6th harmonic current in this case, to the neutral-point of the Y-connected ABC winding. The input armature currents of inverter-1 are responsible for the fundamental-harmonic current of the armature winding that produces the fundamental MMF in the air gap. However, the single-phase 6th harmonic input current of inverter-2 is responsible for generating a high-harmonic current component of the armature winding. This current develops the high-harmonic MMF in the air gap. Although the fundamental and high-harmonic MMF components are produced by the same armature winding, the difference in their frequencies makes them uncoupled. On the rotor side, the rotor has two windings, namely excitation winding and field winding. Both windings are electrically connected by an uncontrolled rectifier. The fundamental MMF component produces the main armature magnetic field, whereas the high-harmonic MMF component induces the harmonic current in the rotor excitation winding. The induced harmonic current is rectified to inject DC to the field winding of the rotor to produce the rotor main field. The produced rotor field interacts with the equal number of stator poles to produce torque. A 4-pole 24-slot (4p24s) machine is utilized to validate the operation of the proposed high-harmonic injection-based WFSM topology. This machine is adopted from [16]. The structure of the employed machine along with its armature and rotor winding configurations are presented in Figure 3. As seen in the figure, the stator core is equipped with 4-pole, 24-slot, double-layered armature winding whereas the rotor is fitted with 4 rotor main teeth and 8 rotor sub-teeth to accommodate the excitation and field windings. The parameters of the armature winding used for the employed machine are presented in Table 1, which shows that the winding factor of the winding is 0.933. The winding factor and the three-phase MMF for the different harmonic numbers generated for armature winding used for the employed machine are presented in Figure 4a,b, respectively. The winding configuration used for the armature of the machine is chosen from the various options available on online tool named as Emetor-Electric motor winding calculator [21].

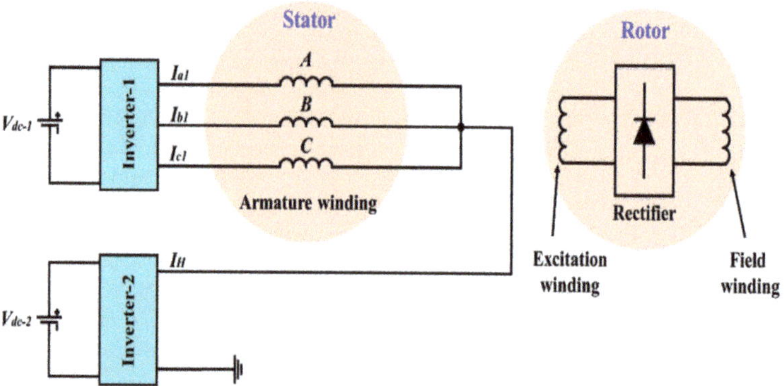

Figure 2. Proposed brushless WFSM topology.

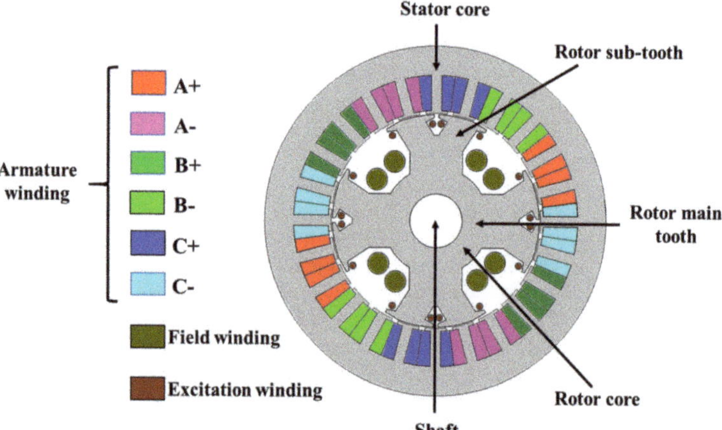

Figure 3. Two-dimensional machine layout and its winding configuration.

Table 1. Winding parameters for the employed machine.

Parameter	Value
Number of poles/slots/layers	4/24/2
Coil span	5 slots
Pole pitch	6 slots
Periodicities	2, 4
Winding factor	0.933

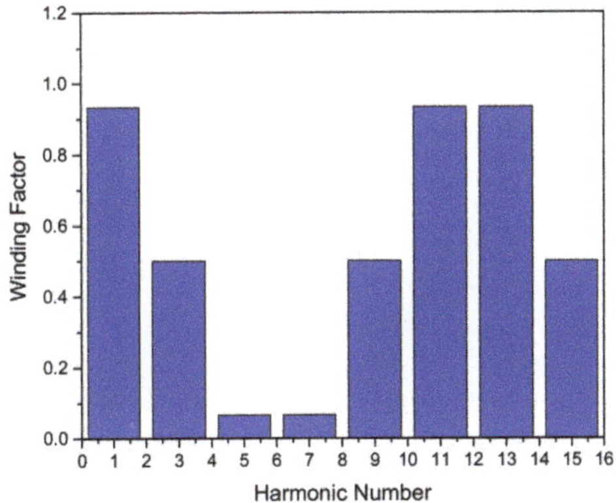

(**a**)

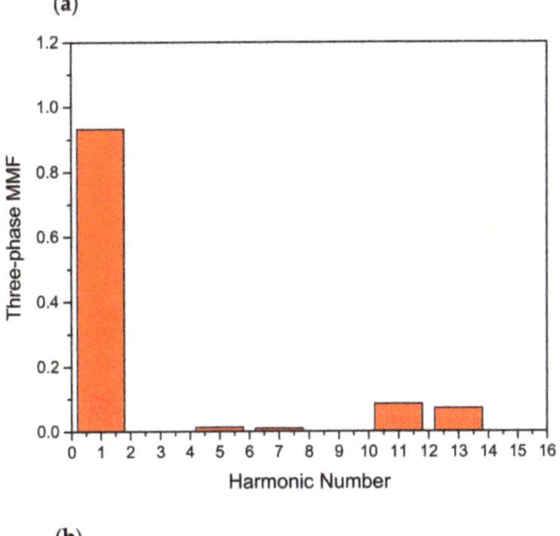

(**b**)

Figure 4. (**a**) Winding factor and (**b**) three-phase MMF for different harmonic numbers generated for the armature winding used for employed machine [21].

2.2. Mathematical Modelling

In order to explain the operation of the proposed high-harmonic injection-based field excitation scheme for the brushless operation of WFSMs, a mathematical model is presented in this sub-section.

For the typical WFSM topology employing 3rd harmonic injection-based field excitation scheme [12], the magnetomotive force (MMF) for each phase is expressed as:

$$
\begin{aligned}
MMF_A &= i_A N_{\varphi 1}\left(\sin\theta_s + \tfrac{1}{3}\sin 3\theta_s\right) \\
MMF_B &= i_B N_{\varphi 1}\left\{\sin(\theta_s - \tfrac{2\pi}{3}) + \tfrac{1}{3}\sin 3\theta_s\right\} \\
MMF_C &= i_C N_{\varphi 1}\left\{\sin(\theta_s + \tfrac{2\pi}{3}) + \tfrac{1}{3}\sin 3\theta_s\right\}
\end{aligned}
\tag{1}
$$

where,

$$
N_{\varphi 1} = \frac{2}{\pi}\,(per\ phase\ number\ of\ turns)
$$

θ_s = electrical angle (spatial), and
ω = angular frequency (electrical).
The armature winding currents can be expressed as:

$$
\begin{aligned}
i_A &= I_1 \sin(\omega t) + I_H \\
i_B &= I_1 \sin(\omega t - \tfrac{2\pi}{3}) + I_H \\
i_C &= I_1 \sin(\omega t + \tfrac{2\pi}{3}) + I_H
\end{aligned}
\tag{2}
$$

where I_1 is the fundamental and I_H is the supplied single-phase harmonic current into the armature winding.

When you bring Equation (2) to Equation (1), and add the MMF of three-phase armature windings A, B, and C, the net MMF of the armature winding is expressed as:

$$
MMF_{ABC}(\theta_s, i) = MMF_A + MMF_B + MMF_C
\tag{3}
$$

$$
MMF_{ABC}(\theta_s, i) =
\begin{bmatrix}
N_{\varphi 1}\left\{\sin(\theta_s) + \tfrac{1}{3}\sin 3\theta_s\right\}\left\{I_1 \sin(\omega t) + I_H\right\} \\
+N_{\varphi 1}\left\{\sin(\theta_s - \tfrac{2\pi}{3}) + \tfrac{1}{3}\sin 3\theta_s\right\}\left\{I_1 \sin(\omega t - \tfrac{2\pi}{3}) + I_H\right\} \\
+N_{\varphi 1}\left\{\sin(\theta_s + \tfrac{2\pi}{3}) + \tfrac{1}{3}\sin 3\theta_s\right\}\left\{I_1 \sin(\omega t + \tfrac{2\pi}{3}) + I_H\right\}
\end{bmatrix}
\tag{4}
$$

$$
MMF_{ABC}(\theta_s, i) = N_{\varphi 1}
\begin{bmatrix}
I_1\left\{\begin{array}{l} \sin(\omega t)\sin(\theta_s) + \sin(\omega t - \tfrac{2\pi}{3})\sin(\theta_s - \tfrac{2\pi}{3}) \\ + \sin(\omega t + \tfrac{2\pi}{3})\sin(\theta_s + \tfrac{2\pi}{3}) \end{array}\right\} \\
+I_H \sin 3\theta_s
\end{bmatrix}
\tag{5}
$$

$$
MMF_{ABC}(\theta_s, i) = \frac{3}{2} I_1 N_{\varphi 1}\cos(\omega t - \theta_s) + I_H N_{\varphi 1}\sin 3\theta_s
\tag{6}
$$

Equation (6) gives the net MMF produced in the air gap, which consists of two parts: (1) the fundamental MMF component, which produces the main armature field of rotating nature, and (2) spatial-location-fixed MMF component that develops a harmonic field in the air gap. This MMF component is utilized by the injected single-phase harmonic current. Since the main armature and the harmonic fields are produced by the currents having different frequencies, both the fields are not coupled.

Assuming that θ_0 is the rotor excitation winding initial position angle, the spatial position of the excitation winding can be calculated as under:

$$
\theta_s = \omega t + \theta_0
\tag{7}
$$

The generated flux of each winding pole will be:

$$
\begin{aligned}
\psi_E &= n_E P_g N_{\varphi 1}\left\{\tfrac{3}{2} I_1 \cos(\omega t - \theta_s) + I_H \sin 3\theta_s\right\} \\
\psi_E &= n_E P_g N_{\varphi 1}\left\{\tfrac{3}{2} I_1 \cos\theta_0 + I_H \sin(3\omega t + 3\theta_0)\right\}
\end{aligned}
\tag{8}
$$

where n_E is the rotor excitation winding number of turns, and P_g is the air gap permeance.
When I_H is the 3rd harmonic current, as in the case of [12], we have:

$$
I_H = I_3 \sin 3(\omega t)
\tag{9}
$$

and the magnitude of the induced EMF in the rotor excitation winding can be calculated as [22,23]:

$$
\begin{aligned}
e_E &= 6\frac{d\psi_E}{dt} \\
e_E &= 6n_E P_g N_{\varphi 1}\{0 + 3I_3\omega \cos 3(\omega t)\sin(3\omega t + 3\theta_0)\} \\
e_E &= 18 n_E P_g N_{\varphi 1} I_3 \omega \sin(6\omega t + 3\theta_0)
\end{aligned}
\tag{10}
$$

As the proposed high-harmonic field excitation scheme is based on the injection of single-phase 6th harmonic current, I_H in this case will become:

$$
I_H = I_3 \sin 6(\omega t)
\tag{11}
$$

and the magnitude of the induced EMF in the rotor excitation winding for the proposed high-harmonic injection-based field excitation scheme can be calculated as:

$$
\begin{aligned}
e_E &= 6\frac{d\psi_E}{dt} \\
e_E &= 6n_E P_g N_{\varphi 1}\left\{ \begin{array}{l} 0 + 6I_3\omega \cos 6(\omega t)\sin(3\omega t + 3\theta_0) \\ +3I_3\omega \sin 6(\omega t)\cos(3\omega t + 3\theta_0) \end{array} \right\}
\end{aligned}
\tag{12}
$$

$$
e_E = 3n_E P_g N_{\varphi 1} I_3 \omega \{9\sin(9\omega t + 3\theta_0) - 3\sin 3(\omega t - 3\theta_0)\}
\tag{13}
$$

The induced EMF in the rotor excitation winding is rectified by an uncontrolled rectifier in order to inject DC to the field winding of the rotor to create the main rotor field. This field, when interacting with the 4-pole armature field, generates torque and archives brushless operation for WFSM.

3. Electromagnetic Analysis

To support the presented theory and to attain the electromagnetic performance of the proposed high-harmonic injection-based WFSM topology, a finite element method (FEM) is utilized in order to carry out a finite element analysis (FEA) in JMAG-Designer ver. 19.1. For the comparative performance analysis, typical brushless WFSM topology based on a 3rd harmonic injection field excitation scheme (as presented in Figure 1) is also implemented. Both topologies are investigated under the same loading conditions and for the same machine (as presented in Figure 3). The parameters of this machine are presented in Table 2. A slide mesh is generated to carry out the FEA. 50H1300 material manufactured by NIPPON STEEL is used for the stator and rotor cores. The FEA parameters along with the attributes associated with the generated mesh are presented in Table 3.

Table 2. Machine specifications.

Parameter	Value
Rated power	3 kW
Machine poles/Stator slots	4/24
Rated speed	1800 rpm
Frequency	60 Hz
Stator outer/inner diameter	130/80 mm
Air gap	0.5 mm
Rotor diameter	79 mm
Rotor main/sub-teeth	4/8
Exciter/Field winding number of turns	25/250
Armature winding number of turns	100
Stack length	90 mm

Table 3. FEA parameters.

Attribute	Value
Boundry conditions	195 mm
Number of divisions	21,600
Time interval/1 step	92.5 µs
Number of steps	21,601
Number of elements (for Mesh)	5508
Number of nodes (for Mesh)	4024
Radial divisions (for Mesh)	7
Circumferential divisions (for Mesh)	360

3.1. No-Load Analysis

To examine the no-load operation of the machine used for the validation of 3rd and 6th harmonic injection-based brushless WFSM topologies, no-load analysis is carried out in JMAG-Designer ver. 19.1. A field current of 2 A is directly supplied to the rotor field winding in order to generate the main rotor field. The rotor of the machine is rotated at 1800 rpm, as in the previous case. The magnetic flux density plot of the machine under such a condition is shown in Figure 5. This flux density plot shows that the operation of the machine is under a saturation level of 1.98 T. A back-EMF of 546.30 V_{rms} is generated. The generated back-EMF of the machine is shown in Figure 6a; however, the FFT plot of this back-EMF is presented in Figure 6b to show its harmonic contents. The peak-to-peak cogging torque of the employed machine is 1.68 Nm. The generated cogging torque is shown in Figure 7.

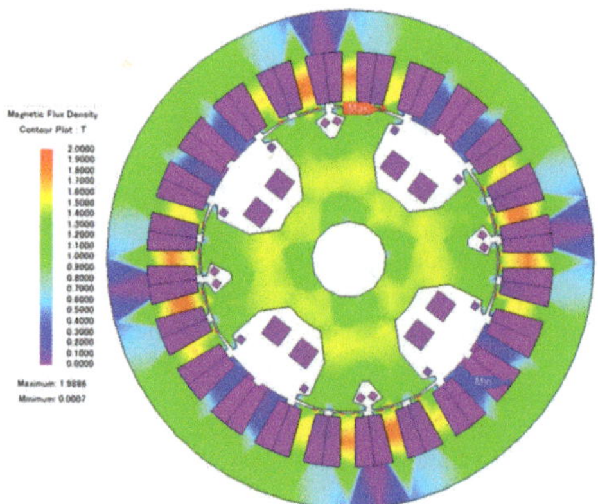

Figure 5. Magnetic flux density plot of the employed machine.

3.2. Full-Load Analysis

The three-phase input armature current of inverter-1 (I_{abc1}) for the 3rd and 6th harmonic injection-based brushless WFSM topologies is 5 A (peak) at a 60 Hz frequency. This current is presented in Figure 8a. In the case of the 3rd harmonic injection-based WFSM topology, the magnitude of the single-phase input armature current for inverter-2 is 1 A (peak) at a 180 Hz frequency, whereas it is 1 A (peak) at a 360 Hz frequency in the case of the proposed high-harmonic injection-based WFSM topology, as presented in Figure 8b,c, respectively. Figure 8d,e show the three-phase armature currents flowing across A, B, and C windings employing the 3rd and 6th harmonic injection-based brushless WFSM topologies. Fast Fourier transform (FFT) plots for phase A of the armature currents are produced to show the amplitude of the fundamental and harmonic current components for the 3rd and 6th harmonic injection-based WFSM topologies. These graphs are presented in Figure 9a,b, respectively.

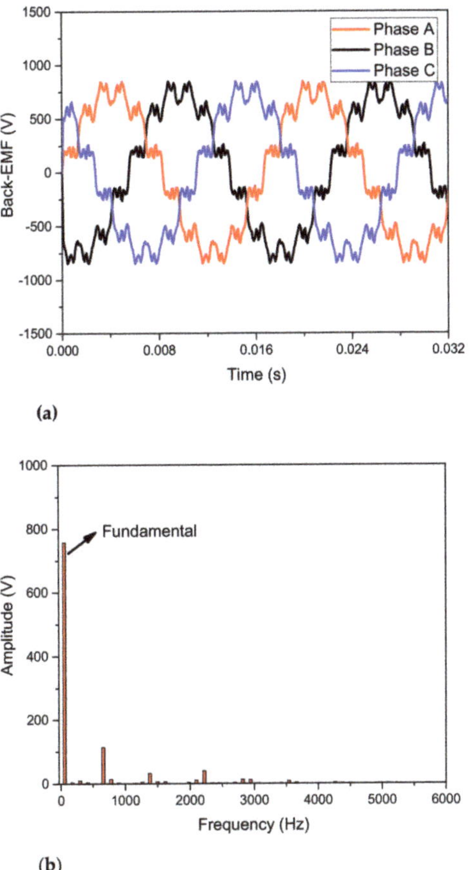

(a)

(b)

Figure 6. (a) Back-EMF of the employed machine, and (b) its FFT plot.

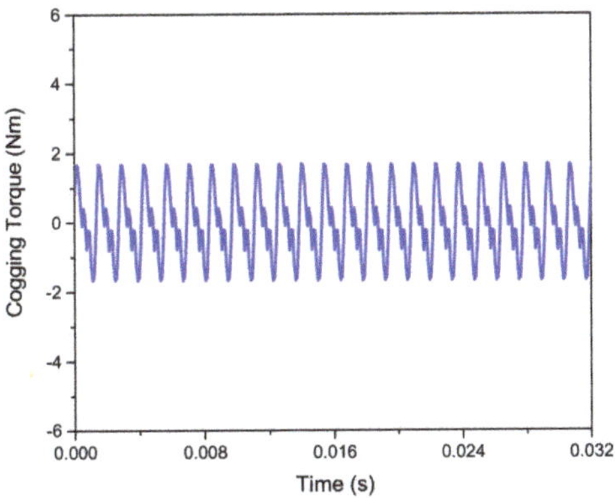

Figure 7. Cogging torque of the employed machine.

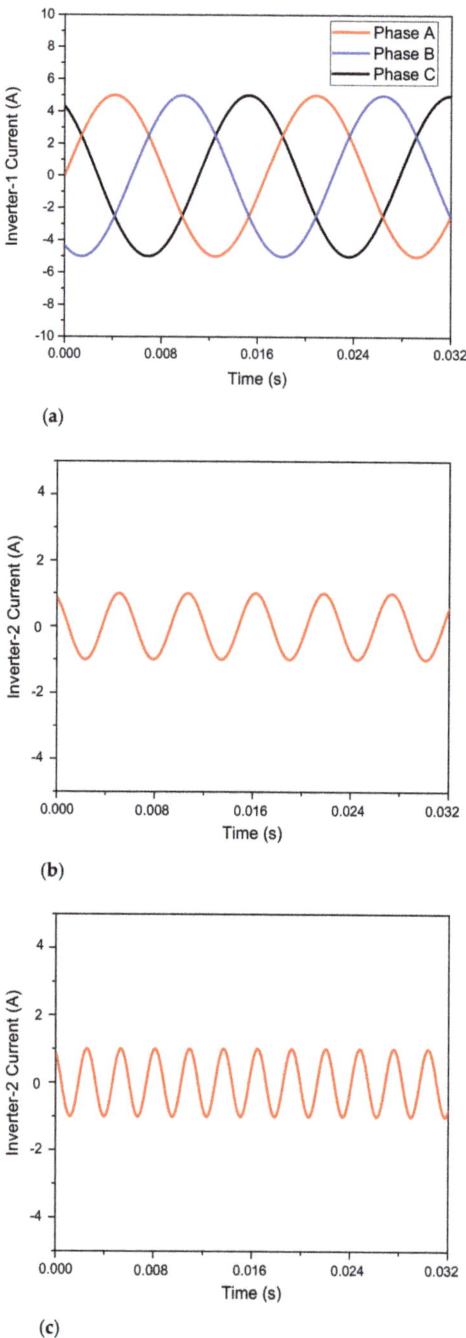

Figure 8. *Cont.*

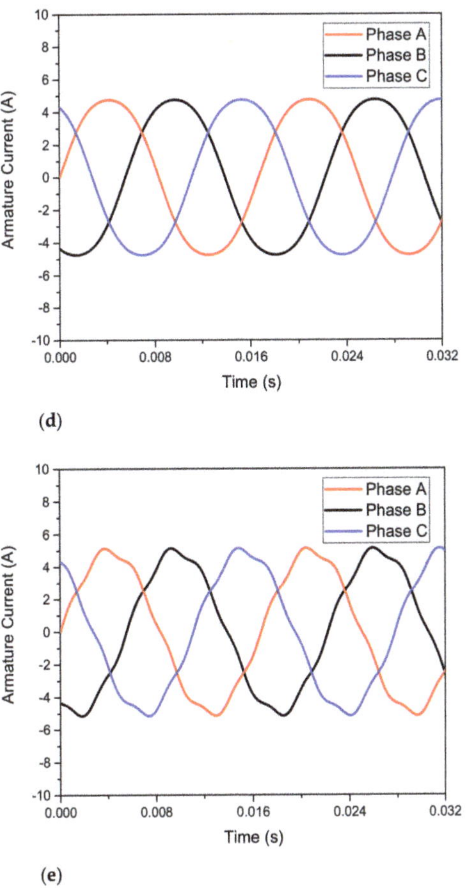

(d)

(e)

Figure 8. (**a**) Inverter-1 current, (**b**) inverter-2 current for the 3rd harmonic injection-based topology, (**c**) inverter-2 current for the 6th harmonic injection-based topology, (**d**) armature currents for the 3rd harmonic injection-based topology, and (**e**) armature currents for the 6th harmonic injection-based WFSM topology.

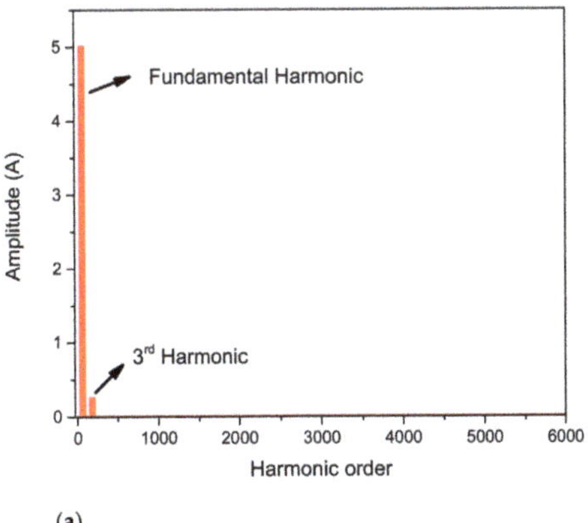

(a)

Figure 9. *Cont.*

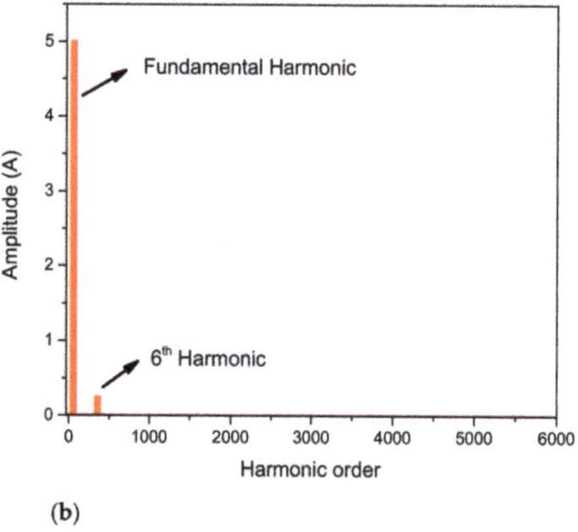

(b)

Figure 9. FFT plot of phase *A* of the armature currents for (**a**) typical brushless WFSM topology based on 3rd harmonic field excitation scheme, and (**b**) proposed high-harmonic injection-based WFSM topology.

The shaft of both machines is rotated at a speed of 1800 rpm, whereas the simulations are performed for 1 s. The flux linkages of the machine using the conventional and proposed WFSM topologies are presented in Figure 10a,b, respectively. Figure 11a,b show the magnetic field density plot of the analyzed machines employing the 3rd and 6th harmonic injection-based brushless WFSM topologies, respectively. These figures show that the operation of the investigated machines is under a saturation level of 2.24 T, and none of its parts become saturated during the operation. The rotor harmonic and field currents for the 3rd and 6th harmonic injection-based brushless WFSM topologies are presented in Figure 12a,b, respectively. As seen in the figures, the magnitude of the average field current for the brushless WFSM topology based on the 3rd harmonic field excitation scheme is around 2.007 A, whereas, for the proposed high-harmonic injection-based WFSM topology its magnitude is 2.08 A, which is 3.63% higher than the 3rd harmonic injection-based topology. Figure 13a,b show the output torque of the analyzed machines. The comparative performance analysis of the 3rd and 6th harmonic injection-based brushless WFSM topologies is presented in Table 4.

Table 4. Comparative performance analysis.

Attribute	3rd **Harmonic Injection-Based Topology**	6th **Harmonic Injection-Based Topology**
Average output torque	12.4814 Nm	12.8048 Nm
Maximum torque	15.1361 Nm	15.3927 Nm
Torque ripple	42%	42.5%
Efficiency	82.91%	85.095%

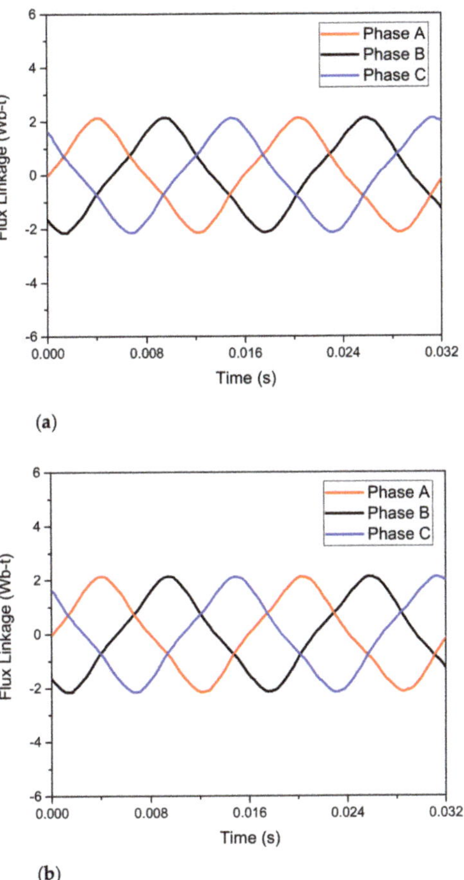

(a)

(b)

Figure 10. Flux linkages for (**a**) 3rd harmonic injection-based, and (**b**) 6th harmonic injection-based WFSM topologies.

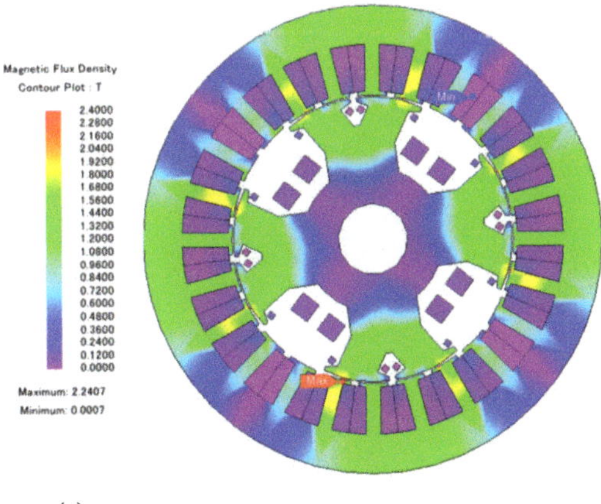

(a)

Figure 11. *Cont.*

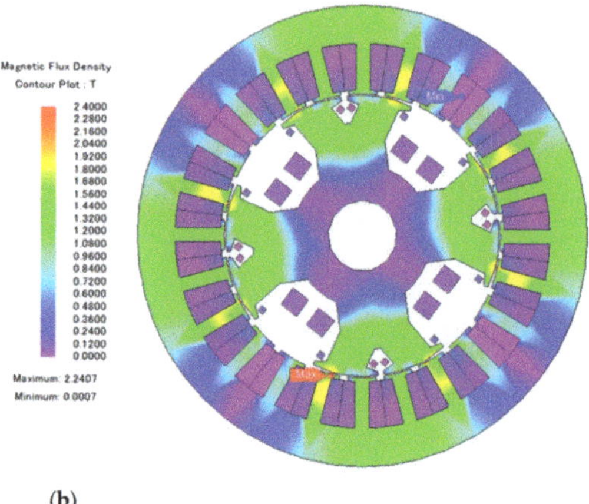

(**b**)

Figure 11. Magnetic flux density graph for (**a**) 3rd harmonic injection-based, and (**b**) 6th harmonic injection-based WFSM topologies.

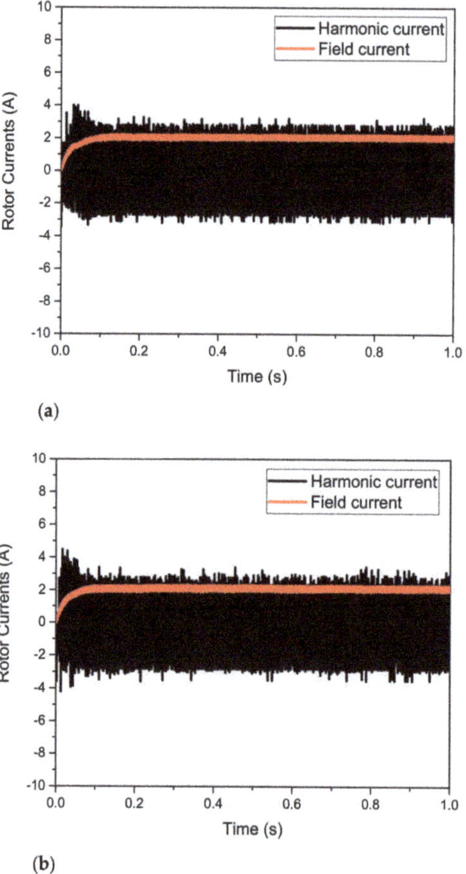

(**a**)

(**b**)

Figure 12. Rotor currents for (**a**) 3rd harmonic injection-based, and (**b**) 6th harmonic injection-based WFSM topologies.

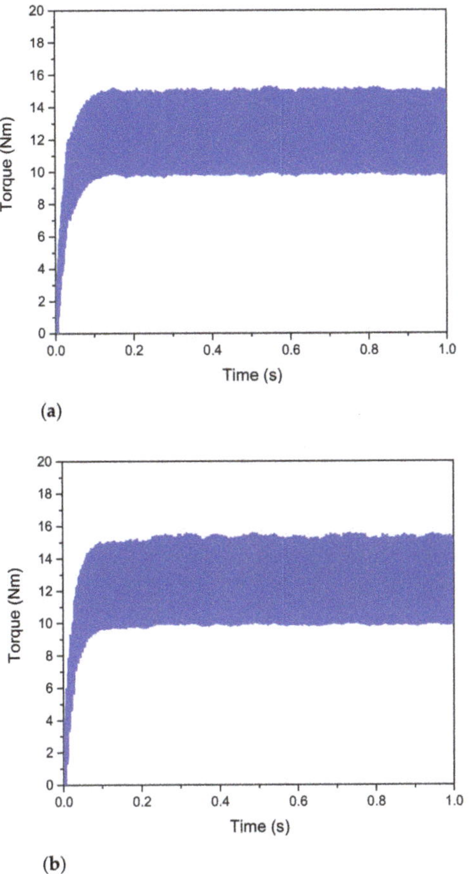

(a)

(b)

Figure 13. Output torque for (**a**) 3rd harmonic injection-based, and (**b**) 6th harmonic injection-based WFSM topologies.

4. Conclusions

A high-harmonic injection-based field excitation scheme for the brushless operation of WFSMs has been discussed in this paper. This scheme utilized two inverters. One inverter provided the regular three-phase current, while the second inverter injected the single-phase 6th harmonic current to the neutral-point of the Y-connected armature winding. The regular three-phase current was used to produce the main armature field, whereas the single-phase 6th harmonic induced the harmonic current in the rotor excitation winding. The induced harmonic current was used to supply DC to the field winding of the rotor after rectification, in order to attain a brushless operation.

A four-pole, 24-slot machine was utilized to validate the operation of the proposed scheme and attain its electromagnetic performance by FEA. The proposed high-harmonic injection-based field excitation scheme attains a better performance for the output average torque, maximum torque, and efficiency. However, the torque ripple of the proposed scheme is 0.5% higher than the typical brushless WFSM topology, based on the 3rd harmonic field excitation scheme.

Author Contributions: Conceptualization, S.S.H.B. and F.H.M.; software, S.S.H.B. and I.S.; formal analysis, S.S.H.B. and Q.A.; writing—original draft preparation, S.S.H.B.; funding acquisition, J.-S.R. All authors have read and agreed to the published version of the manuscript.

Funding: This work was supported in part by the Brain Pool (BP) Program funded by the Ministry of Science and ICT through the National Research Foundation of Korea under Grant 2019H1D3A1A01102 988, and in part by the Basic Science Research Program through the National Research Foundation of Korea funded by the Ministry of Education under Grant 2016R1D1A1B01008058.

Conflicts of Interest: The authors declare no conflict of interest.

References

1. Sun, L.; Gao, X.; Yao, F.; An, Q.; Lipo, T. A new type of harmonic current excited brushless synchronous machine based on an open winding pattern. In Proceedings of the 2014 IEEE Energy Conversion Congress and Exposition (ECCE), Pittsburgh, PA, USA, 14–18 September 2014; pp. 2366–2373.
2. Yao, F.; An, Q.; Sun, L.; Lipo, T.A. Performance Investigation of a Brushless Synchronous Machine with Additional Harmonic Field Windings. *IEEE Trans. Ind. Electron.* **2016**, *63*, 6756–6766. [CrossRef]
3. Inoue, K.; Yamashita, H.; Nakamae, E.; Fujikawa, T. A brushless self-exciting three-phase synchronous generator utilizing the 5th-space harmonic component of magneto motive force through armature currents. *IEEE Trans. Energy Convers.* **1992**, *7*, 517–524. [CrossRef]
4. Yao, F.; An, Q.; Sun, L.; Illindala, M.S.; Lipo, T.A. Optimization design of stator excitation windings in brushless synchronous machine excited with double-harmonic-windings. In Proceedings of the 2017 International Energy and Sustainability Conference (IESC), Farmingdale, NY, USA, 19–20 October 2017; pp. 1–6.
5. Ayub, M.; Hussain, A.; Jawad, G.; Kwon, B. Brushless Operation of a Wound-Field Synchronous Machine Using a Novel Winding Scheme. *IEEE Trans. Magn.* **2019**, *55*, 8201104. [CrossRef]
6. Ali, Q.; Lipo, T.A.; Kwon, B.-I. Design and analysis of a novel brushless wound rotor synchronous machine. *IEEE Trans. Magn.* **2015**, *51*. [CrossRef]
7. Hussain, A.; Kwon, B.-I. A new brushless wound rotor synchronous machine using a special stator winding arrangement. *Electr. Eng.* **2018**, *100*, 1797–1804. [CrossRef]
8. Ali, Q.; Bukhari, S.S.H.; Atiq, S. Variable-speed, sub-harmonically excited BL-WRSM avoiding unbalanced radial force. *Electr. Eng.* **2019**, *101*, 251–257. [CrossRef]
9. Ayub, M.; Atiq, S.; Sirewal, G.J.; Kwon, B. Fault-Tolerant Operation of Wound Field Synchronous Machine Using Coil Switching. *IEEE Access* **2019**, *7*, 67130–67138. [CrossRef]
10. Ayub, M.; Jawad, G.; Kwon, B. Consequent-Pole Hybrid Excitation Brushless Wound Field Synchronous Machine with Fractional Slot Concentrated Winding. *IEEE Trans. Magn.* **2019**, *55*, 8203805. [CrossRef]
11. Khan, S.; Bukhari, S.S.H.; Ro, J. Design and Analysis of a 4-kW Two-Stack Coreless Axial Flux Permanent Magnet Synchronous Machine for Low-Speed Applications. *IEEE Access* **2019**, *7*, 173848–173854. [CrossRef]
12. Bukhari, S.S.H.; Sirewal, G.J.; Chachar, F.A.; Ro, J.-S. Brushless Field Excitation Scheme for Wound Field Synchronous Machines. *Appl. Sci.* **2020**, *10*, 5866. [CrossRef]
13. Bukhari, S.S.H.; Sirewal, G.J.; Chachar, F.A.; Ro, J.-S. Dual-Inverter-Controlled Brushless Operation of Wound Rotor Synchronous Machines Based on an Open-Winding Pattern. *Energies* **2020**, *13*, 2205. [CrossRef]
14. Jawad, G.; Ali, Q.; Lipo, T.A.; Kwon, B.I. Novel brushless wound rotor synchronous machine with zero-sequence third-harmonic field excitation. *IEEE Trans. Magn.* **2016**, *52*, 8106104. [CrossRef]
15. Bukhari, S.S.H.; Sirewal, G.J.; Ro, J.-S. A New Small-Scale Self-Excited Wound Rotor Synchronous Motor Topology. *IEEE Trans. Magn.* **2021**, *57*, 8200205. [CrossRef]
16. Bukhari, S.S.H.; Ahmad, H.; Sirewal, G.J.; Ro, J.-S. Simplified Brushless Wound Field Synchronous Machine Topology Based on a Three-phase Rectifier. *IEEE Access* **2021**, *9*, 8637–8648. [CrossRef]
17. Bukhari, S.S.H.; Sirewal, G.J.; Madanzadeh, S.; Ro, J.-S. Cost-Effective Single-Inverter-Controlled Brushless Scheme for Wound Rotor Synchronous Machines. *IEEE Access* **2020**, *8*, 204804–204815. [CrossRef]
18. Ayub, M.; Bukhari, S.S.H.; Sirewal, G.J.; Arif, A.; Kwon, B.-I. Utilization of Reluctance Torque for Improvement of the Starting and Average Torques of a Brushless Wound Field Synchronous Machine. *Electr. Eng.* **2021**, *103*. [CrossRef]
19. Ayub, M.; Sirewal, G.J.; Bukhari, S.S.H.; Kwon, B.-I. Brushless wound rotor synchronous machine with third-harmonic field excitation. *Electr. Eng.* **2020**, *102*, 259–265. [CrossRef]
20. Ayub, M.; Bukhari, S.S.H.; Kwon, B.-I. Brushless Wound-Field Synchronous Machine with Third-Harmonic Field Excitation using a Single Inverter. *Electr. Eng.* **2019**, *101*, 165–173. [CrossRef]
21. Emetor, A.B. Electric Motor Winding Calculator. Available online: https://www.emetor.com/windings/ (accessed on 25 June 2021).
22. Yao, F.; An, Q.; Gao, X.; Sun, L.; Lipo, T.A. Principle of operation and performance of a synchronous machine employing a new harmonic excitation scheme. *IEEE Trans. Ind. Appl.* **2015**, *51*, 3890–3898. [CrossRef]
23. An, Q.; Gao, X.; Yao, F.; Sun, L.; Lipo, T. The structure optimization of novel harmonic current excited brushless synchronous machines based on open winding pattern. In Proceedings of the 2014 IEEE Energy Conversion Congress and Exposition (ECCE), Pittsburgh, PA, USA, 14–18 September 2014; pp. 1754–1761.

Article

Analysis and Optimization of Axial Flux Permanent Magnet Machine for Cogging Torque Reduction

Hina Usman [1], Junaid Ikram [1], Khurram Saleem Alimgeer [1], Muhammad Yousuf [2], Syed Sabir Hussain Bukhari [3,4] and Jong-Suk Ro [4,5,*]

[1] Electrical and Computer Engineering Department, Islamabad Campus, COMSATS University Islamabad, Islamabad 45550, Pakistan; hinausman93@gmail.com (H.U.); Junaidikram@comsats.edu.pk (J.I.); khurram_saleem@comsats.edu.pk (K.S.A.)
[2] Electrical and Computer Engineering Department, Abbottabad Campus, COMSATS University Islamabad, Islamabad 22060, Pakistan; m.yousaf891@gmail.com
[3] Department of Electrical Engineering, Sukkur IBA University, Sukkur 65200, Pakistan; sabir@iba-suk.edu.pk
[4] School of Electrical and Electronics Engineering, Chung-Ang University, Seoul 06910, Korea
[5] Department of Intelligent Energy and Industry, Chung-Ang University, Seoul 06910, Korea
* Correspondence: jongsukro@gmail.com

Citation: Usman, H.; Ikram, J.; Alimgeer, K.S.; Yousuf, M.; Bukhari, S.S.H.; Ro, J.-S. Analysis and Optimization of Axial Flux Permanent Magnet Machine for Cogging Torque Reduction. *Mathematics* **2021**, *9*, 1738. https://doi.org/10.3390/math9151738

Academic Editors: Vladimir Prakht, Mohamed N. Ibrahim and Aleksey S. Anuchin

Received: 7 July 2021
Accepted: 20 July 2021
Published: 23 July 2021

Abstract: In this paper, a hexagonal magnet shape is proposed to have an arc profile capable of reducing torque ripples resulting from cogging torque in a single-sided axial flux permanent magnet (AFPM) machine. The arc-shaped permanent magnet increases the air-gap length effectively and makes the flux of the air-gap more sinusoidal, which decreases air-gap flux density and hence causes a reduction in cogging torque. Cogging torque is the basic source of vibration, along with the noise in PM machines, since it is the main cause of torque ripples. Cogging torque is independent of the load current and is proportional to the air-gap flux and the reluctance variation. Three-dimensional finite element analysis (FEA) is used in the JMAG-Designer to analyze the performance of the conventional and proposed hexagonal-shaped PM AFPM machines. The proposed shape is designed to reduce cogging torque, and the voltage remains the same as compared to the conventional hexagonal-shaped PM machine. Further, optimization is performed by utilizing an asymmetric overhang. Latin hypercube sampling (LHS) is used to create samples, the kriging method is applied to approximate the model, and a genetic algorithm is applied to obtain the optimum parameters of the machine.

Keywords: Axial flux permanent magnet machine; 3D FEA; Genetic algorithm; hexagonal-shaped PMs; PM overhang

1. Introduction

Permanent magnet machines are divided into three categories—axial, radial, and transverse machines—depending upon the direction of flux through the air-gap. AFMs have several benefits over radial and transverse flux machines. They have high a power-to-weight ratio and easily modifiable air-gaps [1,2]. The axial flux permanent magnet (AFPM) machines are widely used, especially for power generation and electric vehicle (EVs) applications, due to their high power density and their compressed structure [3–9]. Furthermore, AFPM machines perform exceptionally well at a wide range of rotational speeds, which is appropriate for low-speed, high-torque characteristics. A single-sided AFPM machine is discussed in this paper due to its compact size and low cost. However, similar to other machines, single-sided AFPM machines also suffer from torque ripples, mainly due to cogging torque [10].

In PM machine design, cogging torque is a concern since it adds undesirable harmonics to torque. There are various techniques that exist to overcome cogging torque [11,12]. Skewing is the simplest, most common, and most effective technique used in PM machines to decrease the cogging torque. It also decreases the high order harmonics in the back-EMF

199

waveform [13]. Cogging torque reduction is still a considerable issue in PM machines. Investigating methods of cogging torque minimization is of high importance, and various methods are discussed and investigated in other work [14–19].

Another technique of reducing cogging torque is the stator slots-to-rotor poles ratio. In the integral slot machines, components of cogging torque that are produced by all of the magnets are in phase with each other, which results in a high resultant cogging torque. However, in the fractional slot machines, components of the cogging torque are out of phase with each other [11]. Hence, the resultant cogging torque is reduced since some of the cogging components are partially canceled. Generally, it is desirable to use a combination of slot and pole numbers that has a higher least common multiple [20]. There are also different methods for reducing torque ripples; the shifting and stepping of rotor magnets is one of the most effective methods [21]. Three-dimensional FEA is utilized for a more precise performance analysis of the AFPM machine as compared to the analytical methods. To achieve high efficiency, stator-sided iron must not be saturated, and torque ripples should be nearby zero.

In the AFPM machine, the pole length is measured from the inner side to the outer diameter of the rotor back iron. The circumference of the outer side of the rotor is larger than the inner side. Consequently, to increase the active utilization of the surface area of the rotor, a hexagonal shape is more advantageous than a rectangular or circular shape. In practice, circular-shaped magnets may not be used in AFPM machines, especially when a higher number of poles is required [22].

A symmetrical, sinusoidal-shaped radial flux permanent magnet machine (RFPM) was designed, which removes the harmonics of back-EMF more effectively and has been proposed as a method for obtaining reduced cogging torque [23]. In [24], a trapezoidal arc-shaped is introduced to reduce torque ripples and cogging torque. In arc-shaped PM, the effective length of the air-gap is increased, which is not identical across the one magnet; it is at its maximum at the pole edges and reaches its minimum at the center of the pole. Hence, the increased length of an air-gap reduces the magnetic flux between magnets; therefore, cogging torque and torque ripples are reduced.

Generally, permanent magnet motors have an overhang structure in which the length of PM is longer than the length of the stator's stack. The overhang structure is commonly installed in PM machines to increase the air-gap flux density by concentrating magnetic flux [25]. The optimization of rotor overhang variation and PM overhang within the tangential direction is proposed to enhance the performance of AFPM machines.

Multi-objective optimization has received the most attention recently and is used in the optimization of electric machines. To improve the average output torque of the switched reluctance motor, a multi-objective optimization analysis is investigated in [26]. In [27], another optimization technique, based on particle swarm optimization (PSO), is discussed for cogging torque reduction and efficiency enhancement of the brushless DC motor. A 3D Pareto front linear induction motor with finite element model evaluation is optimized and analyzed in [28]. This paper performs the well-known, simple and practical optimization approach due to its suitability for non-linear data, as implemented in [29,30].

In this paper, a single-rotor, single-stator AFPM with a proposed PM shape is introduced. It utilizes the arc profile for the conventional hexagonal-shaped PM. The proposed hexagonal-shaped PM decreases the air-gap flux density, thus decreasing the cogging torque and torque ripple, which improves the performance of the machine. In addition, the optimization of the proposed model, which is obtained through a genetic algorithm (GA), further enhances its output characteristics by using asymmetric overhang. The research methodology, as described in [31], is presented in Figure 1.

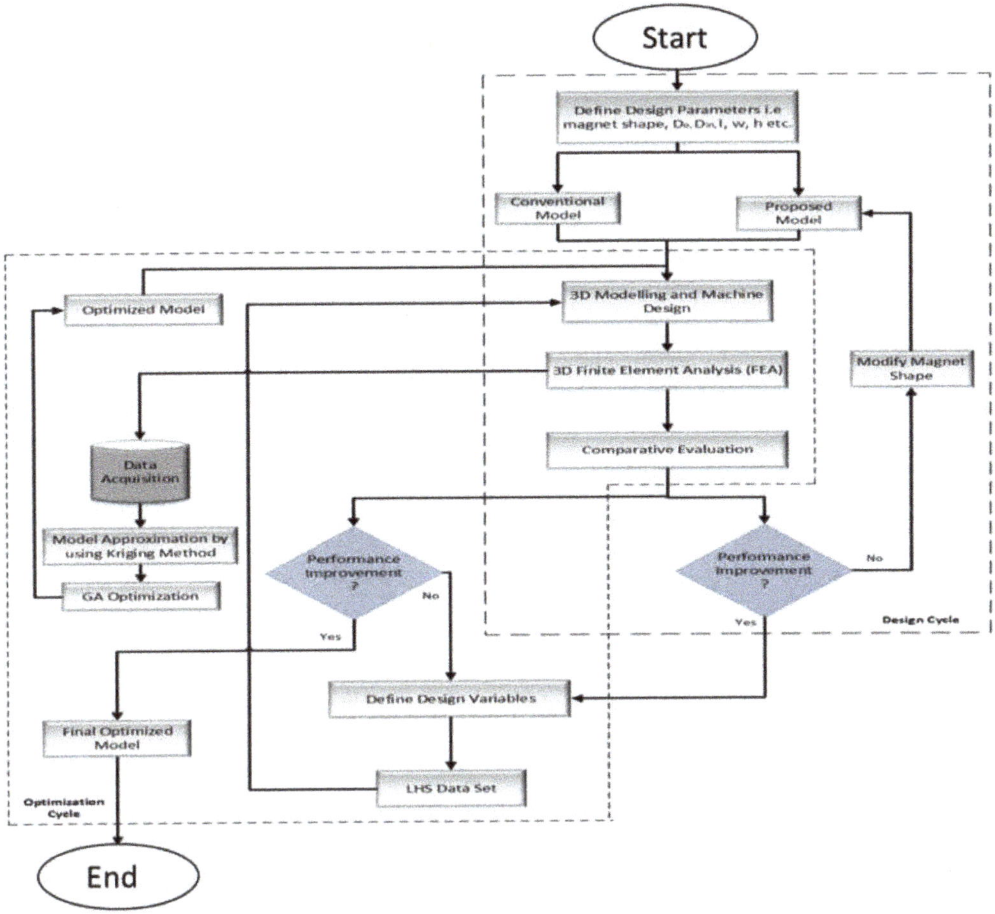

Figure 1. Flow chart of design and optimization process.

2. Machine Topology and Design Process

The single-sided machine is the most fundamental of these topologies for AFPM's. This arrangement contains one stator disc containing coils and one rotor disc containing magnets. Magnets in the rotor alternate between having south and north poles facing the stator, as shown in Figure 2.

In this topology, it is possible to attain a high ratio between the diameter and machine length. The topology normally consists of a compact design and offers low cost. The main disadvantage of single-sided topology is the major force attraction problem; because magnetic flux is coming from only one side, the attractive forces result in the bending of the yokes and high bearing losses. Table 1 shows the parameters of the basic model. The power of this machine is 2.5 kW and the frequency is about 700 Hz. A SRSS machine, with 14 poles mounted on the rotor and 12 slots on the stator having a double-layer winding, is used in this research.

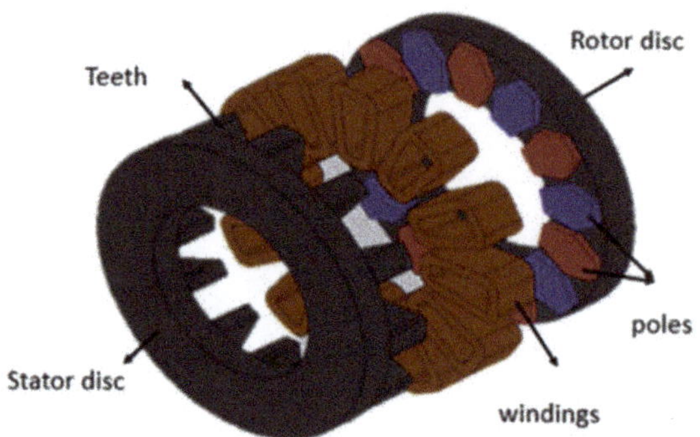

Figure 2. Single-rotor, single-stator axial flux permanent magnet machine.

Table 1. Main dimensions and operating parameters of the conventional and proposed models.

Parameters	Conventional/Proposed	Units
Outer diameter	110	mm
Inner diameter	64	mm
Axial length	40/40.3	mm
Stator slots depth	15	mm
Stator slots width	5	mm
Stator core thickness	10	mm
Rotor core thickness	12	mm
Operating speed	6000	rpm
Number of turns	32	-
Length of magnet (L_m)	22.5	mm
Side length (L_n)	12.5	mm
Width (W_m)	18	mm
Top and bottom width (W_n)	9.5	mm
Magnetic gap	1	mm
Magnet thickness	2.5/2.8	mm
Arc height	NA/1.8	mm
Flat portion (H_m)	NA/1	mm
Height of center (H_n)	NA/1.5	mm

The mathematical discussion of the basic model that is briefly discussed here is the same as the one discussed in [21]. The input power calculations can be made as discussed in (1).

$$P_{in} = 2\frac{m}{T}\int_0^T E_m \sin(\omega t) I_m \sin(\omega t) dt \tag{1}$$

where m, E_m, and I_m represent the phase numbers, the amplitude of back-EMF, and the phase current, respectively. The output power calculations are depicted in (2), where γ represents efficiency [29–31].

$$P_{out} = \gamma m E_m I_m \tag{2}$$

The flux can be computed as in (3)

$$\varphi = \varphi_m \cos(N_p \alpha_r) \tag{3}$$

where φ denotes the magnitude of flux, α_r shows the rotor position, and N_p represents the number of rotor poles, respectively. The back-EMF can be expressed as in (4).

$$e(t) = N_{ph}\omega_r N_P \varphi_m \sin(N_p \alpha_r) \tag{4}$$

where N_{ph} and ω_r show the phase coil turns and the angular speed of the rotor, respectively. By combining (1) and (2), back-EMF is determined as in (5).

$$e(t) = N_{ph}\omega_r N_p k_d k_f B_g \alpha_i \frac{1}{N_s} \frac{\pi}{4}(D_{out}^2 - D_{in}^2) \tag{5}$$

where k_f, k_d, B_g, N_s, D_{out}, and D_{in} represent the air-gap flux density coefficient, the lux leakage coefficient, the air-gap flux density, the slot numbers, the pole arc coefficient, and the outer and inner diameters of the rotor, respectively. The armature current is computed as in (6) [29–31].

$$I_m = \frac{\sqrt{2}A_e \pi D_{in}}{2mN_{ph}} \tag{6}$$

where A_e shows the electrical loading. By putting the (3) and (4) in (2), the input power is given, as shown in (7).

$$P_{in} = \frac{\sqrt{2\pi^3}}{240} \frac{N_p}{N_s} k_d k_f k_{io}(1 - k_{io}^2) A_e B_g \alpha_i D_{out}^3 n_{rpm} \gamma \tag{7}$$

where k_{io} represents the ratio of the inner and outer diameters and n_{rpm} represents the rotor speed. Furthermore, the output torque can be computed as in (8).

$$T_{out} = \frac{\sqrt{2}}{240} \pi^2 \frac{N_p}{N_s} k_d k_f k_{io}(1 - k_{io}^2) A_e B_g \alpha_i D_{out}^3 \gamma \tag{8}$$

It is very clear from (7) and (8) that the power and torque are dependent on the A_e, B_g, and N_p/N_s. From (7), the outer diameter can be represented as in (9).

$$D_{out} = \sqrt[3]{\frac{240 P_{out} N_s}{\sqrt{2\pi^3} N_r k_d k_f k_{io}(1 - k_{io}^2) A_e B_g \alpha_i n_{rpm} \gamma}} \tag{9}$$

3. Proposed and Optimized Model

The proposed design is generated using the creation of an arc in the conventional model and is presented in the next subsection. Later, the proposed design is optimized to further improve the results presented in the subsequent section.

3.1. Hexagonal Conventional and Proposed Model Magnets

The conventional and proposed hexagonal magnets' shapes are shown in Figure 3. The length and width, as shown in Figure 3, are kept the same in both the conventional and proposed models. Moreover, the volume of both of the models is also kept constant. However, the height of the PMs is different in order to keep the volume of the magnets constant. In order to preserve the constant volume, magnetic thickness is increased.

3.2. Optimization of Proposed Model

The asymmetric magnetic overhang concept is utilized to optimize the proposed model. The dimensions of the PM are varied by the inner and outer overhang. Extending the magnet on the upper side is termed as the upper overhang, and that on the lower side is the lower overhang. The volume of the magnet is kept constant. The magnets' inner width, outer width, and height are also varied, as shown in Figure 4. The limits of the variables are shown in Figure 4.

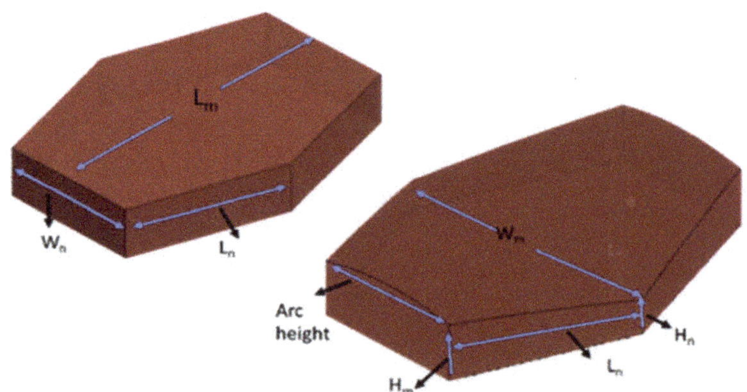

Figure 3. Conventional and proposed hexagonal-shaped magnets for AFPM machines.

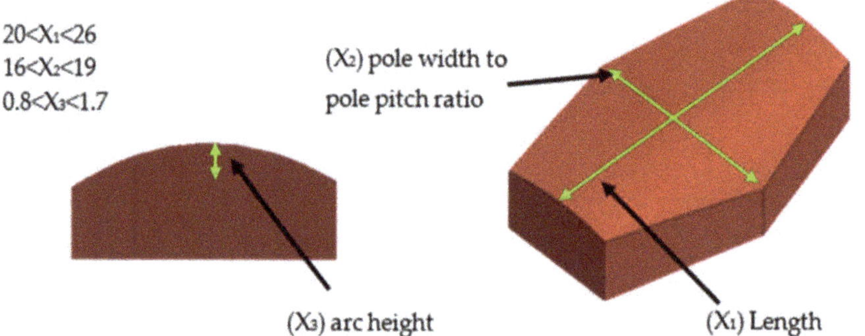

Figure 4. Variables (overhang, height, and width) for optimization of the axial flux permanent magnet machine.

4. Results and Discussion of Conventional vs Proposed Model

In this section, the various types of results are produced using the 3D transient FEA of both the conventional and proposed models. The proposed model is compared with the conventional model to show its improvements. The comparison is performed for cogging torque, back-EMF, three-phase voltage, V_{THD}, torque, torque ripples, and power.

4.1. Comparison of Cogging Torque and Back-EMF of Conventional and Proposed Models

The cogging torque and back-EMF of the conventional and proposed models are shown in the graphs in Figures 5 and 6. The peak-to-peak cogging torque of the conventional model is 0.705 Nm and that of the proposed model is 0.45 Nm. The cogging torque is reduced in the proposed model due to the change in PM shape. There is a significant reduction in the cogging torque due to the proposed magnet shape. The percentage reduction in the cogging torque of the proposed model is 36.17%. This reduction in the cogging torque is due to the large, effective air-gap in the proposed model, which decreases the air-gap flux in the machine. Back-EMF in the proposed model is lower than in the conventional model due to the enhancement of the effective air-gap length. The back-EMF of the proposed and conventional models is 109.361 V_{rms} and 109.092 V_{rms}, respectively. The decrease in back-EMF is 0.245%. As can be seen, the decrease in cogging torque is more significant as compared to the decrease in the back-EMF. This is mainly because cogging torque is proportional to the square of the air-gap flux.

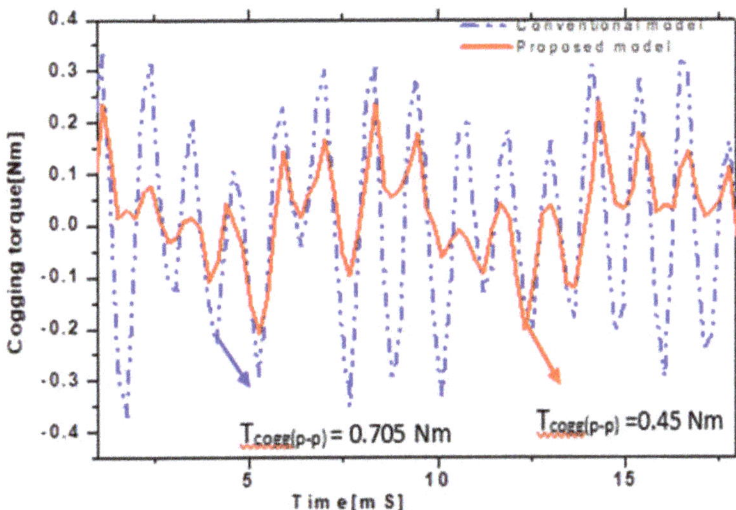

Figure 5. Cogging torque comparison for the conventional and proposed models.

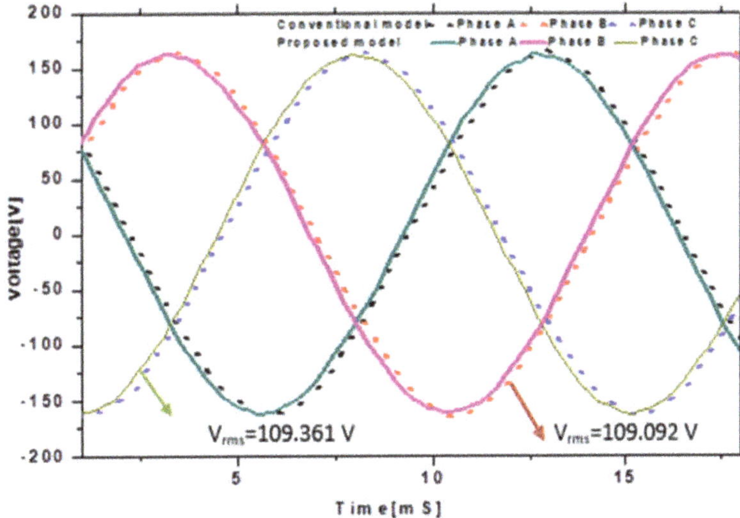

Figure 6. Voltage comparison of the conventional and proposed models.

4.2. Voltage Total Harmonic Distortion (V_{THD}) of the Conventional and Proposed Models

The V_{THD} of both the conventional and proposed models is shown in Figure 7. The V_{THD} of the conventional hexagonal model is 1.33% and the V_{THD} of the proposed hexagonal model is 0.789%. The percentage decrease in V_{THD} is 40.6%. Thus, the proposed model has enhanced the fundamental back-EMF component as compared to the conventional model, due to the arc hexagonal-shaped PM. Thus, with the proposed PM shape's smoothness in the back-EMF is enhanced, which will be helpful in further reducing the torque ripples.

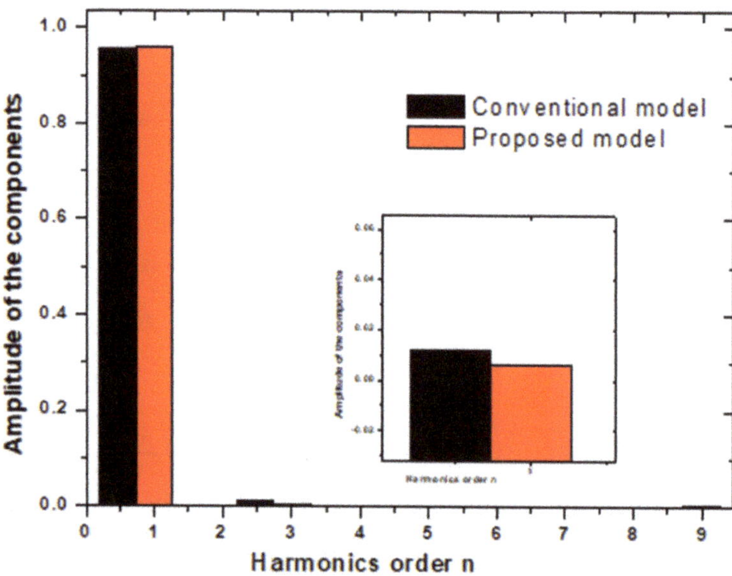

Figure 7. Voltage Total Harmonic Distortion of the conventional and proposed models.

4.3. Torque Analysis of Conventional and Proposed Models

The conventional and proposed models' output torque is shown in Figure 8. The output torque of the conventional and proposed models is 3.5984 Nm and 3.56787 Nm, respectively. The output torque of both of the models is almost equal because both models have the same back-EMF. The torque ripples of the conventional model are 18.89% and those of the proposed model are 12.1%. Hence, with the proposed model, the amount of torque ripple is reduced significantly by 35.94%, and there is no significant change in average output torque.

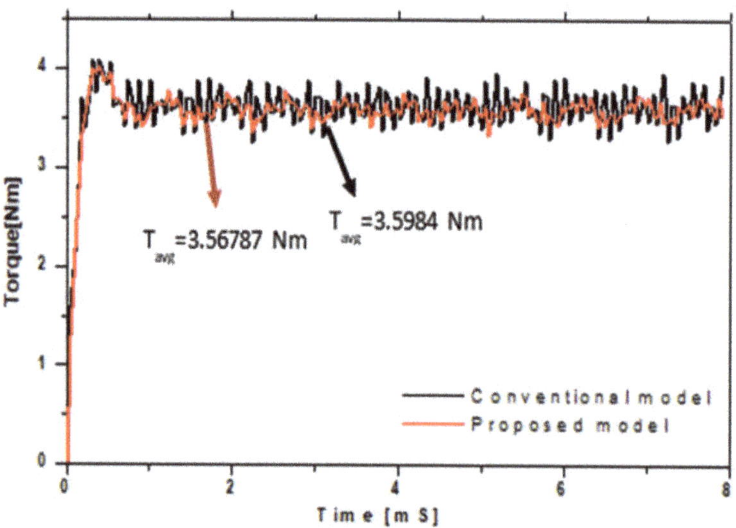

Figure 8. Comparison of output torque for the conventional and proposed models.

4.4. Comparison of Conventional and Proposed Models

The performance comparison of the conventional flat hexagonal and the proposed arc hexagonal magnet models is specified in Table 2. Table 2 shows that the conventional model provides the most V_{rms}, B_g, torque ripples, V_{THD}, and $T_{cogging}$. Table 2 shows that cogging torque and torque ripples are reduced in the proposed model. Further, the air-gap flux density of the proposed model is reduced compared to the conventional model. Moreover, the results show that the reduction in V_{THD} is achieved with the proposed model in comparison to the conventional model. Furthermore, the back-EMF and torque levels in both of the models are almost the same.

Table 2. Performance Comparison of Conventional and Proposed Model.

Model	Conventional	Proposed	Units
Voltage (V_{rms})	109.361	109.092	V
Cogging torque	0.705	0.45	Nm
B_g (rms)	0.548	0.496	T
V_{THD}	1.33	0.789	%
Torque	3.59	3.56	Nm
Torque ripples	18.89	12.1	%

5. Optimized Model Results and Discussion

The asymmetric magnetic overhang concept is utilized to optimize the proposed model, as shown in Figure 9. The dimensions of the PM are varied by the inner and the outer overhang. Extending the magnet on the upper side is termed as the upper overhang, and that on the lower side is the lower overhang. The volume of the magnet is kept constant. The inner width, outer width, and height of the magnets are also varied. Latin Hyper Cube (LHS) sampling is used to create the experiments. The krigging is used to approximate the function, and a genetic algorithm is used for the minimization of the objective function. The volume of the magnet is kept constant. The minimization of the cogging torque and the maximization of the back-EMF are the objective functions. The objective functions, constraints, and limits of the variables are described in (10), (11), and (12), respectively.

$$\begin{aligned} &Maximize\ the\ voltage \\ &Minimize\ the\ cogging\ torque \end{aligned} \tag{10}$$

$$\begin{aligned} &Cogging\ torque < 0.45 Nm \\ &Voltage > 109.092 \end{aligned} \tag{11}$$

$$\begin{aligned} 20 &\leq X_1 \leq 26 \\ 16 &\leq X_2 \leq 19 \\ 0.8 &\leq X_3 \leq 1.7 \end{aligned} \tag{12}$$

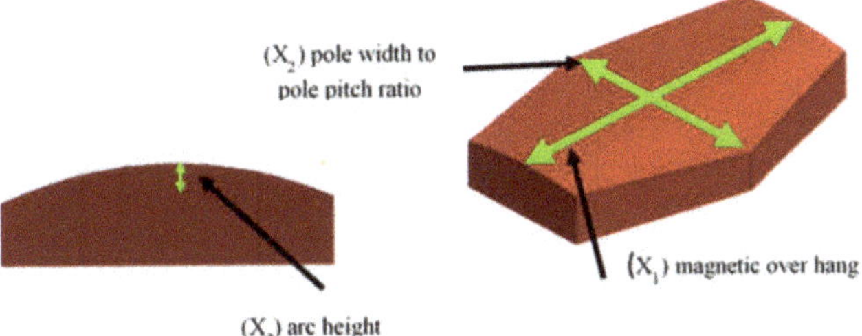

Figure 9. Variables for optimization.

The convergence process of optimization for the objective functions and design variables is shown in Figures 10 and 11, respectively. The cogging torque is reduced from 0.45 to 0.35, and voltage is increased from 109 V to 112.59 V. Then, the results are verified through finite element analysis.

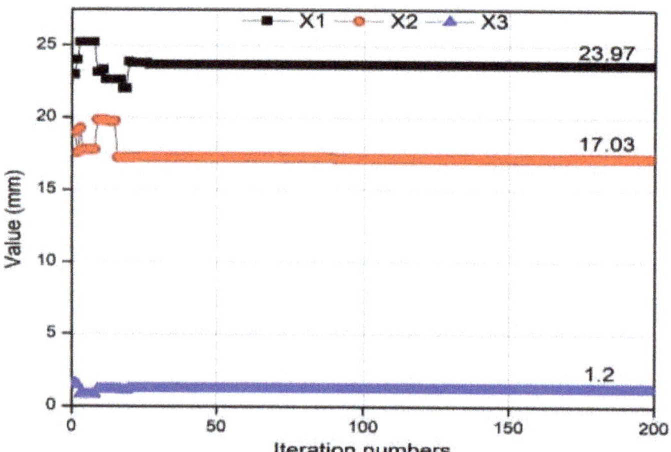

Figure 10. Convergence process for design variables.

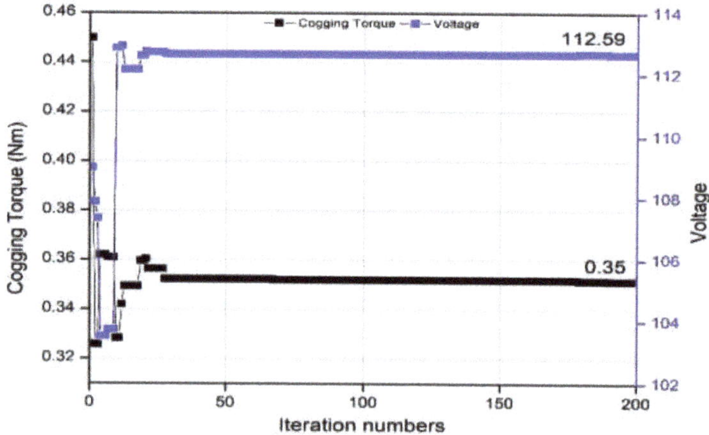

Figure 11. Convergence process for objective functions.

5.1. Voltage and Cogging Torque Improvement in the Optimized Model

Voltage is improved in the optimized model, from 109.361 V_{rms} to 112.592 V_{rms}. The voltage comparison graph of the proposed and optimized models is shown in Figure 12. A considerable reduction in cogging torque is achieved during the optimization process, and the cogging torque is reduced from 0.45 Nm to 0.356 Nm. Figure 13 shows the improvements in cogging torque. The increase in back-EMF is obtained due to the asymmetric overhang effect, and the increase of 3.232 V is achieved while the reduction in cogging torque is 20.88%.

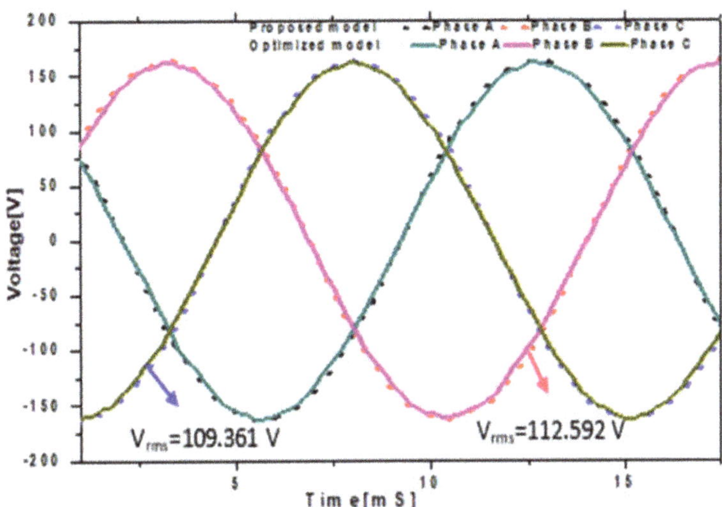

Figure 12. Voltage comparison of the proposed and optimized models.

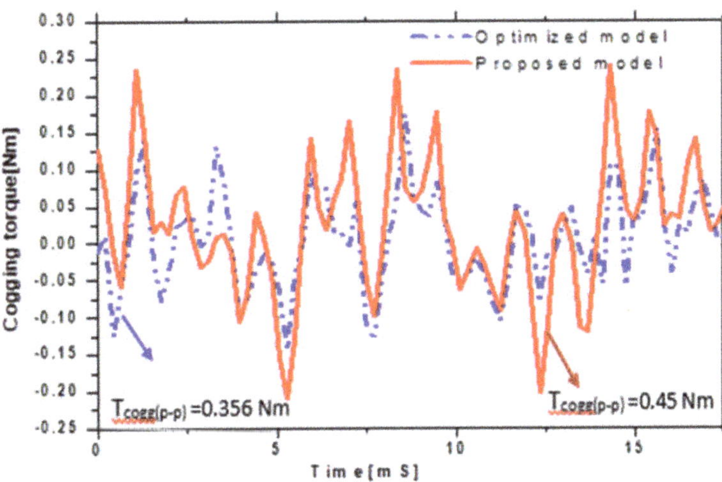

Figure 13. Cogging torque comparison of the proposed and optimized models.

5.2. Power and Torque Analysis of Proposed and Optimized Models

The output power of the machine is improved, through optimization, from 2.26 kW to 2.47 kW. The output torque of the machine is improved from 3.859 Nm to 3.643 Nm peak-to-peak in the proposed model. The enhancement achieved in the output power is 0.21 KW, and 0.2 Nm torque enhancement is achieved in the optimized model. Moreover, a reduction in the torque ripple is achieved, from 12.1% to 9.55%. during the optimization process. Figures 14 and 15 show the output power and torque of the proposed and optimized models.

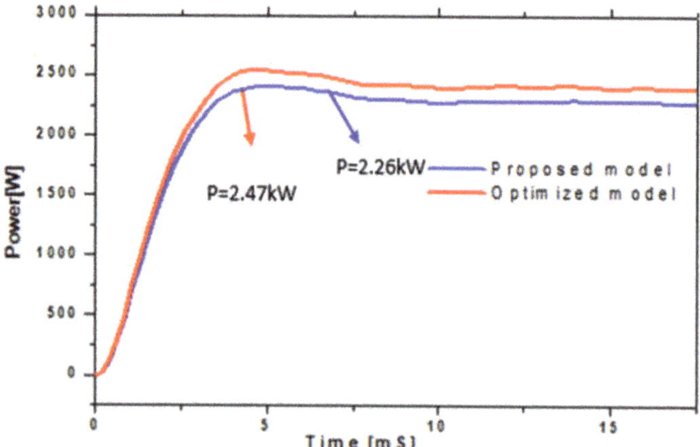

Figure 14. Comparison of power for the proposed and optimized models.

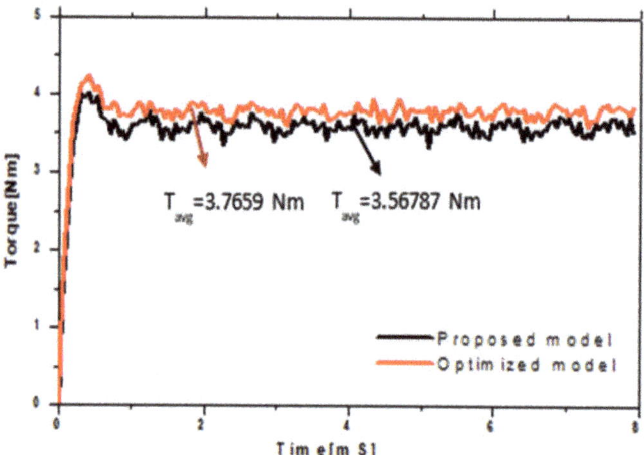

Figure 15. Torque comparison of the proposed and optimized models.

5.3. Comparison of Performance for Proposed and Optimized Models

The performance comparison of the proposed arc hexagonal magnets and the optimized model is provided in Table 3, which includes V_{rms}, B_g, torque ripple, and $T_{cogging}$. The table illustrates that minimum cogging torque, output torque ripples, and maximum voltage are achieved in the optimized model, and all of these performance parameters are summarized in Table 3.

Table 3. Performance Comparison of Conventional and Proposed Models.

Model	Proposed	Optimized	Units
Voltage (V_{rms})	109.092	112.592	V
Cogging torque	0.745	0.356	Nm
B_g (rms)	0.496	0.481	T
V_{THD}	0.789	1.9	%
Torque ripples	12.1	9.55	%

6. Conclusions

An AFPM machine with an arc hexagonal magnet shape is proposed to decrease cogging torque. In the proposed model, cogging torque is reduced, and output torque has been reduced due to an increase in air-gap length as compared to the conventional model. The V_{THD} of the conventional hexagonal-shaped model is 1.33%, and the V_{THD} of the proposed model is 0.789%; hence, the V_{THD} is reduced in the proposed model. The arc hexagonal-shaped proposed model is then further optimized to achieve reduced cogging torque and increased output torque. In the optimized model, cogging torque and torque ripples are reduced, the output torque is improved, and power is also increased as compared with the proposed hexagonal model. The performance characteristics of the optimized model exhibit improved results relative to the proposed and conventional models.

Author Contributions: Conceptualization, J.I.; software, H.U. and M.Y.; formal analysis, H.U., M.Y., and K.S.A.; writing—original draft preparation, S.S.H.B.; funding acquisition, J.-S.R. All authors have read and agreed to the published version of the manuscript.

Funding: National Research Foundation of Korea, 2016R1D1A1B01008058; the Brain Pool (BP) Program, 2019H1D3A1A01102988.

Acknowledgments: This work was supported in part by the Brain Pool (BP) Program, funded by the Ministry of Science and ICT through the National Research Foundation of Korea under Grant 2019H1D3A1A01102988, and in part by the Basic Science Research Program through the National Research Foundation of Korea, funded by the Ministry of Education under Grant 2016R1D1A1B01008058.

Conflicts of Interest: The authors declare no conflict of interest.

References

1. Honsinger, V.B. Performance of polyphase permanent magnet machines. *IEEE Trans. Power Appar. Syst.* **1980**, *4*, 1510–1518. [CrossRef]
2. Huang, S.; Luo, J.; Leonardi, F.; Lipo, T.A. A general approach to sizing and power density equations for comparison of electrical machines. *IEEE Trans. Ind. Appl.* **1998**, *34*, 92–97. [CrossRef]
3. Taran, N.; Heins, G.; Rallabandi, V.; Patterson, D.; Ionel, D.M. Torque Production Capability of Axial Flux Machines with Single and Double Rotor Configurations. In Proceedings of the 2018 IEEE Energy Conversion Congress and Exposition (ECCE), Portland, OR, USA, 23–27 September 2018.
4. Ahmadi Darmani, M.; Hooshyar, H. Optimal Design of Axial Flux Permanent Magnet Synchronous Motor for Electric Vehicle Applications Using GA and FEM. *J. Electr. Comput. Eng. Innov.* **2015**, *3*, 89–97.
5. Lee, C.; Liu, C.; Chau, K. A Magnetless Axial-Flux Machine for Range-Extended Electric Vehicles. *Energies* **2014**, *7*, 1483–1499. [CrossRef]
6. Luo, X.; Niu, S. Maximum Power Point Tracking Sensorless Control of an Axial-Flux Permanent Magnet Vernier Wind Power Generator. *Energies* **2016**, *9*, 581. [CrossRef]
7. Hedlund, M.; Kamf, T.; De Santiago, J.; Abrahamsson, J.; Bernhoff, H. Reluctance Machine for a Hollow Cylinder Flywheel. *Energies* **2017**, *10*, 316. [CrossRef]
8. Kappatou, J.; Zalokostas, G.; Spyratos, D. 3-D FEM Analysis, Prototyping and Tests of an Axial Flux Permanent-Magnet Wind Generator. *Energies* **2017**, *10*, 1269. [CrossRef]
9. Mohamed, A.; Hemeida, A.; Vansompel, H.; Sergeant, P. Parametric Studies for Combined Convective and Conductive Heat Transfer for YASA Axial Flux Permanent Magnet Synchronous Machines. *Energies* **2018**, *11*, 2983. [CrossRef]
10. Amin, S.; Khan, S.; Bukhari, S.S.H. A Comprehensive Review on Axial Flux Machines and Its Applications. In Proceedings of the 2019 2nd International Conference on Computing, Mathematics and Engineering Technologies (iCoMET), Sukkur, Pakistan, 30–31 January 2019.
11. Aydin, M.; Zhu, Z.Q.; Lipo, T.A.; Howe, D. Minimization of cogging torque in axial-flux permanent-magnet machines: Design concepts. *IEEE Trans. Magn.* **2007**, *43*, 3614–3622. [CrossRef]
12. Kumar, P.; Srivastava, R.K. Influence of Rotor Magnet Shapes on Performance of Axial Flux Permanent Magnet Machines. *Prog. Electromagn. Res.* **2018**, *85*, 155–165. [CrossRef]
13. Aydin, M. Magnet skew in cogging torque minimization of axial gap permanent magnet motors. In Proceedings of the 2008 18th International Conference on Electrical Machines, Vilamoura, Portugal, 6–9 September 2008.
14. Henrotte, F.; Hameyer, K. Computation of electromagnetic force densities: Maxwell stress tensor vs. *virtual work principle*. *J. Comput. Appl. Math.* **2004**, *168*, 235–243. [CrossRef]

15. Ishak, D.; Zhu, Z.Q.; Howe, D. High torque density permanent magnet brushless machines with similar slot and pole numbers. *J. Magn. Magn. Mater.* **2004**, *272*, E1767–E1769. [CrossRef]
16. Zieliński, P.; Schoepp, K. Wolnoobrotowe generatory synchroniczne wzbudzane magnesami trwałymi o uzwojeniach skupionych. In *Prace Naukowe Instytutu Maszyn, Napedów i Pomiarów Elektrycznych Politechniki Wrocławskiej, Studia i Materiały*; Oficyna Wydawnicza Politechniki Wrocławskiej: Wrocław, Poland, 2006.
17. Glinka, T. *Maszyny Elektryczne Wzbudzane Magnesami Trwałymi*; Wydawnictwo Politechniki Śląskiej: Gliwice, Poland, 2002.
18. Jonczyk, J.; Kołodziej, J. Modelowanie Zagadnie´n Polowych z Wykorzystaniem MES. Master's Thesis, Politechnika Opolska, Opole, Poland, 2004.
19. Yang, Y.; Wang, X.; Zhang, R.; Zhu, C.; Ding, T. Research of cogging torque reduction by different slot width pairing permanent magnet motors. In Proceedings of the 8th International Electric Machines and Systems Conference, Nanjing, China, 27–29 September 2005; pp. 367–370.
20. Zhu, Z.Q.; Howe, D. Influence of design parameters on cogging torque in permanent magnet machines. *IEEE Trans. Energy Convers.* **2000**, *15*, 407–412. [CrossRef]
21. Cetin, E.; Daldaban, F. Analyzing the profile effects of the various magnet shapes in axial flux PM motors by means of 3D-FEA. *Electronics* **2018**, *7*, 13. [CrossRef]
22. Shokri, M.; Rostami, N.; Behjat, V.; Pyrhönen, J.; Rostami, M. Comparison of performance characteristics of axial-flux permanent-magnet synchronous machine with different magnet shapes. *IEEE Trans. Magn.* **2015**, *51*, 1–6. [CrossRef]
23. Zhao, W.; Lipo, T.A.; Kwon, B.I. Material-efficient permanent-magnet shape for torque pulsation minimization in SPM motors for automotive applications. *IEEE Trans. Ind. Electron.* **2014**, *61*, 5779–5787. [CrossRef]
24. Ikram, J.; Khan, N.; Khaliq, S.; Kwon, B.I. Reduction of Torque Ripple in an Axial Flux Generator Using Arc Shaped Trapezoidal Magnets in an Asymmetric Overhang Configuration. *J. Magn.* **2016**, *21*, 577–585. [CrossRef]
25. Jo, W.; Cho, Y.; Chun, Y.; Koo, D. Characteristic analysis on overhang effect in axial flux PM synchronous motors with slotted winding. In Proceedings of the 2006 CES/IEEE 5th International Power Electronics and Motion Control Conference, Shanghai, China, 14–16 August 2006.
26. Lebensztajn, L.; Marretto, C.A.; Perdiz, F.A.; Costa, M.C.; Nabeta, S.I.; Dietrich, Á.B.; Chabu, I.E.; Cavalcanti, T.T.; Cardoso, J.R. A multi-objective analysis of a special switched reluctance motor. *COMPEL-Int. J. Comput. Math. Electr. Electron. Eng.* **2005**, *24*, 931–941. [CrossRef]
27. Nagalingam, U.; Mahadevan, B.; Vijayarajan, K.; Loganathan, A.P. Design optimization for cogging torque mitigation in brushless DC motor using multi-objective particle swarm optimization algorithm. *COMPEL Int. J. Comput. Math. Electr. Electron. Eng.* **2015**, *34*, 1302–1318. [CrossRef]
28. Gong, J.; Berbecea, A.C.; Gillon, F.; Brochet, P. Multi-objective optimization of a linear induction motor using 3D FEM. *COMPEL-Int. J. Comput. Math. Electr. Electron. Eng.* **2012**, *31*, 958–971. [CrossRef]
29. Bilal, M.; Ikram, J.; Fida, A.; Bukhari, S.S.H.; Haider, N.; Ro, J.S. Performance Improvement of Dual Stator Axial Flux Spoke Type Permanent Magnet Vernier Machine. *IEEE Access* **2021**, *9*, 64179–64188. [CrossRef]
30. Baig, M.A.; Ikram, J.; Iftikhar, A.; Bukhari, S.S.H.; Khan, N.; Ro, J.S. Minimization of cogging torque in axial field flux switching machine using arc shaped triangular magnets. *IEEE Access* **2020**, *8*, 227193–227201. [CrossRef]
31. Yousuf, M.; Khan, F.; Ikram, J.; Badar, R.; Bukhari, S.S.H.; Ro, J.S. Reduction of Torque Ripples in Multi-Stack Slotless Axial Flux Machine by Using Right Angled Trapezoidal Permanent Magnet. *IEEE Access* **2021**, *9*, 22760–22773. [CrossRef]

 *mathematics*

Article

Novel Single Inverter-Controlled Brushless Wound Field Synchronous Machine Topology

Syed Sabir Hussain Bukhari [1,2], Ali Asghar Memon [3], Sadjad Madanzadeh [2], Ghulam Jawad Sirewal [4], Jesús Doval-Gandoy [5] and Jong-Suk Ro [2,6,*]

[1] Department of Electrical Engineering, Sukkur IBA University, Sukkur 65200, Sindh, Pakistan; sabir@iba-suk.edu.pk
[2] School of Electrical and Electronics Engineering, Chung-Ang University, Seoul 06974, Korea; s.madanzadeh@gmail.com
[3] Department of Electrical Engineering, Mehran University of Engineering and Technology, Jamshoro 76062, Sindh, Pakistan; ali.asghar@faculty.muet.edu.pk
[4] Department of Electrical Engineering Technology, The Benazir Bhutto Shaheed University of Technology and Skill Development, Khairpur Mir's 66020, Pakistan; jawadsirewal1@gmail.com
[5] Applied Power Electronics Technology Research Group, University of Vigo, 36310 Vigo, Spain; jdoval@uvigo.es
[6] Department of Intelligent Energy and Industry, Chung-Ang University, Seoul 06974, Korea
* Correspondence: jsro@cau.ac.kr

Citation: Bukhari, S.S.H.; Memon, A.A.; Madanzadeh, S.; Sirewal, G.J.; Doval-Gandoy, J.; Ro, J.-S. Novel Single Inverter-Controlled Brushless Wound Field Synchronous Machine Topology. *Mathematics* **2021**, *9*, 1739. https://doi.org/10.3390/math9151739

Academic Editor: Vladimir Prakht

Received: 4 July 2021
Accepted: 21 July 2021
Published: 23 July 2021

Publisher's Note: MDPI stays neutral with regard to jurisdictional claims in published maps and institutional affiliations.

Abstract: This paper proposes a novel brushless excitation topology for a three-phase synchronous machine based on a customary current-controlled voltage source inverter (VSI). The inverter employs a simple hysteresis-controller-based current control scheme that enables it to inject a three-phase armature current to the stator winding which contains a dc offset. This dc offset generates an additional air gap magneto-motive force (MMF). On the rotor side, an additional harmonic winding is mounted to harness the harmonic power from the air gap flux. Since a third harmonic flux is generated in this type of topology, the machine structure is also modified to accommodate the third harmonic rotor winding to have a voltage induced as the rotor rotates at synchronous speed. Specifically, four-pole armature and field winding patterns are used, whereas the harmonic winding is configured for a twelve-pole pattern. A diode rectifier is also mounted on the rotor between the harmonic and field windings. Therefore, the generated voltage on the harmonic winding feeds the current to the field winding for excitation. A 2D-finite element analysis (FEA) in JMAG-Designer was carried out for performance evaluation and verification of the topology. The simulation results are consistent with the proposed theory. The topology could reduce the cost and stator winding volume compared to a conventional brushless machine, with good potential for various applications.

Keywords: brushless topology; third harmonic flux; dc offset; wound field synchronous machines

1. Introduction

High cost of rare earth magnets and flux control method complexity in permanent magnet synchronous machines (PMSMs) are serious problems in making the PMSM suitable for many applications [1]. However, alternatives such as wound field synchronous machines (WFSMs) can be used in a variety of applications ranging from small capacity motors to large capacity power generation applications [1–3]. In [2], it was specifically investigated for use in automotive driving compared to permanent magnet excited machines. Given the potential in WFSM, due to its low-price benefit, an inherent problem of its assembly of brushes and slip rings must be solved for comparison. In small capacity applications, operating it without brushes and slip rings will be convenient [3–7].

In [8], utilizing space harmonics power, a brushless WFSM was designed with an additional winding on the rotor. The additional rotor winding in that case will retrieve the space harmonics power and feed a current to the field winding of the machine as a voltage source, removing the need for any external dc power source for excitation. A similar approach has been explored in [9] to design a brushless WFSM using space harmonics for excitation.

In [10], a self-excited brushless synchronous generator was designed and investigated to utilize the fifth space harmonic of armature MMF to excite main field winding in a cylindrical rotor design.

In [11,12], brushless excitation was achieved by injecting a third harmonic in the same armature windings to create a harmonic air gap flux for rotor excitation. In a similar attempt, an additional winding on the stator was used for more controllability of the harmonics in [13].

In [14], to reduce the components, a thyristor-generated harmonic current was used to feed and control the current, employing an additional harmonic winding on the stator with a similar machine structure as in [13]. However, these brushless machines have some disadvantages, such as high stator/rotor winding or core volume compared to a benchmark machine. Otherwise, the machine performance is not comparable to a conventional PMSM or WFSM.

However, recently new topologies have been designed and investigated to solve the remaining problems in the brushless excitation of WFSMs [15–18].

In [15], stator armature winding was connected in such a way that the three phase armature winding terminals were connected to an exciter winding mounted on the stator through diode rectifier. This topology will generate two types of fluxes in the air gap creating both the fundamental and harmonic flux through single power supply. However, it is limited to small scales due to load insensitivity and unwanted harmonics.

In [16,17], stator armature winding was connected in such a way that the ampere-turns resulting from supplying current were unequal in two portions of the machine structure. This attempt results in sub-harmonic air gap flux in addition to the fundamental flux.

In all these types of machines, the additional flux is generally created to be induced in a rotor harmonic winding which acts as power source for the field winding excitation. However, changing the conventional winding connection patterns generally results in unwanted harmonics along with the desired harmonic frequencies.

Recently, a brushless WFSM based on a single rectifier is proposed in [18]. This topology is presented in Figure 1. The proposed WFSM is based on two power sources. One is directly connected to the main armature winding, whereas the second power source is connected to the three-phase rectifier whose output is connected to the neutral point of the armature winding which is Y-connected. This results in a dc offset for the armature currents which generates a third harmonic air gap flux. The rotor harmonic winding harnesses the third harmonic power from the air gap flux for brushless operation.

In this paper, a dc offset for the armature winding currents, previously realized using two power sources in [18], is produced using a single customary current-controlled voltage source inverter (CCVSI). The inverter operation is based on a simple and easy to implement current control scheme. This control scheme involves a hysteresis controller that enables it to generate three-phase currents for the stator winding which contains a dc offset for each phase. The magnitude of the dc offset can be varied by varying the reference currents of the controller. This dc offset generates an additional air gap MMF which produces a third harmonic air gap flux. The generated flux is intercepted by the rotor harmonic winding wound along with the field winding for brushless excitation. In the proposed WFSM topology, the conventional armature winding is exploited to reduce any additional unwanted harmonics which can reduce the machine performance. In addition, the proposed topology requires a single customary CCVSI employing a simple control scheme, which makes it cost-effective compared to the brushless WFSM topology presented in [18]. The proposed inverter topology and its operating principle are discussed in subsequent sections. Finite element analysis (FEA) in JMAG-Designer 19.1 is employed to validate the proposed topology and achieve its electromagnetic performance.

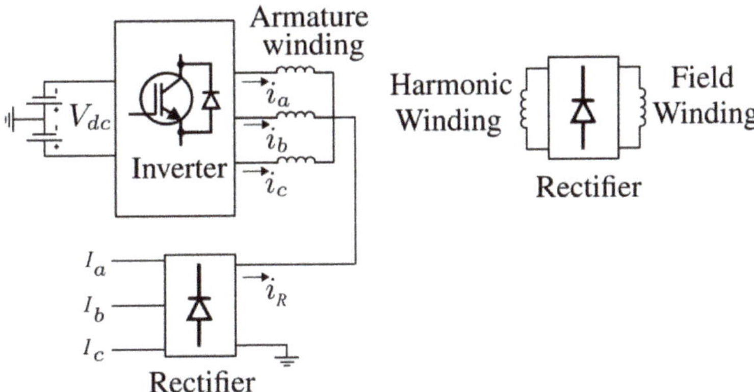

Figure 1. Conventional three-phase rectifier-based brushless WFSM topology.

2. Proposed Inverter Topology

A simplified representation of the proposed brushless WFSM topology, in which the inverter generates an armature current with a dc offset for the stator winding of the machine, is presented in Figure 2. The armature winding in the proposed brushless WFSM is connected to the customary three-phase, two-level, current-controlled voltage source inverter. The rotor of the machine has harmonic and field windings connected through a rotating diode rectifier. The comprehensive illustration of the proposed topology is shown in Figure 3a, in which the inverter uses a control scheme for the required commutation of the inverter which switches based on a typical hysteresis controller. The employed hysteresis controller controls the phase currents of the armature winding with a specific hysteresis band over the given reference current signals. The controlled output currents with the required dc offset, which essentially contain the fundamental and the bias/dc current components, are given to the armature winding of the machine. Consequently, the proposed inverter topology involves two dc sources connected in series as shown in Figures 2 and 3a. The coupling point of the dc source is connected to the neutral point of the Y-connected armature winding of the machine. The switching potentials of the inverter switches are used to decide the bandwidth of the current controllers. The reference current signals i^*_a, i^*_b, and i^*_c used for the employed hysteresis current control scheme are generated through the following equation:

$$
\begin{aligned}
i^*_a &= I\sin(\omega t) + I_{bias} \\
i^*_b &= I\sin\left(\omega t - \frac{2\pi}{3}\right) + I_{bias} \\
i^*_c &= I\sin\left(\omega t + \frac{2\pi}{3}\right) + I_{bias}
\end{aligned}
\tag{1}
$$

Figure 3b shows the reference and controlled inverter output currents for phase A of the proposed inverter topology. In the given figure, the black color waveform represents the reference signal, whereas the red color waveform denotes the inverter output current. Figure 3c illustrates the three-phase input armature currents, which produces the following neutral current:

$$
I_N = i_a + i_b + i_c = 3I_{bias}
\tag{2}
$$

The generalized voltage equation for the armature winding is as follows:

$$
v_x = Ri_x + L\frac{di_x}{dt}
\tag{3}
$$

where $x \in \{a, b, c\}$, v_x, and i_x represent the proposed CCVSI output voltage and current, respectively. R and L represent the armature winding resistance and inductance of the machine.

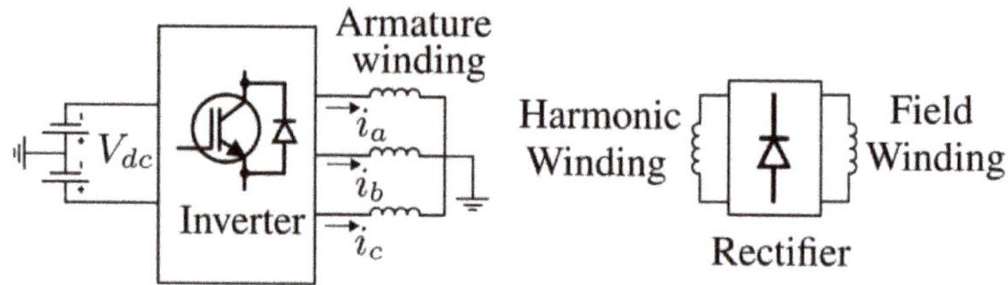

Figure 2. Simplified diagram of the proposed brushless WFSM drive topology.

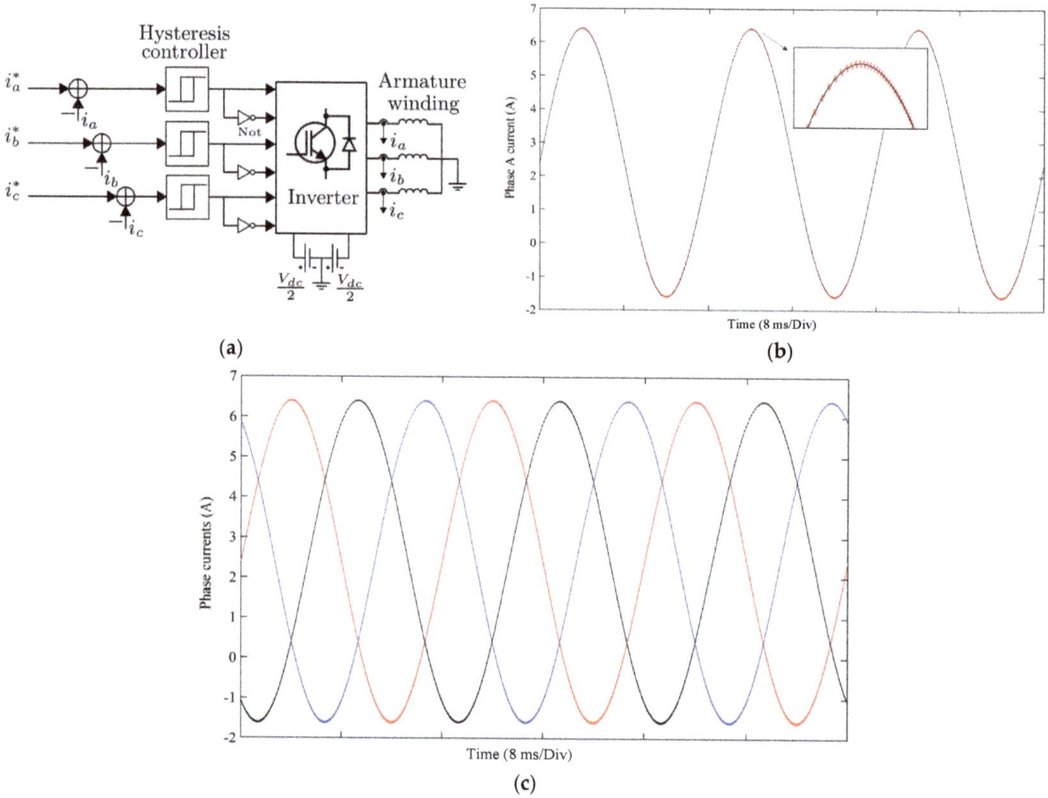

Figure 3. (**a**) Proposed inverter topology, (**b**) reference and controlled output current for phase A, and (**c**) controlled three-phase output currents of the inverter.

3. Machine Topology and Working Principle

The input armature currents (i_{abc}) generated through the method discussed in Section 2 are given to the machine's armature winding. A 4-pole, 42-slot (4p42s) machine with a concentrated, double-layered armature winding, which has a winding factor of 0.932, is employed to validate the proposed brushless WFSM topology. The employed machine along with its stator and rotor winding configurations are shown in Figure 4a,b, respectively. As seen from Figure 4b, the rotor of the machine has four main teeth to accommodate the four-pole rotor field winding, whereas each main tooth is further altered to have two sub-teeth to house the rotor harmonic winding. The rotor harmonic winding is based on a twelve-pole winding configuration to harness the harmonic power generated in the air gap flux. The detailed winding specifications are presented in Table 1.

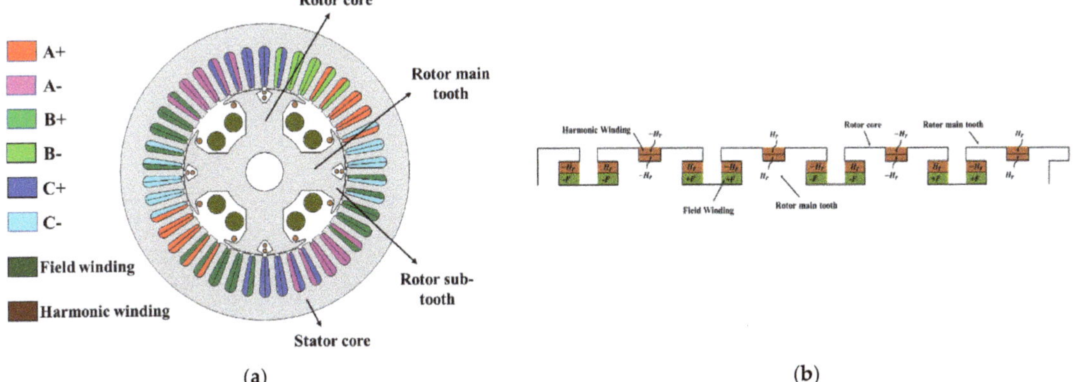

(a)

(b)

Figure 4. (a) Machine model with stator winding and (b) rotor winding configurations.

Table 1. Winding parameters for the employed machine.

Parameter	Value
Number of poles/slots/layers	4/42/2
Coil span	9 slots
Pole pitch	10.5 slots
Periodicities	2
Winding factor	0.932

The operation of the proposed brushless WFSM topology is investigated considering four different cases, each is based on the supply of input armature currents (i_{abc}) with a different magnitude of dc offset/bias. In case 1 and 2, the armature winding is supplied with currents having a dc offset of 0.6 A and 1.2 A for each phase, respectively. However, a dc offset of 1.8 A and 2.4 A is achieved for the stator armature currents of the employed machine in case 3 and 4, respectively. The input armature currents during all operating conditions, i.e., case 1 to 4, are presented in Figure 5a–d.

These currents produce a magnetomotive force (F) for each phase of the armature currents, as given under:

$$F_a = i_a N_{\varphi 1}\left(\sin\theta_s + \tfrac{1}{3}\sin 3\theta_s\right)$$
$$F_b = i_b N_{\varphi 1}\left\{\sin(\theta_s - \tfrac{2\pi}{3}) + \tfrac{1}{3}\sin 3\theta_s\right\}$$
$$F_c = i_c N_{\varphi 1}\left\{\sin(\theta_s + \tfrac{2\pi}{3}) + \tfrac{1}{3}\sin 3\theta_s\right\} \tag{4}$$

In the above equation
$N_{\varphi 1} = \tfrac{2}{\pi}(per\ phase\ number\ of\ turns)$
θ_s = electrical angle (spatial), and
ω = angular frequency (electrical).
The controlled armature winding currents can be expressed as:

$$i_a = I_1 \sin(\omega t) + I_{bias}$$
$$i_b = I_1 \sin(\omega t - \tfrac{2\pi}{3}) + I_{bias}$$
$$i_c = I_1 \sin(\omega t + \tfrac{2\pi}{3}) + I_{bias} \tag{5}$$

where I_1 is the fundamental and I_{bias} is the magnitude of the dc offset for the armature winding currents for each phase.

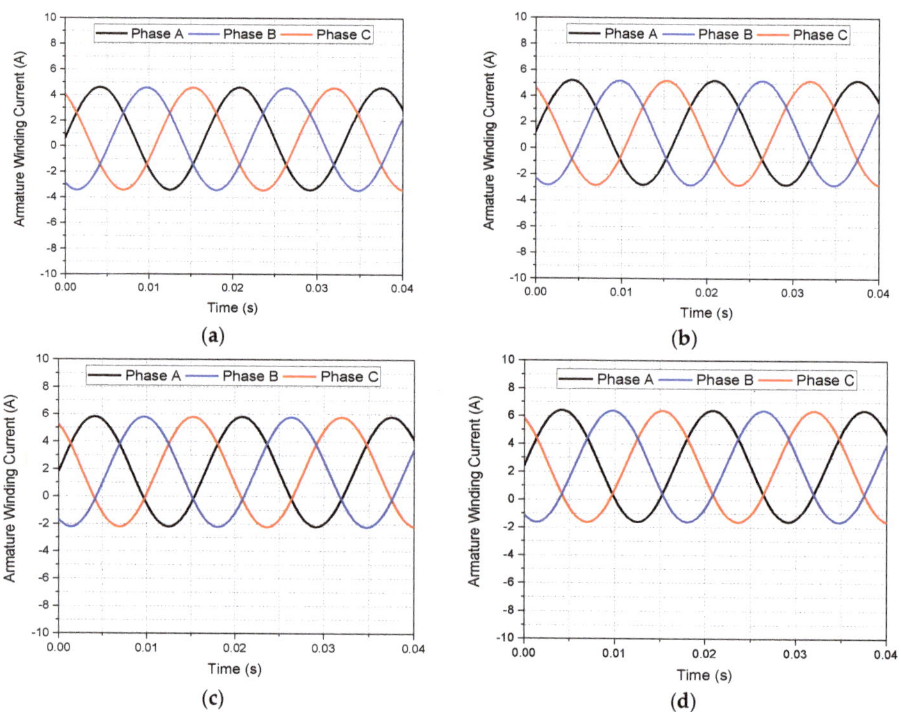

Figure 5. Armature winding currents with (**a**) 0.6 A, (**b**) 1.2 A, (**c**) 1.8 A, and (**d**) 2.4 A dc offset for each phase.

Bring Equation (5) to Equation (4), and add the MMF of three-phase armature windings a, b, and c. The net MMF (F_{abc}) of the armature winding is expressed as:

$$F_{abc}(\theta_s, i) = F_a + F_b + F_c \tag{6}$$

$$F_{abc}(\theta_s, i) = \begin{bmatrix} N_{\varphi 1}\left\{\sin(\theta_s) + \frac{1}{3}\sin 3\theta_s\right\}\{I_1 \sin(\omega t) + I_{bias}\} \\ +N_{\varphi 1}\left\{\sin\left(\theta_s - \frac{2\pi}{3}\right) + \frac{1}{3}\sin 3\theta_s\right\}\left\{I_1 \sin\left(\omega t - \frac{2\pi}{3}\right) + I_{bias}\right\} \\ +N_{\varphi 1}\left\{\sin\left(\theta_s + \frac{2\pi}{3}\right) + \frac{1}{3}\sin 3\theta_s\right\}\left\{I_1 \sin\left(\omega t + \frac{2\pi}{3}\right) + I_{bias}\right\} \end{bmatrix} \tag{7}$$

$$F_{abc}(\theta_s, i) = N_{\varphi 1}\begin{bmatrix} I_1\left\{\begin{matrix} \sin(\omega t)\sin(\theta_s) + \sin\left(\omega t - \frac{2\pi}{3}\right)\sin\left(\theta_s - \frac{2\pi}{3}\right) \\ + \sin\left(\omega t + \frac{2\pi}{3}\right)\sin\left(\theta_s + \frac{2\pi}{3}\right) \end{matrix}\right\} \\ +3I_{bias} \end{bmatrix} \tag{8}$$

$$F_{abc}(\theta_s, i) = \frac{3}{2}I_1 N_{\varphi 1}\cos(\omega t - \theta_s) + 3I_{bias} \tag{9}$$

The above equation shows that F_{abc} consists of the normal fundamental MMF rotating at synchronous speed and the spatial-location-fixed MMF generated by the dc offset component of the armature currents. These two fields are not coupled due to the difference of frequencies.

If the rotor rotates at synchronous speed, the fundamental component of MMF will not produce any *EMF* in the harmonic winding of the rotor as the speed of the rotor and the fundamental MMF is same; however, the stationary MMF component caused by the dc offset component of the armature currents will induce a rotating *EMF* and transformer action-based *EMF* in the harmonic winding of the rotor.

Assuming that θ_0 is the rotor excitation winding initial position angle, the spatial position of the excitation winding can be calculated as:

$$\theta_s = \omega t + \theta_0 \tag{10}$$

The generated flux of each winding pole will be:

$$\psi_h = n_h P_g N_{\varphi 1} \left\{ \frac{3}{2} I_1 \cos(\omega t - \theta_s) + 3 I_{bias} \right\} \tag{11}$$

where n_h is the rotor excitation winding number of turns, and P_g is the air gap permeance.

The magnitude of the induced *EMF* in the rotor harmonic winding can be calculated as:

$$e_h = 6 \frac{d\psi_h}{dt}$$
$$e_h = 18 n_h P_g N_{\varphi 1} I_{bias} \omega \cos(3\omega t + 3\theta_0) \tag{12}$$

From Equation (12), it can be seen that the induced *EMF* in the harmonic winding of the rotor is three times as much as the synchronous angular frequency. The induced *EMF* (e_h) in the rotor harmonic winding is rectified by a rotating rectifier to supply dc to the rotor field winding to archive brushless operation for WFSM [19,20].

4. Finite Element Analysis

To validate the proposed single inverter-controlled brushless WFSM topology, finite element analysis (FEA) was carried out in JMAG-Designer. The electromagnetic performance of the proposed topology was achieved by developing a 4-pole, 42-slot machine as presented in Figure 4. The machine was investigated under four different operating conditions i.e., case 1, case 2, case 3, and case 4. In case 1 and 2, the armature currents had a dc offset of 0.6 A and 1.2 A for each phase, respectively. However, a dc offset of 1.8 A and 2.4 A was achieved for the armature currents in case 3 and 4. The input armature currents under these cases are shown in Figure 5a–d and the machine specifications are presented in Table 2.

Table 2. Machine specifications.

Parameter	Value
Rated power	1 kW
Machine poles/Stator slots	4/42
Rated speed	1800 rpm
Frequency	60 Hz
Stator outer/inner diameter	88.5/50 mm
Air gap	0.5 mm
Rotor diameter	49.5 mm
Rotor main/sub-teeth	4/8
Harmonic/Field winding number of turns	9/150
Armature winding number of turns	20
Stack length	80 mm

The machine was operated at a speed of 1800 rpm. The simulations of the proposed brushless WFSM were carried out for 0.6 s. The flux linkages of the machine under all four cases are shown in Figure 6a–d. Fast Fourier transform (FFT) plots for phase *A* of these flux linkages were carried out to show its harmonic contents. The FFT plots for the flux linkages of the machine under the investigated operating cases are presented in Figure 7a–d. These figures show that a considerable magnitude of third harmonic is present in the flux linkages produced by rotating the shaft of the machine at synchronous speed and the armature currents generated through the proposed CCVSI. Figure 8a–d show the magnetic field density plot of the machine under the investigated operating cases. These plots show that the operation of the machine is under the saturation level of 1.6 T in case 1, 1.7 T in case 2 and 3, and 1.8 T in case 4.

The third harmonic flux induces the harmonic current in the twelve-pole rotor harmonic winding, which is rectified through a diode rectifier to excite the rotor field winding. The induced harmonic and rectified field currents of the employed machine under investigated operating cases are shown in Figure 9a–d.

A four-pole rotor field can get locked with the four-pole main stator field and develop torque. In case 1 and 2, the magnitude of the average torque generated through the proposed brushless WFSM topology is 3.198 Nm and 5.252 Nm, respectively. However, in case 3 and 4, the magnitude of the average torque is 6.4478 Nm and 7.2367 Nm. The output torque of the machine under the investigated operating cases is shown in Figure 10a–d. The magnitude of the generated output torque and its torque ripple during the investigated operating conditions are presented in Table 3. From the table, it can be seen that as the magnitude of the dc offset increases, the average torque of the machine increases. However, the torque ripple also increases. It is because the increase in the magnitude of the dc offset increases the additional harmonics in the machine air gap.

The torque ripple of the machine can be minimized by using parametric optimization techniques and skewing.

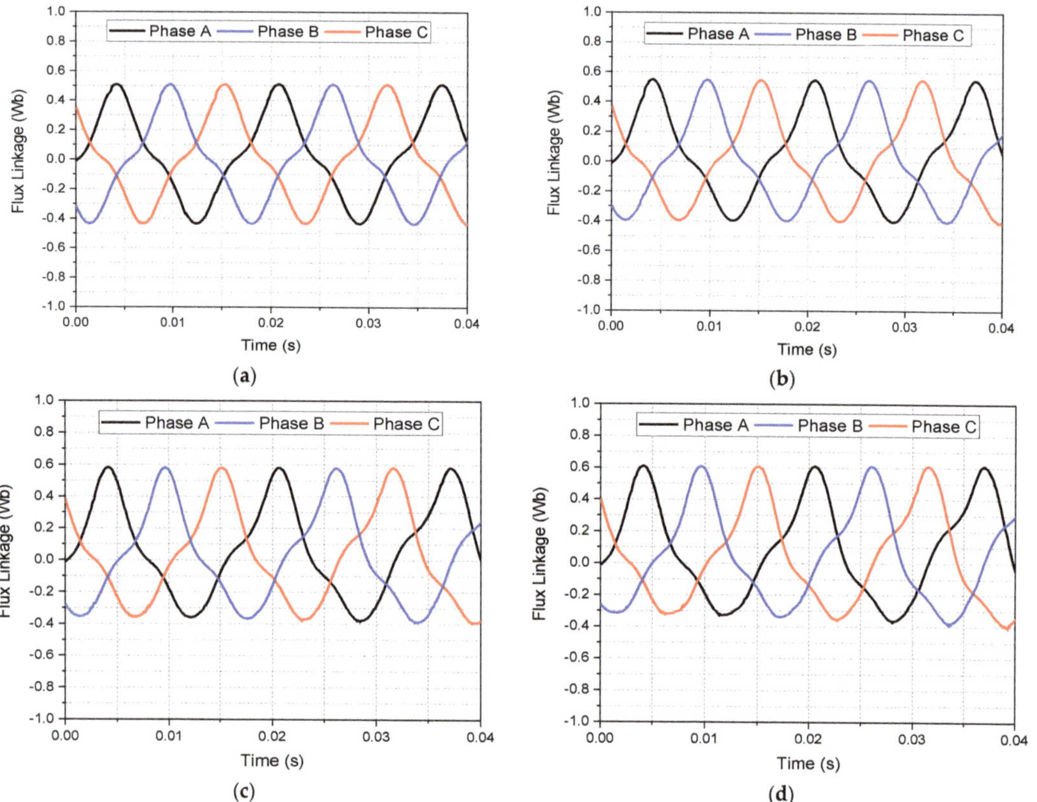

Figure 6. Flux linkages of the machine having (**a**) 0.6 A, (**b**) 1.2 A, (**c**) 1.8 A, and (**d**) 2.4 A dc offset for each phase of armature currents.

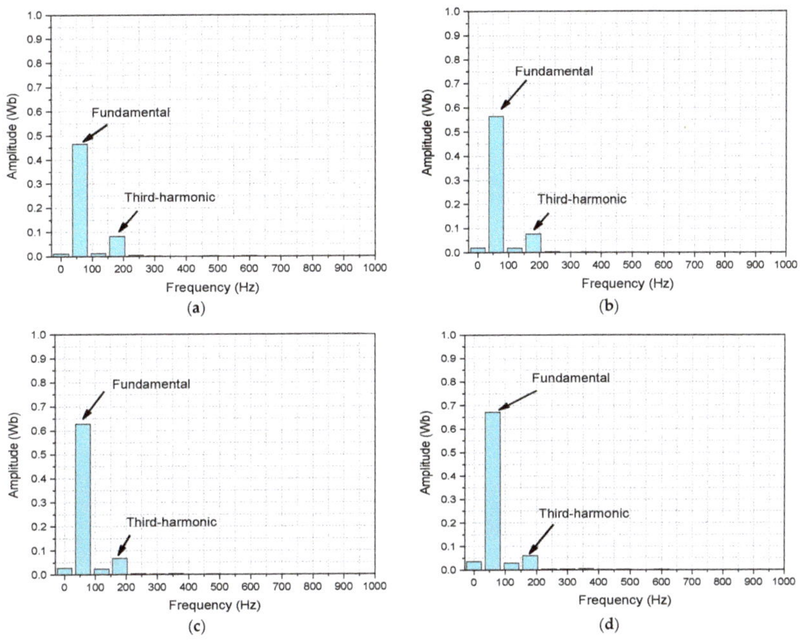

Figure 7. FFT of flux linkages of the machine having (**a**) 0.6 A, (**b**) 1.2 A, (**c**) 1.8 A, and (**d**) 2.4 A dc offset for each phase of armature currents.

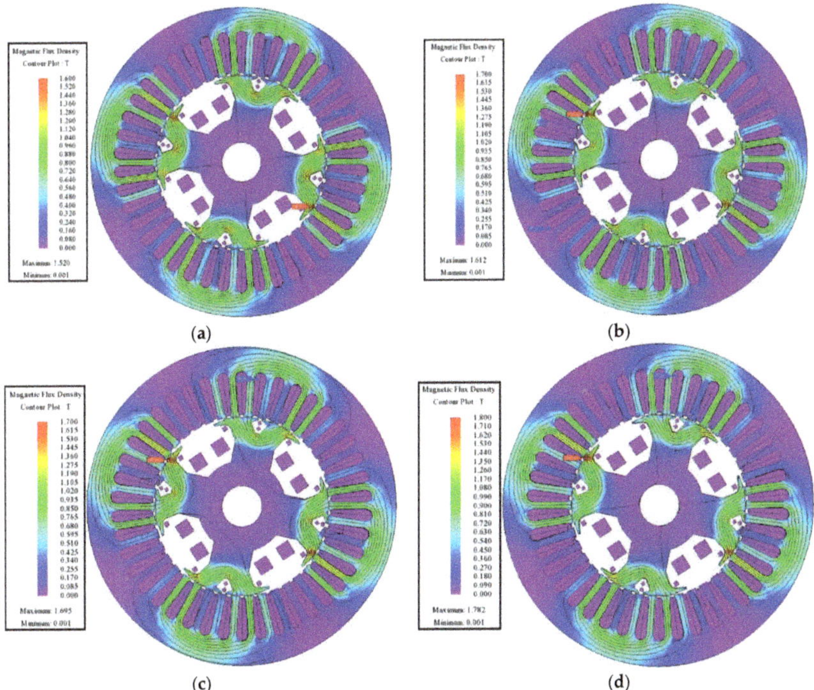

Figure 8. Magnetic field density plots of the machine having (**a**) 0.6 A, (**b**) 1.2 A, (**c**) 1.8 A, and (**d**) 2.4 A dc offset for each phase of armature currents.

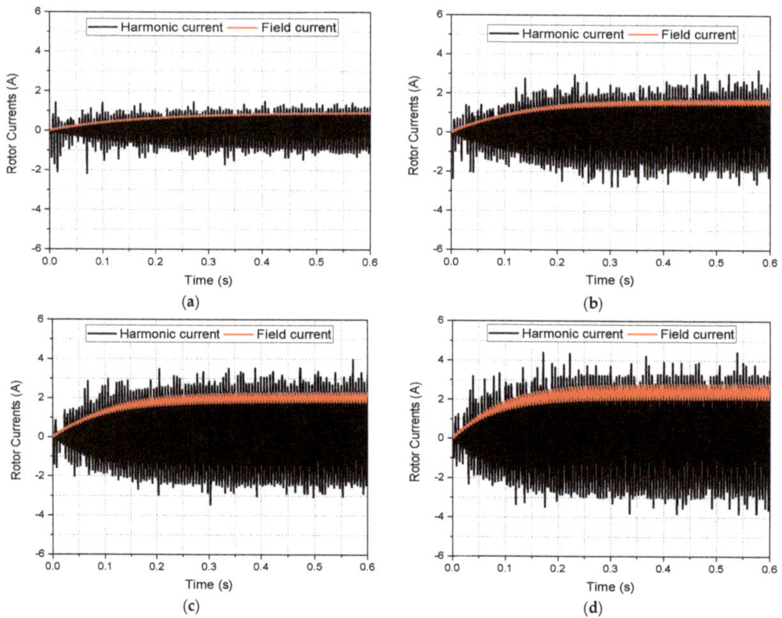

Figure 9. Rotor currents of the machine having (**a**) 0.6 A, (**b**) 1.2 A, (**c**) 1.8 A, and (**d**) 2.4 A dc offset for each phase of armature currents.

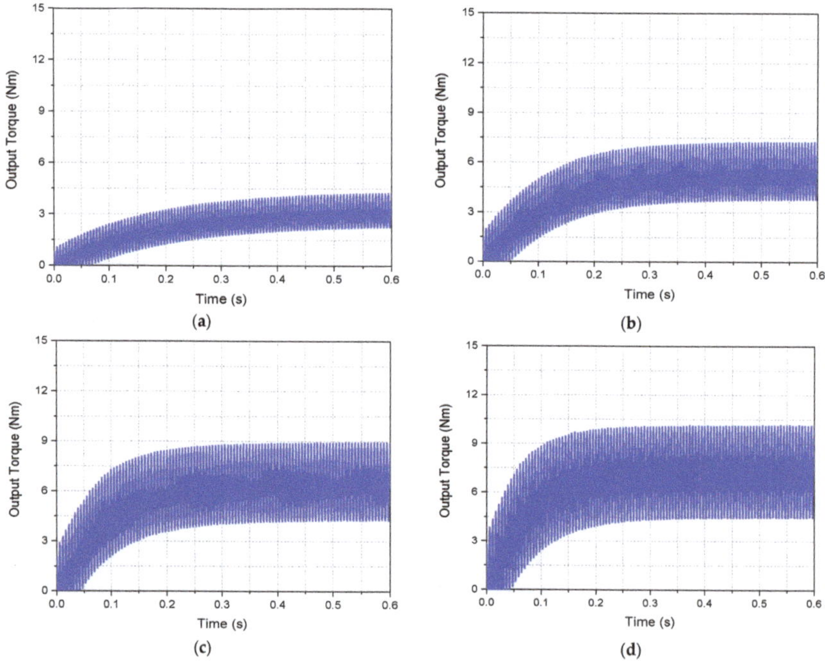

Figure 10. Output torque of the machine having (**a**) 0.6 A, (**b**) 1.2 A, (**c**) 1.8 A, and (**d**) 2.4 A dc offset for each phase of armature currents.

Table 3. Comparative performance analysis.

Attribute	Case 1	Case 2	Case 3	Case 4
Average output torque (in Nm)	3.1980	5.2520	6.4478	7.2367
Torque ripple (in %)	60.97	65.689	71.342	76
Maximum torque (in Nm)	4.36	7.25	8.85	10.15

No-Load Analysis

To examine the operation of the machine used to validate the proposed single inverter-controlled brushless WFSM topology under no-load condition, no load analysis of the machine was carried out in JMAG-Designer. The machine was provided with a field current of 1 A dc and was operated at 1800 rpm. The flux linkage of the machine under such conditions is presented in Figure 11a, whereas the magnetic flux density plot is presented in Figure 11b. A back-EMF of 76.671 V_{rms} was generated in the stator winding of the machine and is presented in Figure 12a. To show the harmonics present in the induced back-EMF, a FFT plot of the back-EMF was generated and is shown in Figure 12b. The cogging torque of the machine is 0.04 Nm (peak-to-peak). The generated cogging torque is presented in Figure 13. The no-load analysis results are presented in Table 4.

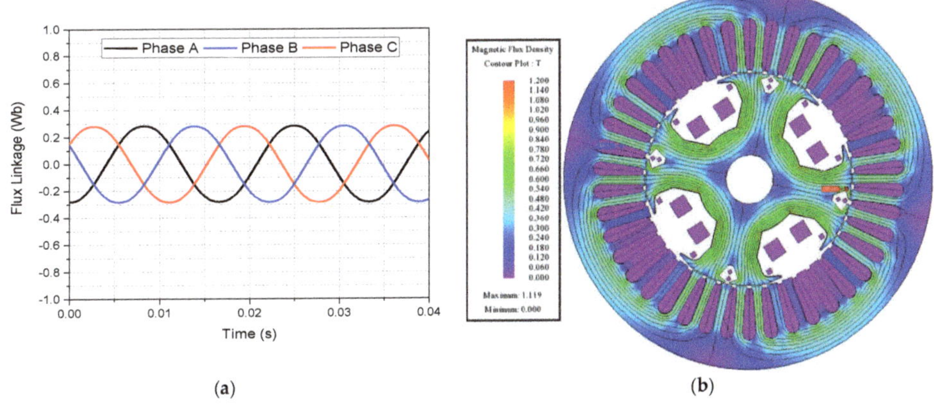

(a)

(b)

Figure 11. (**a**) Flux linkage and (**b**) magnetic flux density plot.

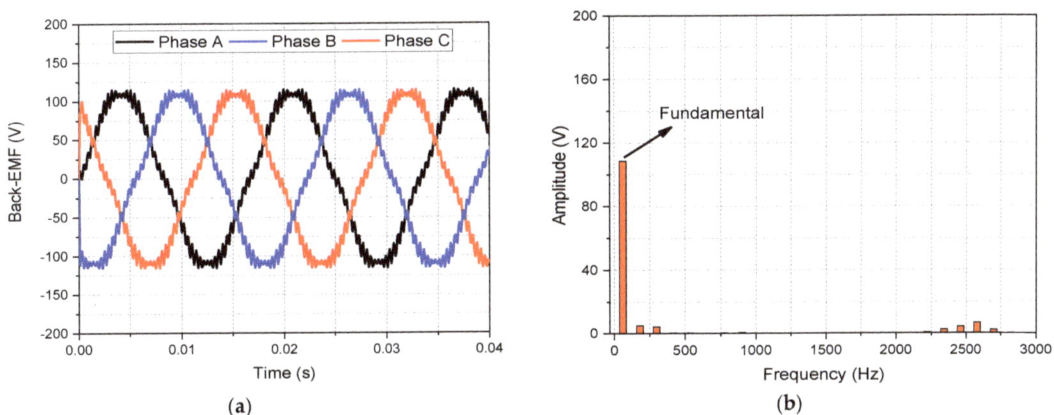

(a)

(b)

Figure 12. (**a**) Back-EMF of the employed machine and (**b**) its FFT plot.

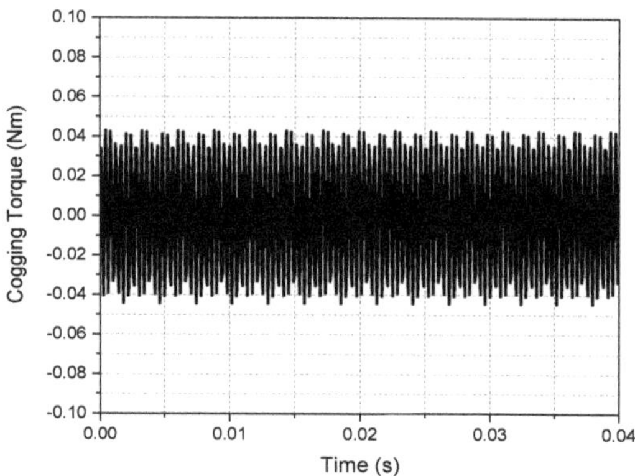

Figure 13. Cogging torque.

Table 4. No-load analysis.

Attribute	Case 4
Back-EMF (in V_{rms})	76.671
Cogging torque (in peak-to-peak Nm)	0.04

5. Conclusions

This paper proposes and investigates a single inverter-controlled brushless WFSM topology in which the inverter injects a three-phase armature current having a dc offset. In this arrangement, the volume of the machine can be reduced compared to a conventional brushless WFSM where an additional winding is used on the stator for excitation purpose. In addition, the proposed topology uses a single inverter and a simple current control strategy which makes it cost-effective compared to the brushless WFSM topologies of same kind. In particular, a 4-pole, 42-slot machine was used to simulate in JMAG for performance evaluation and verification of the working principle. The results show that a significant amount of torque is produced with the proposed brushless WFSM topology. This conclusion is, however, only based on the verification of the topology with reference to its working principle. The performance can be enhanced with an improved machine structure and optimized for various applications.

The limitation of the proposed single inverter controlled brushless WFSM topology includes the selection of the magnitude of dc offset for armature currents within a limit which may never cause any saturation for the stator and rotor cores.

Author Contributions: Conceptualization, S.S.H.B. and J.D.-G.; software, S.S.H.B. and G.J.S.; formal analysis, S.S.H.B. and S.M.; writing—original draft preparation, S.S.H.B. and A.A.M.; funding acquisition, J.-S.R. All authors have read and agreed to the published version of the manuscript.

Funding: This work was supported by the Brain Pool (BP) Program funded by the Ministry of Science and ICT through the National Research Foundation of Korea (2019H1D3A1A01102988), the Basic Science Research Program through the National Research Foundation of Korea funded by the Ministry of Education (2016R1D1A1B01008058), and the Human Resources Development (No.20204030200090) of the Korea Institute of Energy Technology Evaluation and Planning (KETEP) grant funded by the Korea government Ministry of Trade, Industry and Energy.

Institutional Review Board Statement: Not applicable.

Informed Consent Statement: Not applicable.

Data Availability Statement: Not applicable.

Conflicts of Interest: The authors declare no conflict of interest.

References

1. Lipo, T.A.; Du, Z.S. Synchronous motor drives-a forgotten option. In Proceedings of the 2015 International Aegean Conference on Electrical Machines & Power Electronics (ACEMP), 2015 International Conference on Optimization of Electrical & Electronic Equipment (OPTIM) & 2015 International Symposium on Advanced Electromechanical Motion Systems (ELECTROMOTION), Side, Turkey, 2–4 September 2015; pp. 1–5.
2. Dorrell, D. Are wound-rotor synchronous motors suitable for use in high efficiency torque-dense automotive drives? In Proceedings of the IECON 2012—38th Annual Conference on IEEE Industrial Electronics Society, Montreal, QC, Canada, 25–28 October 2012; pp. 4880–4885.
3. Zhu, S.; Liu, C.; Wang, K.; Hu, Y. The novel brushless excitation methods for wound rotor excitation synchronous generators. In Proceedings of the 2016 Eleventh International Conference on Ecological Vehicles and Renewable Energies (EVER), Monte Carlo, Monaco, 6–8 April 2016; pp. 1–7.
4. Kano, Y. Design optimization of brushless synchronous machines with wound-field excitation for hybrid electric vehicles. In Proceedings of the 2015 IEEE Energy Conversion Congress and Exposition (ECCE), Montreal, QC, Canada, 20–24 September 2015; pp. 2769–2775.
5. Di Gioia, A.; Brown, I.P.; Nie, Y.; Knippel, R.; Ludois, D.C.; Dai, J.; Hagen, S.; Alteheld, C. Design of a wound field synchronous machine for electric vehicle traction with brushless capacitive field excitation. In Proceedings of the 2016 IEEE Energy Conversion Congress and Exposition (ECCE), Milwaukee, WI, USA, 18–22 September 2016; pp. 1–8.
6. An, Q.; Gao, X.; Yao, F.; Sun, L.; Lipo, T. The structure optimization of novel harmonic current excited brushless synchronous machines based on open winding pattern. In Proceedings of the 2014 IEEE Energy Conversion Congress and Exposition (ECCE), Pittsburgh, PA, USA, 14–18 September 2014; pp. 1754–1761.
7. Sun, L.; Gao, X.; Yao, F.; An, Q.; Lipo, T. A new type of harmonic current excited brushless synchronous machine based on an open winding pattern. In Proceedings of the 2014 IEEE Energy Conversion Congress and Exposition (ECCE), Pittsburgh, PA, USA, 14–18 September 2014; pp. 2366–2373.
8. Aoyama, M.; Noguchi, T. Rare-earth free motor with field poles excited by space harmonics current phase-torque characteristics of self-excitation synchronous motor. In Proceedings of the 2013 International Conference on Renewable Energy Research and Applications (ICRERA), Madrid, Spain, 20–23 October 2013; pp. 149–154.
9. Dajaku, G.; Gerling, D. New self-excited synchronous machine with tooth concentrated winding. In Proceedings of the 3rd International Electric Drives Production Conference (EDPC-2013), Erlangen-Nürnberg, Germany, 29–30 October 2013.
10. Inoue, K.; Yamashita, H.; Nakamae, E.; Fujikawa, T. A brushless self-exciting three-phase synchronous generator utilizing the 5th-space harmonic component of magneto motive force through armature currents. *IEEE Trans. Energy Convers.* **1992**, *7*, 517–524. [CrossRef]
11. Yao, F.; An, Q.; Gao, X.; Sun, L.; Lipo, T.A. Principle of operation and performance of a synchronous machine employing a new harmonic excitation scheme. *IEEE Trans. Ind. Appl.* **2015**, *51*, 3890–3898. [CrossRef]
12. Jawad, G.; Ali, Q.; Lipo, T.A.; Kwon, B.-I. Novel brushless wound rotor synchronous machine with zero-sequence third-harmonic field excitation. *IEEE Trans. Magn.* **2016**, *52*, 1–4. [CrossRef]
13. Yao, F.; An, Q.-T.; Sun, L.; Lipo, T.A. Performance investigation of a brushless synchronous machine with additional harmonic field windings. *IEEE Trans. Ind. Electron.* **2016**, *63*, 6756–6766. [CrossRef]
14. Sirewal, G.J.; Ayub, M.; Atiq, S.; Kwon, B.-I. Analysis of a brushless wound rotor synchronous machine employing a stator harmonic winding. *IEEE Access* **2020**, *8*, 151392–151402. [CrossRef]
15. Bukhari, S.S.H.; Sirewal, G.J.; Ayub, M.; Ro, J.-S. A New small-scale self-excited wound rotor synchronous motor topology. *IEEE Trans. Magn.* **2021**, *57*, 1–5. [CrossRef]
16. Ayub, M.; Jawad, G.; Kwon, B.-I. Consequent-pole hybrid excitation brushless wound field synchronous machine with fractional slot concentrated winding. *IEEE Trans. Magn.* **2019**, *55*, 1–5. [CrossRef]
17. Ayub, M.; Hussain, A.; Jawad, G.; Kwon, B.-I. Brushless operation of a wound-field synchronous machine using a novel winding scheme. *IEEE Trans. Magn.* **2019**, *55*, 1–4. [CrossRef]
18. Bukhari, S.S.H.; Ahmad, H.; Sirewal, G.J.; Ro, J.-S. Simplified brushless wound field synchronous machine topology based on a three-phase rectifier. *IEEE Access* **2021**, *9*, 8637–8648. [CrossRef]
19. Bukhari, S.S.H.; Ahmad, H.; Chachar, F.A.; Ro, J.-S. Brushless field-excitation method for wound-rotor synchronous machines. *Int. Trans. Electr. Energy Syst.* **2021**, *31*, e12961.
20. Ayub, M.; Bukhari, S.S.H.; Sirewal, G.J.; Arif, A.; Kwon, B.-I. Utilization of reluctance torque for improvement of the starting and average torques of a brushless wound field synchronous machine. *Electr. Eng.* **2021**, *103*.

MDPI

Article

Feasibility Study of Direct-on-Line Energy-Efficient Motors in a Pumping Unit, Considering Reactive Power Compensation

Vadim Kazakbaev, Safarbek Oshurbekov, Vladimir Prakht and Vladimir Dmitrievskii *

Department of Electrical Engineering, Ural Federal University, Yekaterinburg 620002, Russia;
vadim.kazakbaev@urfu.ru (V.K.); safarbek.oshurbekov@urfu.ru (S.O.); va.prakht@urfu.ru (V.P.)
* Correspondence: vladimir.dmitrievsky@urfu.ru; Tel.: +7-343-375-45-07

Abstract: The paper compares the economic effect of using capacitors in fixed speed drives of a pumping station when using energy-efficient motors of various types. Induction motors of IE2 and IE3 energy efficiency classes, a direct-on-line synchronous motor with a permanent magnet in the rotor, and a direct-on-line synchronous reluctance motor are considered. The comparison takes into account not only the efficiency of the motors, but also their power factor, on which the losses in the cable and transformer depend. The possibility of using static capacitors to compensate for the reactive power of motors and reduce the losses is also considered. The feasibility analysis takes into account that the motors have different initial costs. The cost of capacitors is also taken into consideration. The analysis shows that the use of static capacitors can have a significant impact on the comparison between different motors in this application. Without considering capacitors, the permanent magnet motor has the shortest payback period, otherwise the synchronous reluctance motor has the shortest payback period.

Keywords: centrifugal pump; direct-on-line permanent magnet synchronous motor; direct-on-line synchronous reluctance motor; energy efficiency; induction motor; permanent magnet motor; reactive power compensation

Citation: Kazakbaev, V.; Oshurbekov, S.; Prakht, V.; Dmitrievskii, V. Feasibility Study of Direct-on-Line Energy-Efficient Motors in a Pumping Unit, Considering Reactive Power Compensation. *Mathematics* **2021**, *9*, 2196. http://doi.org/10.3390/math9182196

Academic Editor: Alessandro Niccolai

Received: 29 July 2021
Accepted: 6 September 2021
Published: 8 September 2021

Publisher's Note: MDPI stays neutral with regard to jurisdictional claims in published maps and institutional affiliations.

1. Introduction

The high energy intensity of modern industry makes it necessary to improve the energy efficiency of production processes. About 70% of the electricity generated worldwide is consumed by electric motors, the most significant part of which is powered directly from the electrical grid [1]. Electric motors connected directly to the AC mains consume both real and reactive power. Reactive power does not produce any useful work, but a reactive current creates additional losses in supply cables and transformers. Therefore, to reduce the power consumption of an electric drive, the reactive power must be compensated [2]. Many studies have been devoted to the analysis of the feasibility of reactive power compensation for electric motors powered directly from the grid. The following methods for improving the power factor have been proposed [2,3]:

— reduction in the motor voltage at partial load operation (Figure 1a);
— the use of a double motor winding, one section of which is connected to the grid, and capacitors are connected to the second. Capacitors can be connected in series (Wanlass connection, Figure 1b) or in parallel to the winding (Roberts' connection, Figure 1c);
— the use of static capacitors at the motor terminals (Figure 1d);
— the use of semiconductor devices for reactive power compensation.

A particular and most common case of voltage reduction schemes is switching the winding connection from triangle to star (Δ/Y) [4]. However, such a solution provides an effective increase in the power factor only when the motor is running at low loads. Therefore, it will not be effective in most applications [2].

Existing dual-winding solutions (Figure 1b,c) can effectively improve the power factor of the motor, however, the overall motor losses increase significantly. In addition, the motor cost increases, and the reliability deteriorates. Therefore, motors with double windings have not found wide application [2,3]. The Roberts' connection was also proposed for direct-on-line synchronous motors, however, such solutions also did not find wide practical application [3,5,6].

Another possible way to increase the power factor is to use a semiconductor device connected in parallel with the motor, or an induction machine with a wound rotor, the stator of which is directly connected to the mains, and the wound rotor is connected to two static inverters with a common DC-link (doubly fed induction machine). This method is effective, but not suitable for low-power motors due to its high cost [7,8].

Static capacitors have long been used for reactive power compensation and are the most effective method of the above [2]. Non-switchable and switchable capacitor banks can be used [9]. However, the feasibility analysis of the use of capacitors for various types of motors, including modern DOL-SynRM and DOL-PMSM, is still poorly covered in the literature.

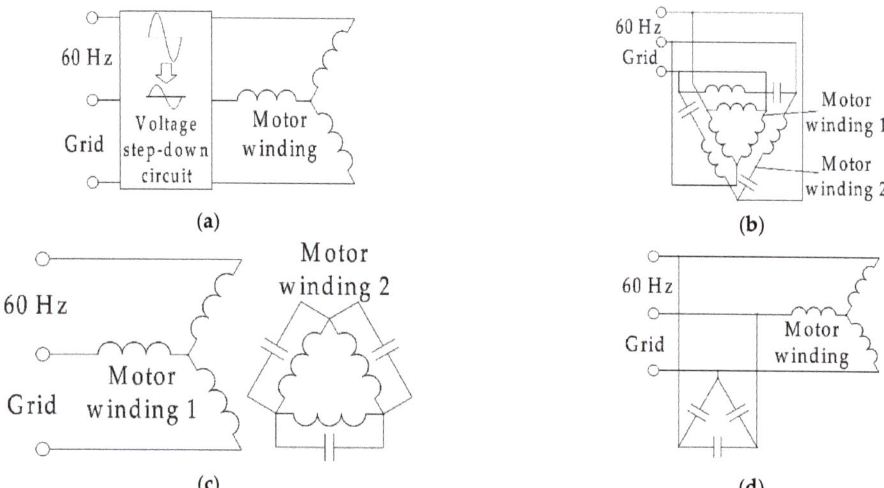

Figure 1. Reactive power compensation methods for low power AC motors: (**a**) reducing the voltage at the motor terminals; (**b**) Wanlass connection motor winding; (**c**) Roberts' connection motor winding; (**d**) Static capacitors at the motor terminals.

A number of studies are devoted to the analysis of energy savings in a pumping station due to an increase in the power factor of electric motors. In [10], energy savings are calculated when using capacitors to compensate for the reactive power of a pumping station with eight pumps equipped with induction motors. A different number of simultaneously operating pumps is considered. It has been shown that when the station is fully loaded, the capacitors will have a very short payback period: about 2 weeks. However, the absolute value of losses in the transmission line, taking into account the parameters of the cable and transformer, is not calculated. Only the percentage reduction in losses in the transmission line and an approximate reduction in electricity costs according to the formula for large consumers in Egypt are estimated, which is not universal and is not suitable for the case of small consumers.

In [11], the energy savings of the pump drive were estimated, taking into account the losses in the cable without taking into account the losses in the transformer. The article compares annual energy savings and lifetime energy savings for induction motors (IM) of different efficiency classes, direct-on-line permanent magnet synchronous motor (DOL PMSM) and direct-on-line synchronous reluctance motor (DOL SynRM) (Figure 2).

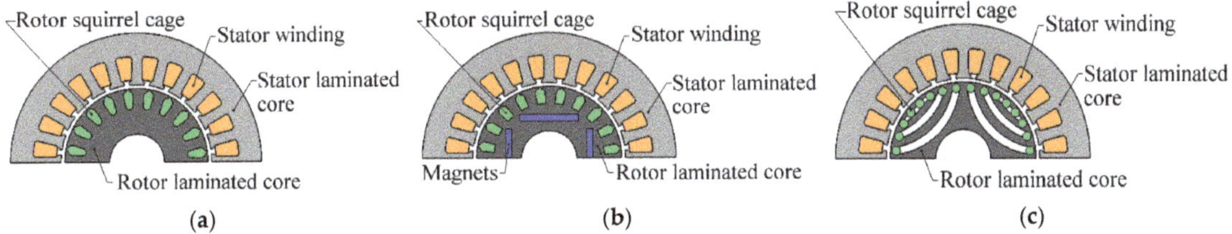

Figure 2. Schematic representation of the motor design (**a**) induction motor (IM); (**b**) direct-on-line permanent magnet synchronous motor (DOL-PMSM); (**c**) direct-on-line synchronous reluctance motor (DOL-SynRM).

All three types of motors under consideration have approximately the same stator design, but different rotor designs. IM operates in an asynchronous steady-state mode and has significant electrical losses in the rotor. DOL PMSM and DOL SynRM usually have a higher efficiency than IM due to the absence of electrical losses in the rotor from the first (fundamental) current harmonic when operating in a synchronous steady-state mode. In [11], it is shown that the power factor of the motor has a significant effect on the cable losses, but the losses in the transformer are not taken into account. However, transformer losses also significantly depend on the reactive component (power factor) of the load current.

Paper [12] also discusses a comparison of the energy consumption of IM, DOL PMSM and DOL SynRM in a pumping application. In this case, the influence of the motor power factor on the losses in the cable and transformer is taken into account. It has been shown that the increased power factor can significantly increase energy savings, shorten the payback period and make the use of DOL PMSM most profitable after several years, despite its higher cost compared to IM and DOL SynRM and its lower efficiency compared to DOL SynRM.

This article, in contrast to [11,12], evaluates the energy savings when using static capacitors for reactive power compensation of various types of motors (IM of energy efficiency classes IE2 and IE3, DOL PMSM, DOL SynRM) in a pumping application. The energy savings when using motors of 3.7 kW, 60 Hz, four poles of various types in a pumping station and their payback period are assessed. Various characteristics of the motors are taken into account, including their cost, efficiency and power factor. The cost of capacitors and their effect on cable and transformer losses are also taken into account.

2. Evaluating Pump Station Power Consumption

The article compares the power consumption of a pumping station when using different motors with a cyclical law of change in flow, shown in Figure 3a. It is shown in [13] that this flow-time diagram is typical for pumps without a variable speed drive (VSD). It is assumed that 100% flow demand corresponds to the best efficiency point (BEP) of the pump. This choice is justified by the fact that if the pump is chosen in such a way that its BEP corresponds to this flow rate, then the pump efficiency will be maximum, and the wear of the pump components will be minimal [14]. Figure 3b shows a diagram of the losses in a pumping unit. To calculate the electrical power P_1 consumed from the mains, it is necessary to calculate the mechanical power P_{mech} on the motor shaft at a certain flow rate Q, as well as the corresponding value of the power losses in the motor, cable and transformer.

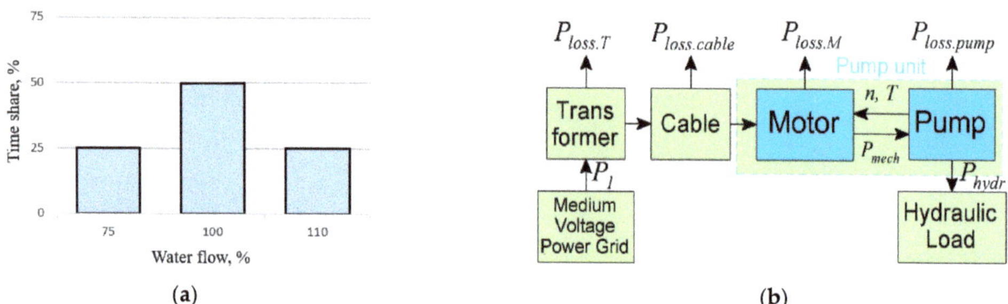

Figure 3. Accessing a pump unit energy consumption: (**a**) variation of the pump flow during operation; (**b**) diagram of power losses in a pump unit.

Figure 4 shows the diagram of the pumping station. Five pumps are connected to the mains through one transformer and cable. Each of the motors has the rated power of 3.7 kW, the rated frequency of 60 Hz and four poles.

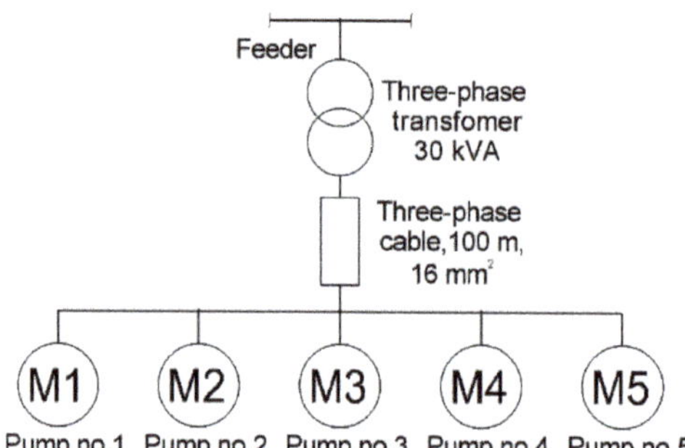

Figure 4. Diagram of connecting the motors of the pumping station to the grid.

The mechanical power of the motor is calculated using the dependence $P_{mech}(Q)$ provided in the catalog of the pump manufacturer. For the calculation, a pump of the HV105 type, 1800 rpm is considered (Figure 5) [15].

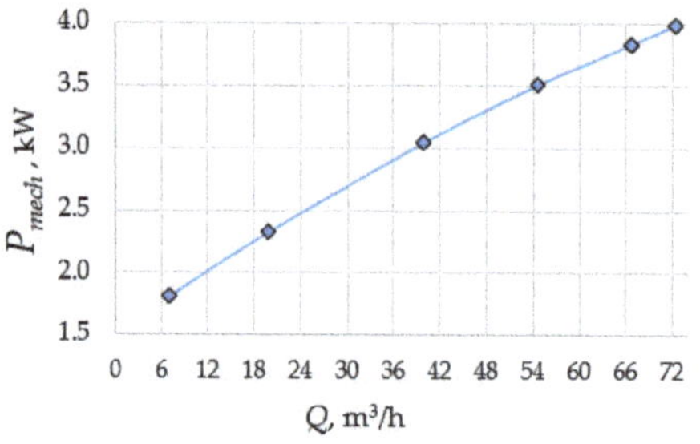

Figure 5. Dependence of the mechanical power on the motor shaft on the flow rate.

The real electrical power consumed by the pumping station from the grid can be calculated as:

$$P_{1\Sigma} = N_1 \cdot P_{mech}/\eta_{motor} + p_{cable} + p_T = N_1 \cdot P_1 + p_{cable} + p_T, \tag{1}$$

where P_{mech} is the input mechanical power of the pump; η_{motor} is the motor efficiency; p_{cable} is the electrical loss in the cable; p_T is the electrical loss in the transformer; $N_1 = 5$ is the number of motors of the pumping station; P_1 is the real power consumed by an individual motor of the pumping station.

The motor efficiency is determined by polynomial interpolation of the catalog data depending on the mechanical power. The losses p_{cable} and p_T depend on the load current:

$$p_{cable} = 3 \cdot R_{cable} \cdot I_{load}^2, \tag{2}$$

$$p_T = A + B \cdot (I_{load} / I_{T\,rate})^2, \tag{3}$$

where I_{load} is the loading current of the cable and the transformer; $R_{cable} = 0.12$ Ohm is the cable phase resistance; $I_{T\,rate} = 43.4$ A is the rated phase current of the transformer; $A = 121$ W and $B = 568$ W are determined based on the value of the transformer losses at $I_{load} = 0$ and $I_{load} = I_{T\,rate}$ A (121 and 689 W, correspondingly, according to [16]).

For the calculation, a cable with a cross section of 16 mm^2 and a length of 100 m and a transformer with a power rating of 30 kVA were considered (Figure 4). The total load current is the sum of the currents of all motors of the pumping station:

$$I_{load} = N_1 \cdot I_{motor}. \tag{4}$$

The single motor current without capacitive compensation is calculated as:

$$I_{motor} = P_{mech} / (\sqrt{3} \cdot V_{motor} \cdot \cos\varphi \cdot \eta_{motor}). \tag{5}$$

3. Reactive Power Compensation Using Capacitors

The paper evaluates the reduction in power consumption of motors when using static capacitors to compensate for reactive power. The capacitors are assumed to be delta connected (Figure 1d). The reactive power generated by the capacitors is calculated by the formula:

$$Q_C = m \cdot \omega \cdot C \cdot V^2, \tag{6}$$

where $m = 3$ is the number of phases of the capacitor bank; $\omega = 2 \cdot \pi \cdot f$ rad/s is the angular electric frequency; $f = 60$ Hz is the electric frequency; C is the line-to-line capacity of the capacitor bank; $V = 400$ V is the linear voltage.

The reactive power generated by the motor at a certain load is calculated using the formula:

$$Q_1 = \sqrt{S_1^2 - P_1^2} = \sqrt{\left(\sqrt{3} \cdot I_{motor} \cdot V\right)^2 - P_1^2}. \tag{7}$$

Therefore, the capacitance C required for full compensation of the reactive power at a certain motor load can be calculated as:

$$C = Q_1 / (3 \cdot \omega \cdot V^2). \tag{8}$$

The current with which the motor loads the cable and the transformer, taking into account the capacitance compensation, is calculated by the formula:

$$I_{motor} = \sqrt{P_1^2 + (Q_1 - Q_c)^2} / \left(V \cdot \sqrt{3}\right). \tag{9}$$

4. Motor Performances in the Pump Operating Cycle

For the calculation, the characteristics of four different motors were considered: DOL-PMSM (manufacturer WEG [17]), IE3-IM (manufacturer WEG [18]), IE2-IM (manufacturer WEG [19]) DOL-SynRM (characteristics taken from the article [20]) (Figure 2). Tables 1 and 2 and Figure 6 show the characteristics of the considered motors.

Table 1. Motor characteristics.

Type of Motor	Rated Mechanical Power, kW	Poles	Frame Size	Frame Material	Weight, kg	Rated Voltage, V
DOL SynRM	3.7	4	IEC 112	No data	No data	400
DOL PMSM	3.7	4	IEC 112	Cast iron	47.2	400
IE2 IM	3.7	4	IEC 112	Cast iron	43.2	400
IE3 IM	3.7	4	IEC 112	Cast iron	44.0	400

Table 2. Motor characteristics.

Type of Motor	Motor Efficiency, %			Motor Power Factor		
	50% Load	75% Load	100% Load	50% Load	75% Load	100% Load
DOL SynRM	91.0	92.1	92.1	0.564	0.658	0.709
DOL PMSM	88.5	90.7	91.6	0.74	0.86	0.92
IE2 IM	86.5	87.5	87.5	0.6	0.72	0.80
IE3 IM	88.1	89.3	89.5	0.61	0.74	0.80

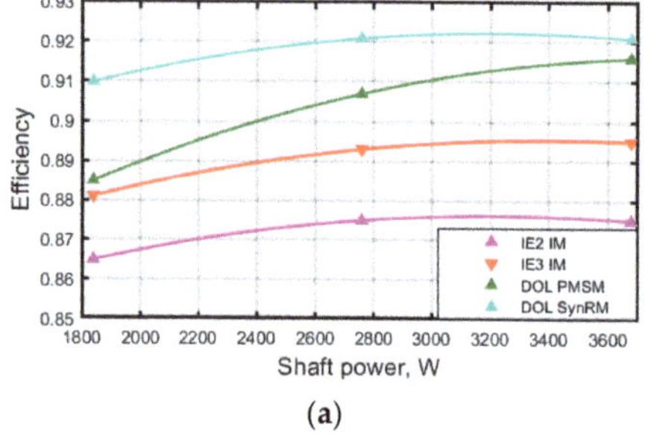

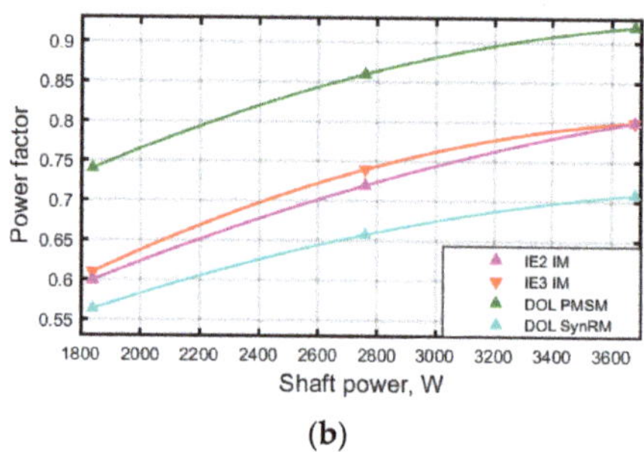

Figure 6. The motor catalogue characteristic comparison (**a**) efficiency; (**b**) power factor.

The characteristics of motors for a certain value of mechanical power are determined using polynomial interpolation of the data shown in Figure 6. The characteristics of the motors calculated in this way at the considered three operating points of the pump (Figure 3a) are shown in Table 3.

Table 3. Interpolated motor performance under various pump load conditions according to Figure 3a.

Q, %	Q, m³/h	P_{mech}, W	Motor Efficiency				Motor Power Factor				Motor Current			
			DOL SynRM	DOL PMSM	IE2 IM	IE3 IM	DOL SynRM	DOL PMSM	IE2 IM	IE3 IM	DOL SynRM	DOL PMSM	IE2 IM	IE3 IM
110	66	3816	0.92	0.916	0.874	0.894	0.712	0.924	0.809	0.803	8.41	6.51	7.79	7.67
100	60	3654	0.921	0.916	0.875	0.895	0.708	0.919	0.798	0.799	8.09	6.27	7.55	7.38
75	45	3204	0.921	0.913	0.876	0.895	0.683	0.897	0.764	0.778	7.35	5.65	6.91	6.65

When using delta-connected capacitors at the motor terminals (Figure 1d), the efficiency of the motor does not change, but the magnitude of the current with which the motor loads the cable and transformer, and, consequently, losses in these elements change, which also affects the total energy consumption. The phase capacitance of the capacitor bank is calculated according to the formula (8) in order to fully compensate for the reactive power of the motor at $Q = 100\%$ (the longest loading condition, according to the diagram in Figure 3a). For the IE3 IM case, the phase capacitance of the compensating device (the capacitance of a separate capacitor) is 16 µF. For the IE2 IM case, the capacitance is 17 µF. For DOL SynRM the capacitance is 20.5 µF. For the DOL PMSM case the capacitance is 9.55

µF. At $Q = 75\%$ and $Q = 110\%$, a slight undercompensation or overcompensation of reactive power is obtained. Table 4 shows a comparison of the current with which the motor loads the cable and transformer, with and without capacitors.

Table 4. Comparison of the current loading the cable and the transformer for different motors.

Q, %	P_{mech}, W	I_{motor}, A							
		DOL SynRM	DOL PMSM	IE3 IM	IE2 IM	DOL SynRM (+capacitors)	DOL PMSM (+capacitors)	IE3 IM (+capacitors)	IE2 IM (+capacitors)
110	3815.7	8.41	6.51	7.67	7.79	6.01	6.01	6.17	6.30
100	3653.9	8.09	6.27	7.38	7.55	5.74	5.76	5.90	6.03
75	3204.2	7.35	5.65	6.64	6.91	5.02	5.07	5.17	5.28

5. Cable and Transformer Losses

When calculating the losses of a station of five pumps (Figure 4), the losses in the cable and transformer are taken into account. For the calculation, the parameters of a 30 kVA transformer from the catalog [16] and the parameters of a cable with a cross-section of 16 mm^2 and a length of 100 m were selected. The cross-section of the cable was selected according [21]. The losses in the cable p_{cable} were calculated using formula (2). The losses in the transformer p_T were calculated using the formula (3). Figure 7 shows the results of calculating the cable and transformer losses. Total loss in the pumping station is also shown.

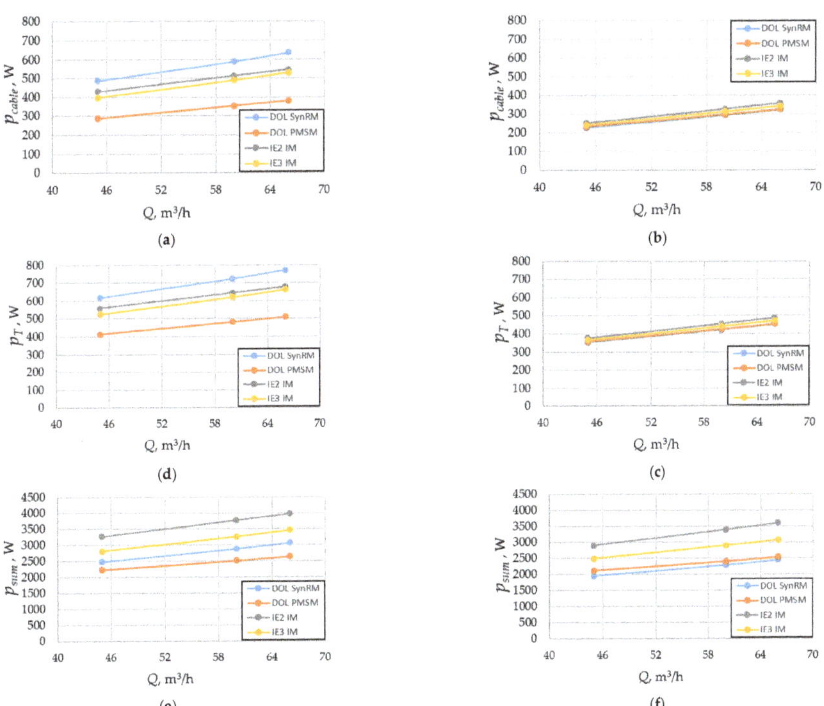

Figure 7. Comparison of losses when using different motors: (**a**) cable loss without using capacitors; (**b**) cable loss using capacitors; (**c**) transformer loss without using capacitors; (**d**) transformer loss using capacitors; (**e**) total pumping station loss without using capacitors; (**f**) total pumping station loss using capacitors.

Figure 7a,c compare cable and transformer losses for various motor types without capacitive compensation. Figure 7b,d compare the cable and transformer losses for various motor types using capacitive compensation. It can be seen that the use of capacitive compensation significantly reduces these losses for the induction motors and for the DOL SynRM, and that the losses in the cable and transformer become approximately the same for all types of motors when using capacitors. Since the DOL PMSM initially has a high power factor, capacitive compensation does not reduce p_{cable} and p_T significantly in the case of the DOL PMSM.

Figure 7e,f compare the total losses of the pumping station (five motors ($N_1 = 5$), cable and transformer) $p_{sum} = P_{1\Sigma} - N_1 \cdot P_{mech}$ with and without using capacitors. It can be concluded that the use of capacitive compensation significantly reduces the total losses of the pumping station in the case of the induction motors and DOL SynRM. When using capacitors, the lowest losses are provided by using the DOL SynRM.

Considering all types of losses, the power $P_{1\Sigma}$ consumed by the pumping station from the grid can be calculated using the formula (1).

6. Lifetime Energy Savings Using Different Motors

Based on the calculated results on the power $P_{1\Sigma}$ consumed from the medium voltage network by the pumping station and the power P_1 consumed by an individual motor, a comparison was carried out for the lifetime electricity savings for various considered variants of the electric drive. Daily electricity consumption is calculated using the formula:

$$E_{day} = t_\Sigma \cdot \sum_{i=1}^{3} (P_{1i} \cdot t_i / t_\Sigma). \tag{10}$$

where $i = 1 \dots 3$ is the index of a loading point; P_{1i} is the eclectic power P_1 in i-th loading point; t_i is the operation time of a loading point; t_Σ is the whole time period (24 h).

Then the annual energy consumption can be obtained as:

$$E_{year} = E_{day} \cdot 365. \tag{11}$$

The cost of electricity consumed (in Euro), considering the applied grid tariffs $GT = 0.2036$ €/kW·h for non-household consumers [22] for Germany in the second half of 2019, was calculated as follows:

$$C_{year} = E_{year} \cdot GT. \tag{12}$$

The expected lifetime of a pump is often evaluated to be about 20 years [23]. In this section, the energy cost is estimated for a service life of $n = 20$ years, excluding maintenance costs and the initial cost of the motors. The net present value (NPV) of the lifecycle cost was obtained as follows:

$$C_{LCC} = \sum_{j=1}^{n} \left(C_{year\,j} / [1 + (y - p)]^j \right), \tag{13}$$

where $C_{year\,j}$ is the energy cost of jth year; y is the interest rate ($y = 0.04$); p is the expected annual inflation ($p = 0.02$); n is the lifetime of the pump unit ($n = 20$ years) [23].

Lifecycle cost savings S_{LCC} for a given motor is calculated as:

$$S_{LCC} = C_{LCC} - C_{LCCIE2}, \tag{14}$$

where C_{LCC} is the lifecycle electricity cost of the considered motor; C_{LCCIE2} is the lifecycle electricity cost of the IE2 IM without capacitors.

S_{LCC} percentage is calculated as:

$$S_{LCC} = 100\% \cdot (C_{LCC} - C_{LCCIE2}) / C_{LCCIE2}, \tag{15}$$

Table 5 and Figure 8 show the results of the calculation of the lifecycle energy savings for various motor types. The savings are calculated compared to IE2 IM motor without capacitors.

Table 5. Comparison of motors 3.7 kW, 4 poles when operating in the pumping unit, taking into account losses in the cable and transformer.

Parameter	DOL SynRM	DOL PMSM	IE3 IM	IE2 IM	DOL SynRM (+capacitors)	DOL PMSM (+capacitors)	IE3 IM (+ capacitors)	IE2-IM (+capacitors)
P_1, W ($Q = 110\%$)	4406	4330	4487	4591	4291	4310	4419	4522
P_1, W ($Q = 100\%$)	4208	4143	4287	4389	4101	4123	4222	4320
P_1, W ($Q = 75\%$)	3682	3639	3751	3840	3587	3619	3693	3774
E_{day}, kW·hour	99	98	96	103	101	97	99	102
E_{year}, kW·hour	36,146	35,601	35,214	37,689	36,819	35,422	36,257	37,092
Annual energy savings, kW·hour	1543	2088	871	–	2475	2267	1432	597
Annual energy savings, %	4.1	5.5	2.3	–	6.6	6.0	3.8	1.6
Annual cost savings Cy.m, EUR	314	425	177	–	504	462	292	122
Life cycle energy cost C_{LCC}, kEUR (per 20 years)	120.3	118.5	122.6	125.5	117.2	117.9	120.7	123.5
Life cycle cost savings S_{LCC}, kEUR (per 20 years)	5.1	7.0	2.90	–	8.2	7.5	4.8	2.0
Life cycle cost savings S_{LCC}, % (per 20 years)	4.1	5.5	2.3	–	6.6	6.0	3.8	1.6

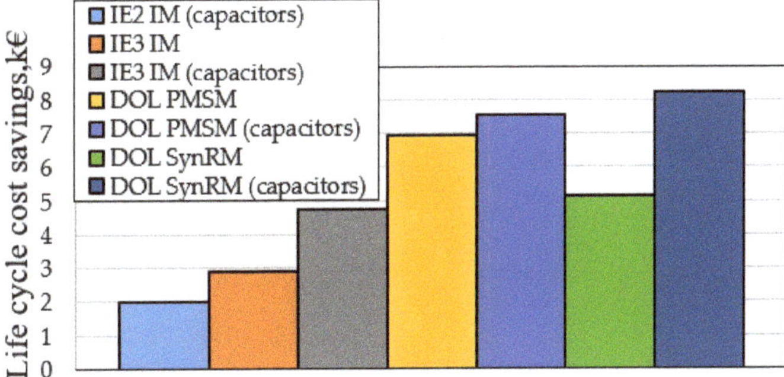

Figure 8. Energy cost savings over 20 years for different motors compared to IE2 IM without capacitors.

It can be concluded that the DOL SynRM without capacitors consumes more energy than the DOL PMSM. At the same time, the energy savings of the DOL SynRM without capacitors are significantly higher than that of the IE3 IM without capacitors. If using the capacitors, the power consumption of the DOL SynRM is lower than that of the DOL PMSM. The savings when using IE3 IM with capacitors are almost the same as when using the DOL SynRM without capacitors. The savings when adding capacitors to the IE2 IM are slightly less than those of the IE3 IM without capacitors.

7. Payback Period of the Motors and Capacitors

Since different motors under consideration have different initial costs, it is necessary to compare not only the energy savings they provide, but also their payback periods. The payback period is calculated for all motors in case of replacement of the IE2 IM without capacitors. For this purpose, the data on the market prices of IE2 IM [24] and AC capacitors of various capacities [25] were used.

Studies [26,27] show that the difference in the market value of the IMs of neighboring energy efficiency classes is usually in the range of 15–30%. A comparison of market price information for specific IM models confirms these findings. For this calculation, we will assume that the IE3 IM price is 22.5% higher than the IE2 IM price. Let us also assume that the IE4 IM price is 22.5% higher than the IE3 IM price.

In the literature, there are various estimates of the increase in the cost of the DOL PMSM in comparison to the IE3 IM. Thus, in [26] it is said the increase in cost is about 100%. However, the authors of this paper see no objective reason for such a large increase in cost. Comparison of information on market prices for specific models, as a rule, leads to a difference in the price of IE3 IMs and DOL PMSMs in the range of 30–40%. For this calculation, we will assume that the price of the DOL PMSM is 35% higher than the price of the IE3 IM. Many studies point out that there are no objective reasons for a significant difference in the cost of DOL SynRMs and IE3 IMs [26,28,29]. For this calculation, we will assume that the DOL SynRM price is equal to the IE3 IM price (see Table 6).

Table 6. Initial costs of motors and capacitors.

Motor	Motor Price, €	C, uF	$N_1 \cdot C$, uF	Capacitor Bank Price (Case Figure 9a), €	Capacitor Bank Price (Case Figure 9b), €	Price: Motor + Capacitors (Case Figure 9a), €	Price: Motor + Capacitors (Case Figure 9b), €
IE2 IM	398.3	17	85	50.94	20.94	449.2	419.2
IE3 IM	487.9	16	80	50.94	20.94	538.8	508.8
DOL SynRM	487.9	21	105	54.00	29.40	541.9	517.3
DOL PMSM	658.6	10	50	41.94	11.99	700.5	670.6

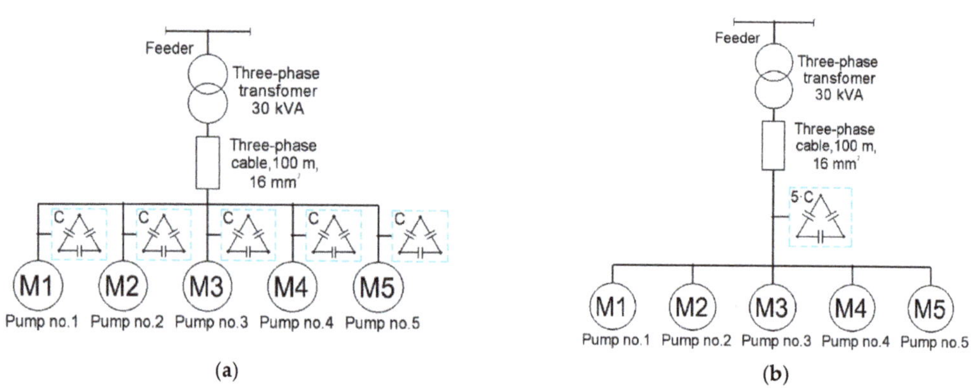

Figure 9. Methods of installing capacitor banks: (**a**) installation of an individual battery with linear capacity C on each motor; (**b**) installation of a battery with linear capacity $N_1 \cdot C$ at the common connection point of the motors.

To compensate for reactive power, capacitors can be installed either individually for each motor (Figure 9a) or connected to the common connection point for all motors (Figure 9b). Capacitor banks (Figure 9a) connected individually to the motor terminals have the advantage that individual pump units can be taken out of operation without generating excess capacitive power ("reactive power overcompensation"). In the considered application, this need can appear if the flow rate of the pumping station changes significantly over time [30].

Similarly interesting is the case of connecting one large capacitor bank with linear capacitance $N_1 \cdot C$ to the common connection point of the motors (Figure 9b). In this case, the total current in the cable and transformer will be the same as when using individual capacitor banks on each of the N_1 motors with linear capacitance C. However, the final cost of the capacitor bank will be lower, because as the rated capacitance of a capacitor increases, its cost per capacitance unit decreases, based on market price analysis [25]. This makes it possible to shorten the payback period, in comparison with the case of installing capacitors on each motor, if the pumping station has an approximately constant flow rate, as in the case under consideration.

Table 6 also shows the prices for capacitor banks for the cases under consideration. For ease of comparison, all prices are for one motor, that is, in the case shown in Figure 9b, the capacitor bank price is divided by the number of motors. For the case of Figure 9b, the cost of the capacitor bank in terms of one motor turns out to be much less, which also makes it possible to significantly reduce the cost of the entire electric drive of the pumping station.

Based on the initial cost of motors and capacitors, as well as the annual energy savings (Table 5), the payback period was calculated for different types of motors for the pumping station drive (Table 7, Figure 10). The results are shown for both the case without capacitors (Figure 4) and for the two cases with capacitors (Figure 9).

Table 7. Motor payback period.

Value	IE2 IM + Capacitors (Case Figure 9a)	IE3 IM	IE3 Motor + Capacitors (Case Figure 9a)	DOL SynRM	DOL SynRM + Capacitors (Case Figure 9a)	DOL PMSM	DOL PMSM + Capacitors (Case Figure 9a)
Annual cost savings, EUR (per 20 years)	122	177	292	314	504	425	462
Payback period (new pump unit commissioning), years	0.419	0.505	0.482	0.285	0.285	0.612	0.655
Payback period (replacing the motor in an exploiting pump unit), years	0.419	2.752	1.848	1.553	1.075	1.549	1.518

Value	IE2 IM + Capacitors (Case Figure 9b)	IE3 motor + Capacitors (Case Figure 9b)	DOL SynRM + Capacitors (Case Figure 9b)	DOL PMSM + Capacitors (Case Figure 9b)
Annual cost savings, EUR (per 20 years)	122	292	504	462
Payback period (new pump unit commissioning), years	0.172	0.379	0.236	0.59
Payback period (replacing the motor in an exploiting pump unit), years	0.172	1.745	1.026	1.453

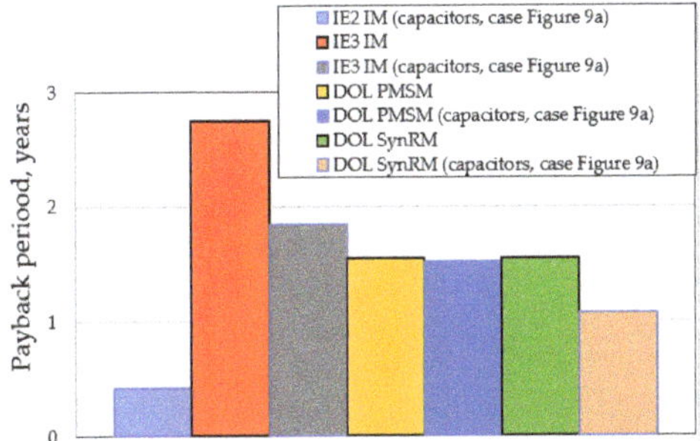

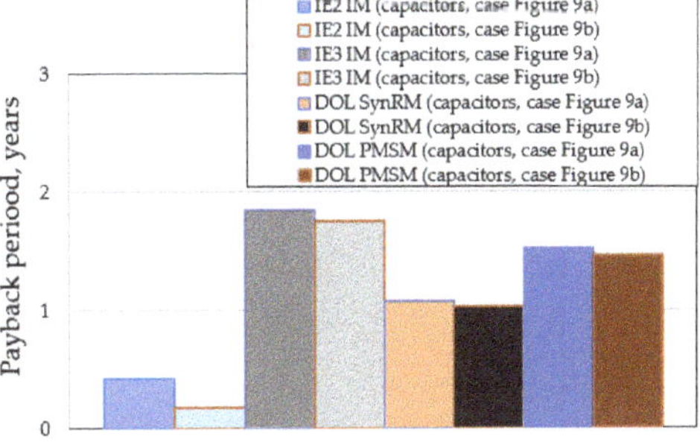

Figure 10. Comparison of payback periods when replacing IE2 IM without capacitors in a pumping unit in service.

Without considering the possibility of installing capacitors, the most profitable solution is the DOL PMSM, which in this case provides the greatest savings and the shortest payback period, despite its highest initial cost. Due to its lower initial cost, the DOL SynRM without capacitors has approximately the same payback period as the DOL PMSM without capacitors, but provides significantly less savings over the lifetime (Figure 8).

The IE3 IM without capacitors has the longest payback period of 2.75 years, which is significantly higher than DOL PMSM and DOL SynRM without capacitors. Comparing the options for installing capacitors, it can be concluded that the solution with installing capacitors on IE2 IM has the shortest payback period. With a limited upgrade budget, this is the most cost-effective solution.

However, in the long term, the most profitable is the use of the DOL SynRM with capacitors, providing the most energy savings. When using capacitors, the payback periods of the IE3 IM and DOL SynRM are significantly reduced. The payback period of the DOL PMSM varies little with using capacitors.

8. Discussion

This article investigated the impact of using static capacitive compensation on energy consumption and payback period for energy efficient electric motors in a pumping application. Compared to the IE2 induction motor, motors such as the IE3 induction motor, direct-on permanent magnet synchronous motor and direct-on-line synchronous reluctance motor are considered.

The comparison considers not only the efficiency of the motors, but also their power factor, on which the losses in the cable and transformer supplying the pumping station depend. The possibility of installing static capacitors to compensate for the reactive power of motors is also taken into account. The analysis takes into account that the motors have different initial costs, and also takes into account the cost of capacitors.

Without taking into account the possibility of using capacitors, the DOL PMSM has the shortest payback period (1.549 years), despite the highest cost, due to its high efficiency and high power factor, which can significantly reduce losses in the cable and transformer. The payback period of the DOL SynRM (1.553 years) is approximately equal to the payback period of DOL PMSM due to its higher efficiency and lower initial cost. At the same time, the payback periods of the DOL PMSM and DOL SynRM are significantly lower than that of IE3 IM (2.75 years).

The analysis shows that considering the possibility of installing static capacitors can significantly affect the results of comparing different motors in the application under consideration. Capacitors have a low initial cost compared to the price of motors, however, they allow you to compensate for the reactive component of the motor current, eliminate losses from this component in the cable and transformer, and therefore significantly reduce total losses. When using capacitors, the DOL SynRM has the shortest payback period (1.07 years).

The use of static capacitors will shorten the payback period of all the motors under consideration. When installing them, the payback period for the IE3 IM decreases most significantly (from 2.75 to 1.75 years). The DOL SynRM payback decreases from 1.553 to 1.07 years. The payback period of the DOL PMSM also decreases, but only slightly (from 1.549 to 1.52 years), since the DOL PMSM has a high power factor even without capacitive compensation.

If a pumping station has an approximately constant flow rate without the need for frequent shutdown of individual pump units, then installing one capacitor bank at the common connection point of the motors is more profitable than installing separate batteries of smaller capacity for each motor.

It is also shown that in the absence of the possibility of replacing the IE2 motor with more energy efficient ones, installing static capacitors on the terminals of the IE2 motor can be a good energy saving solution with a short payback period. In this case, the payback period of the capacitors is only 0.38 years.

The results of this study can be applied not only to pumps, but also to other mechanisms in which electric motors are powered directly from the AC mains and operate for a long time with little changing load, for example, fans, blowers, compressors, mixers, etc.

Author Contributions: Conceptual approach, V.K. and V.P.; data curation, S.O. and V.D.; software, S.O. and V.K.; calculations and modeling, S.O., V.K. and V.P.; writing—original draft, S.O., V.D., V.K. and V.P.; visualization, V.D. and V.K.; review and editing, S.O., V.D., V.K. and V.P. All authors have read and agreed to the published version of the manuscript.

Funding: The work was partially supported by the Ministry of Science and Higher Education of the Russian Federation (through the basic part of the government mandate, Project No. FEUZ-2020-0060).

Institutional Review Board Statement: Not applicable.

Informed Consent Statement: Not applicable.

Data Availability Statement: All data are contained within the article.

Acknowledgments: The authors thank the editors and reviewers for their careful reading and constructive comments.

Conflicts of Interest: The authors declare no conflict of interest.

References

1. Stoffel, B. Assessing the energy efficiency of pumps and pump units. In *Background and Methodology*; Elsevier: Amsterdam, The Netherlands, 2015.
2. Spee, R.; Wallace, A. Comparative evaluation of power factor improvement techniques for squirrel cage induction motors. *IEEE Trans. Ind. Appl.* **1992**, *28*, 381–386. [CrossRef]
3. Obe, E. Steady-state performance of a line-start synchronous reluctance motor with capacitive assistance. *Electr. Power Syst. Res.* **2010**, *80*, 1240–1246. [CrossRef]
4. Ferreira, F.; de Almeida, A. Method for in-field evaluation of the stator winding connection of three-phase induction motors to maximize efficiency and power factor. *IEEE Trans. Energy Convers.* **2006**, *21*, 370–379. [CrossRef]
5. Tola, O.; Obe, E.; Anih, L. Modeling and analysis of dual stator windings permanent magnet synchronous motor. In Proceedings of the 2017 IEEE 3rd International Conference on Electro-Technology for National Development (NIGERCON), Owerri, Nigeria, 7–10 October 2017; pp. 861–871.
6. Ogunjuyigbe, A.; Jimoh, A.; Nicolae, D.; Obe, E. Analysis of synchronous reluctance machine with magnetically coupled three-phase windings and reactive power compensation. *IET Electr. Power Appl.* **2010**, *4*, 291–303. [CrossRef]
7. Yao, Y.; Cosic, A.; Sadarangani, C. Power factor improvement and dynamic performance of an induction machine with a novel concept of a converter-fed rotor. *IEEE Trans. Energy Convers.* **2016**, *31*, 769–775. [CrossRef]
8. Ad'doweesh, K.; Smiai, M.; Haque, S. Power factor improvement of induction motor using microprocessor controlled FC-TCR compensator. *J. King Saud Univ.-Eng. Sci.* **1990**, *2*, 33–41. [CrossRef]
9. Pereira, M.; Fitiwi, D.; Santos, S.; Catalão, J. Managing RES uncertainty and stability issues in distribution systems via energy storage systems and switchable reactive power sources. In Proceedings of the 2017 IEEE International Conference on Environment and Electrical Engineering and 2017 IEEE Industrial and Commercial Power Systems Europe (EEEIC/I&CPS Europe), Milan, Italy, 6–9 June 2017; pp. 1–6.
10. Mohamad, M.; El-gawad, A.; Ramadan, H. Power factor improvement for pumping stations using capacitor banks. *Int. J. Emerg. Electr. Power Syst.* **2016**, *17*, 597–605. [CrossRef]
11. Kazakbaev, V.; Prakht, V.; Dmitrievskii, V.; Oshurbekov, S.; Golovanov, D. Life cycle energy cost assessment for pump units with various types of line-start operating motors including cable losses. *Energies* **2020**, *13*, 3546. [CrossRef]
12. Kazakbaev, V.; Prakht, V.; Dmitrievskii, V.; Golovanov, D. Feasibility study of pump units with various direct-on-line electric motors considering cable and transformer losses. *Appl. Sci.* **2020**, *10*, 8120. [CrossRef]
13. Extended Product Approach for Pumps, Europump. 2014. Available online: http://europump.net/uploads/Extended%20Product%20Approach%20for%20Pumps%20-%20A%20Europump%20guide%20(27OCT2014).pdf (accessed on 29 July 2021).
14. Barringer, H.P. A life cycle cost summary. In Proceedings of the International Conference Maintenance Societies, Perth, WA, Australia, 20–23 May 2003; pp. 20–23.
15. Neptuno Oumps, Engineered Pump Catalog 60 Hz Perfomance Curves, Neptuno Fluid Technology Ltd., 026016 Edition, Document Number NP-PPC-6016R3. Available online: https://neptunopumps.com/wp-content/uploads/2021/02/neptuno-pumps-60-hz-vtp-perfomance-curves.pdf (accessed on 29 July 2021).
16. Design Guide DG009001EN, Dry-Type Distribution Transformers—General Purpose, Eaton, February 2020. Available online: https://www.eaton.com/content/dam/eaton/products/design-guides---consultant-audience/eaton-dtdt-general-purpose-design-guide-dg009001en.pdf (accessed on 29 July 2021).

17. Wquattro Super Premium Efficiency 5 HP 4P 182/4T 3Ph 230/460 V 60 Hz IC411-TEFC. Product Details. Available online: https://www.weg.net/catalog/weg/HT/en/Electric-Motors/AC-Motors---NEMA/General-Purpose-ODP-TEFC/TEFC-Cast-Iron/WQuattro-IE4-Super-Premium/Wquattro-Super-Premium-Efficiency-5-HP-4P-182-4T-3Ph-230-460-V-60-Hz-IC411---TEFC---Foot-mounted/p/13044694 (accessed on 29 July 2021).
18. W22 IE3 5 HP 4P 112M 3Ph 220/440 V 60 Hz IC411-TEFC. Product Details. Available online: https://www.weg.net/catalog/weg/HT/en/Electric-Motors/AC-Motors---IEC/General-Purpose/Cast-Iron-Frame/TEFC-W22-IE2/W22-IE3-5-HP-4P-112M-3Ph-220-440-V-60-Hz-IC411---TEFC---B14L%28D%29/p/13063142 (accessed on 29 July 2021).
19. W22 High Efficiency 5 HP 4P 182/4T 3Ph 230/460//380 V 60 Hz IC411-TEFC. Product Details. Available online: https://www.weg.net/catalog/weg/HR/en/Electric-Motors/Low-Voltage-NEMA-Motors/General-Purpose-ODP-TEFC/Cast-Iron-TEFC-General-Purpose/W22/W22-High-Efficiency-5-HP-4P-182-4T-3Ph-230-460-380-V-60-Hz-IC411---TEFC---Foot-mounted/p/11605508 (accessed on 29 July 2021).
20. Liu, H.; Lee, J. Optimum design of an IE4 line-start synchronous reluctance motor considering manufacturing process loss effect. *IEEE Trans. Ind. Electron.* **2018**, *65*, 3104–3114. [CrossRef]
21. IEC. *Electrical Installations in Buildings—Part. 5-52: Selection and Erection of Electrical Equipment—Wiring Systems, Is the IEC Standard Governing Cable Sizing*; IEC 60364-5-52; IEC: Geneva, Switzerland, 2009; Available online: https://webstore.iec.ch/publication/1878 (accessed on 29 July 2021).
22. Eurostat Data for the Industrial Consumers in Germany. Available online: http://ec.europa.eu/eurostat/statistics-explained/index.php/Electricity_price_statistics#Electricity_prices_for_industrial_consumers (accessed on 29 July 2021).
23. Waghmode, L.; Sahasrabudhe, A. A comparative study of life cycle cost analysis of pumps. In Proceedings of the ASME 2010 International Design Engineering Technical Conferences and Computers and Information in Engineering Conference (ASME 2010), Montreal, QC, Canada, 15–18 August 2010; pp. 491–500.
24. Hindustan 5HP 3.7KW 4 Pole 1500 RPM B5 Flange MTG FR 112M 400V 60HZ IE2 Motor. Product Details. Available online: https://vashielectricals.com/p/hindustan-5hp-3-7kw-4-pole-1500-rpm-b5-flange-mtg-fr-112m-400v-60hz-ie2-motor/ (accessed on 29 July 2021).
25. CAPACITOR 400V 450V 500V 25 Micro Farad Capacitor. Product Details. Available online: https://eclats-antivols.fr/en/condenser/1688-capacitor-400v-450v-500v-25-micro-farad-capacitor-5412810204809.html (accessed on 31 August 2021).
26. Almeida, A. Motor Systems Technology Developments. In Proceedings of the 8th Motor Summit for Energy Efficient Motor Driven Systems Powered by Impact Energy, Zurich, Switzerland, 14–15 November 2018.
27. European Commission Staff Working Document, Brussels, Impact Assessment Accompanying the Directive 2009/125/EC, 1.10.2019 SWD (2019) 343 Final. Available online: https://ec.europa.eu/transparency/regdoc/rep/10102/2019/EN/SWD-2019-343-F1-EN-MAIN-PART-1.PDF (accessed on 29 July 2021).
28. Kersten, A.; Liu, Y.; Pehrman, D.; Thiringer, T. Rotor design of line-start synchronous reluctance machine with round bars. *IEEE Trans. Ind. Appl.* **2019**, *55*. [CrossRef]
29. Hu, Y.; Chen, B.; Xiao, Y.; Shi, J.; Li, L. Study on the Influence of design and optimization of rotor bars on parameters of a line-start synchronous reluctance motor. *IEEE Trans. Ind. Appl.* **2020**, *56*, 1368–1376. [CrossRef]
30. Koor, M.; Vassiljev, A.; Koppel, T. Optimal pump count prediction algorithm for identical pumps working in parallel mode. *Procedia Eng.* **2014**, *70*, 951–958. [CrossRef]

 *mathematics*

Article

Comparative Study of Energy Consumption and CO$_2$ Emissions of Variable-Speed Electric Drives with Induction and Synchronous Reluctance Motors in Pump Units

Victor Goman [1], Vladimir Prakht [2], Vadim Kazakbaev [2,*] and Vladimir Dmitrievskii [2]

[1] Nizhniy Tagil Technological Institute, Ural Federal University, 622000 Nizhniy Tagil, Russia; v.v.goman@urfu.ru
[2] Department of Electrical Engineering, Ural Federal University, 620002 Yekaterinburg, Russia; va.prakht@urfu.ru (V.P.); vladimir.dmitrievsky@urfu.ru (V.D.)
* Correspondence: vadim.kazakbaev@urfu.ru

Abstract: This study carried out a comparative analysis of indicators of electricity consumption and CO$_2$ emissions for four-pole induction motors (IMs) of efficiency classes IE3 and IE4 with a rated power of 2.2–200 kW in a variable speed pump unit. In addition, innovative IE4 converter-fed synchronous reluctance motors (SynRMs) were evaluated. The comparison was derived from the manufacturer's specifications for the power drive systems (PDSs) at various rotational speeds and loads. The results showed that the emission indicators for IE3 class motors were significantly worse compared with IE4 class motors for low power ratings, which make up the vast majority of electric motors in service. This justifies expanding the mandatory power range for IE4 motors to at least 7.5–200 kW or even 0.75–200 kW, as it will dramatically contribute to the achievement of the new ambitious goals for reducing greenhouse gas emissions. In addition, the operational advantages of IE4 SynRMs over IE4 IMs were demonstrated, such as their simpler design and manufacturing technology at a price comparable to that of IE3 IMs.

Keywords: carbon dioxide emissions; climate change mitigation; electric motors; energy conversion; energy efficiency; energy efficiency class; energy policy and regulation; energy saving; sustainable utilization of resources

Citation: Goman, V.; Prakht, V.; Kazakbaev, V.; Dmitrievskii, V. Comparative Study of Energy Consumption and CO$_2$ Emissions of Variable-Speed Electric Drives with Induction and Synchronous Reluctance Motors in Pump Units. *Mathematics* **2021**, *9*, 2679. https://doi.org/10.3390/math9212679

Academic Editors: Paolo Mercorelli and Nicu Bizon

Received: 3 August 2021
Accepted: 20 October 2021
Published: 22 October 2021

1. Introduction

Recently, the European Union has set new ambitious targets to achieve a climate-neutral economy by 2050 according to the European Green Deal plan [1]. In addition, the Council of Europe has raised the target for reducing greenhouse gas emissions by 2030 from the 1990 level of 40% to 55% [2]. The main greenhouse gas is carbon dioxide (CO$_2$). The most modern methods of electricity generation result in CO$_2$ emissions. Therefore, the CO$_2$ emissions can be limited by reducing the electricity consumption of the most significant consumers [3,4]. Electric motors are the largest consumers of electricity. Reducing their consumption will have a significant positive impact on climate change mitigation. According to [5], electric motors consume about half of all electricity in the European Union, which is about 2000 TWh per year; the number of electric motors is about 8 billion; and the volume of associated annual greenhouse gas emissions is 800 MtCO$_2$-eq. It is also known that electric motors account for approximately 53% of all electric power generated worldwide, or 10,700 TWh per year [6]. As a result, ever more stringent legal requirements for the energy efficiency of mains [7] and frequency-converter-powered motors [8] are being adopted.

In accordance with the European Commission Regulations 2009 and 2014 [9] of 1 January 2017, in the European Union, every direct-on-line motor with a power rating of 0.75–375 kW must be at least as efficient as the IE3 energy efficiency class, and the converter-fed motors must not be less efficient than the IE2 class. In 2019, these requirements were updated: in accordance with European Commission Regulation 2019 [10] of 1 July 2021; both motors with the mains supply and converter-fed motors from 0.75 to 1000 kW with the number of poles from 2 to 8 must correspond to at least IE3 class. In many other countries, the IE3 efficiency class has already been adopted as mandatory for newly installed motors, as Table 1 suggests [11,12]. Moreover, in accordance with [10], the European Union plans to adopt the IE4 class as mandatory for motors with a power rating from 75 to 200 kW with the number of poles from 2 to 6.

Table 1. Adopting national regulations for the mandatory IE3 class.

Country	Power Range	First Year of Implementation	Country	Power Range	First Year of Implementation
Switzerland, Turkey	0.75–375 kW	2017	Japan	0.75–375 kW	2014
USA	0.75–200 kW	2007	Saudi Arabia	0.75–375 kW	2018
USA	0.18–2.2 kW	2015	Brazil	0.75–185 kW	2017
Canada	0.75–150 kW	2017	Taiwan	0.75–200 kW	2016
Mexico	0.75–375 kW	2010	Singapore	0.75–375 kW	2013
South Korea	0.75–200 kW	2017	-	-	-

It is known that electrically driven fluid-processing mechanisms such as pumps, fans, and compressors use about 70% of electric power consumed by all electric motors [6]. In this regard, reducing the energy consumption of pumping units becomes critical. There are many ways to reduce the energy consumption of a pump unit. For instance, it is well known that the reduction of their energy consumption is possible with the help of variable-speed drives (VSDs) [13]. A further improvement in the energy efficiency of a pump unit equipped with a VSD can be achieved through the use of a more energy-efficient electric motor.

2. Literature Review and Novelty of the Study

The energy efficiency of three-phase AC electric motors such as induction motors (IMs, Figure 1a), permanent magnet synchronous motors (PMSMs), and synchronous reluctance motors (SynRMs, Figure 1b) has been extensively compared in previous studies.

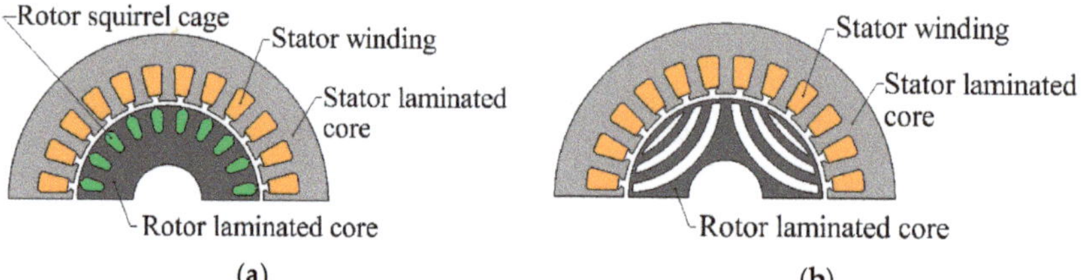

Figure 1. Machine general design representations: (**a**) IM; (**b**) SynRM.

Some studies were not devoted to the analysis of pumps driven by VSDs, but to the analysis of pumps with motors powered directly from the mains. For instance, in [14], a comparison of the efficiency and power factor of the IE1-IE3 induction motors and a direct-on-line permanent magnet motor of the IE4 class was provided. It was concluded that direct-on-line PMSMs could be competitively capable with IMs in the power range of ≤15 kW due to significantly higher efficiency, despite their significantly higher cost. The article claimed that PMSMs cost 2.2–38 times more than IE2 IMs. However, it should be noted that the paper was published more than 7 years ago, motor manufacturers have further mastered PMSM production, and the market price of PMSMs has dropped significantly. In any case, it is shown that PMSMs can have a payback period of fewer than 3 years, even if their cost is 2.3 times higher than IE2 IMs.

In [15], the energy consumption of induction and direct-on-line permanent magnet motors of the IE3 and IE4 energy-efficiency classes and with a power rating of 2.2 kW was assessed for a pump application. A number of energy-efficiency metrics were compared for these motors, such as annual energy costs and lifetime energy costs, as well as cost savings. The study demonstrated that for maximum energy savings, it was necessary to ensure high motor efficiency not only under maximum flow conditions, but also under reduced flow conditions, which have the longest duration in the typical duty cycle of pumping systems with VSDs. In [16], the energy consumption and payback period of IE1 and IE2 induction motors with a rated power of 15 kW were compared. However, CO_2 emissions were not evaluated in [12–16].

A number of papers were devoted to accessing the energy consumption of pump units controlled by VSDs. In [17], the dependencies of efficiency on speed and torque were determined for a pump drive based on 15 kW IMs of IE3 class, 15 and 18.5 kW SynRMs of IE4 class, and 15 kW PMSM of IE5 class. A typical pump load profile with a variable flow rate was considered. Annual energy indicators such as electricity consumption and cost, as well as electricity savings, were assessed. Paper [18] compared experimental data on the energy consumption of IE1 and IE2 IMs and an IE3 SynRM with a power rating of 45 kW in a pumping application. In [19], an IE2 IM and an IE5 SynRM with a power rating of 0.75 kW were compared in the case of a VSD pump application. In [20], an IM and SynRM with a power rating of 1.1 kW were considered in a pump unit with an approximately constant flow rate. The annual energy consumption, annual electricity costs, and annual cost savings were determined. In [21], efficiency maps for an 11 kW IM and SynRM were experimentally obtained and compared.

Paper [22] was devoted to the comparison of the service life of induction motors of various IE classes in various conditions; for example, with an asymmetric and distorted power supply. Paper [23] showed the effect of an overvoltage and undervoltage imbalance on the temperature and performance of IMs of the classes IE2, IE3, and IE4. Paper [24] presented issues related to motor protection, and compared a 7.5 kW IM and PMSM under various operating conditions: thermal steady state at full and partial load, starting, locked rotor, voltage unbalance, and undervoltage. Paper [25] developed and validated simple energy models for pump systems with a VSD drive. In [26], an estimate of energy savings for single pump systems for a storage fill application was developed using various control strategies.

All the previous studies mentioned above [15–26] did not assess CO_2 emissions. In addition, in [15–24], the analyses were carried out only for one specific power rating of the pump and PDS, which did not allow the authors to draw conclusions about the energy savings, cost savings, and CO_2 emissions for the wide range of power ratings commonly used in the industry.

Based on the literature overview presented above, it can be concluded that the energy efficiency metrics and CO_2 emissions for IMs and SynRMs of the IE3 and IE4 classes over a wide range of power outputs in the application of variable speed pumps have not yet been investigated in previous studies.

This paper presents a comparative analysis of the main metrics of energy efficiency, cost savings, and CO_2 emissions of PDSs with four-pole IMs of the IE3 and IE4 classes and SynRMs of the IE4 class with rated outputs of 2.2, 15, 75, 200 kW in a variable-speed pump unit. An analysis of PMSM metrics was not included in this study, as they are rarely used in variable speed pumps due to their higher cost compared to IMs and SynRMs. The main advantage of PMSMs is their reduced weight and dimensions, but this is not in demand in the considered application [20]. The authors believe that the converter-powered PMSMs are more suitable for use in applications such as traction and servo drives. The power ratings of 75 and 200 kW were selected for the analysis, as they are the upper and lower limits of the mandatory range of IE4 motors according to the directive [10].

For the comparative analysis, data from catalogues of pump manufacturers [27–30] and a typical flow-time diagram for the application under consideration were used.

The pump characteristics from the catalogues were interpolated by the second-order polynomial, which was most often used in previous works [31–33]. The semianalytical pump model based on interpolation of catalogue data at the rated rotational speed is commonly used for pumping system analysis [15,33]. When analyzing pumping systems, methods of multicriteria optimization [34,35], statistical methods of reliability analysis [36], and neural networks [37] are also often used.

Typical pump operating conditions and flow-time diagrams are well described in the literature [38,39]. The flow-time diagram with the approximately constant flow rate given in [38,39] was selected for analysis as the most suitable for considered power range. As shown in Figure 2, it was considered: 75% flow for a quarter of the total duty cycle (below this case marked as $i = 1$, where i is the index of the duty cycle load point); 100% flow for half of the total duty cycle ($i = 2$); and 110% flow for another quarter of the total duty cycle ($i = 3$).

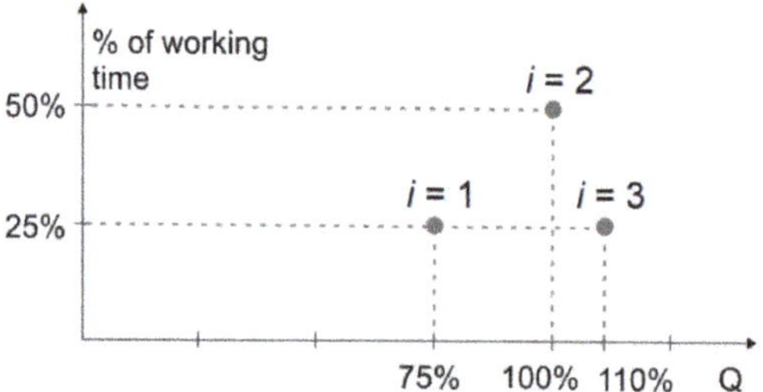

Figure 2. Flow versus time for the constant flow pump units [38,39].

The PDS efficiency depends on the torque and rotational speed. We considered the PDS as a power drive system in terms of the standard IEC 61800-9-2: 2017 [40]. It consisted of a complete drive module (CDM) and an electric motor. The CDM included a frequency converter and protection devices. Nowadays, sensorless field-oriented control is commonly used as a method for controlling AC motors in applications that do not require high-precision control, such as pumps and fans [41]. According to [40], the PDS power loss measurement must be provided at eight loading points (Figure 3a). Values of the motor efficiency under variable speed operations and partial load conditions can be calculated, for example, according to [42], but this does not provide values for the entire PDS efficiency. Some manufacturers already provide such information; therefore, their experimental data were used for analysis as the most reliable source. Information on PDS loss at these eight points is available in the online application Sinasave [43]. Since the range of the power rating of the Siemens SynRM was limited to 45 kW, a PDS based on ABB's SynRMs were considered for 75 and 200 kW ratings. Efficiency data for the ABB PDSs at 16 loading points (shown in Figure 3b) were provided in the manufacturer's statements [44,45].

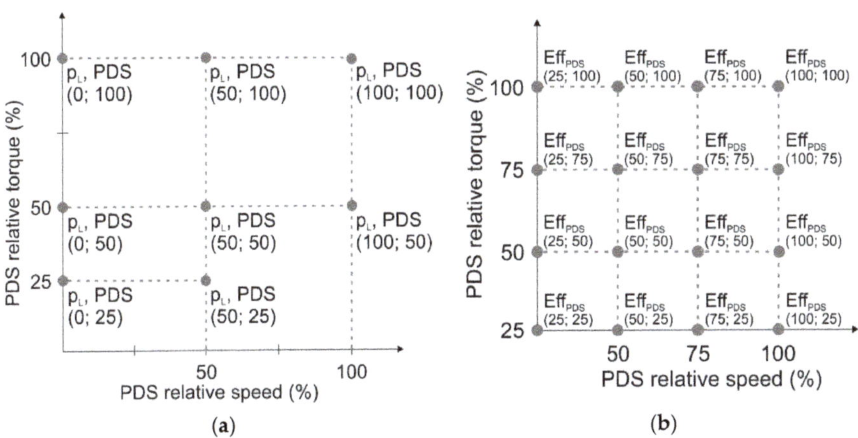

Figure 3. Data formats for PDS efficiency interpolation: (**a**) Siemens (losses according to IEC 61800-9-2:2017 [40]; source: Sinasave software [43]); (**b**) ABB (efficiency; source: manufacturer's statements [44,45]).

3. Evaluation of Pump and PDS Performances

Data from the Grundfos NB 50-200/210 [27], NB 80-315/305 [28], NB 150-400/375 [29], and NB 250-500/445 [30] pumps with the rated powers $P_{RATE.pump}$ consequently of 2.2, 15, 75, and 200 kW were considered in the analysis. All considered pumps had a rated rotational speed of 1450 rpm. The key parameters of these pumps at the best efficiency point (BEP) are specified in Table 2, where H is the pump head, Q is the pump flow rate, and η is the pump efficiency. Let us evaluate the power consumption of the PDS in combination with a pump of a certain power rating. First, a polynomial interpolation of the head-flow curve for the rated rotational speed was carried out according to Equation (1):

$$H(Q) = a_2 \cdot Q^2 + a_1 \cdot Q + a_0. \tag{1}$$

Table 2. Nameplate data of pumps.

No. of Pump	Type	$P_{RATE.pump}$, kW	$n_{RATE.pump}$, rpm	Q_{BEP}, m³/h	H_{BEP}, m	η_{BEP}, %
1	NB 50-200/210	2.2	1450	35.2	13.7	69.8
2	NB 80-315/305	15	1450	125.0	29.0	76.8
3	NB 150-400/375	75	1450	441.3	45.0	84.4
4	NB 250-500/445	200	1450	852.5	60.1	81.5

At nonrated rotational speed n_{pump}, according to the hydraulic affinity laws [33], which were applicable for the considered flow variation range (75 … 110%), the *H-Q* curve is described as:

$$H(Q) = a_2 \cdot Q^2 + a_1 \cdot \frac{n_{pump}}{n_{RATE.pump}} \cdot Q + a_0 \cdot \left(\frac{n_{pump}}{n_{RATE.pump}} \right)^2 \tag{2}$$

The rotational speeds required to provide 75% and 110% of the rated flow were determined according to the reference control curve *C(Q)* [38], which was a straight line (3) passing through points $(0, H_{BEP}/2)$ and (Q_{BEP}, H_{BEP}), where Q_{BEP} and H_{BEP} are the flow and head of the pump in the BEP:

$$C(Q) = k \cdot Q + b. \tag{3}$$

The rotational speed $n_{pump.i}$ (in rpm) for each required value of a flow was determined by jointly solving Equations (2) and (3). Using the results of the pump operating performances (n_i, Q_i, H_i) and the manufacturer's ISO efficiency curves [27–30], the pump efficiency $\eta_{pump.i}$ was evaluated at each operating point considered (75, 100, and 110% of the rated flow). According to Equations (4) and (5), the motor mechanical power and torque were evaluated at i-th loading point:

$$P_{mech.i} = \rho \cdot g \cdot Q_i \cdot H_i / \eta_{pump.i}. \tag{4}$$

$$T_{pump.i} = 9.55\, P_{mech.i} / n_{pump.i} \tag{5}$$

where $g = 9.81$ m/s^2; $\rho = 1000$ is the water density, kg/m^3; and $i = 1, 2, 3$ (see Figure 1).

The rated rotational speed $n_{RATED.motor}$ and rated torque $T_{RATED.motor}$ of the motors employed in the considered PDSs are shown in Table 3.

Table 3. Catalogue data of motors.

m	Type of Motor, IE Class	$n_{RATED.motor}$, rpm				$T_{RATED.motor}$, N·m			
	Rated Output, kW	**2.2**	**15**	**75**	**200**	**2.2**	**15**	**75**	**200**
1	SynRM, IE4	1500	1500	1500	1500	14.0	95.0	478	1272
2	IM, IE4	1465	1480	1490	1490	14.3	97.0	480	1280
3	IM, IE3	1465	1475	1485	1488	14.3	97.0	480	1280

In the case of the data format shown in Figure 3a, the loss of the m-th PDS at the i-th load point $P_{loss.i.m}$ was calculated using two-dimensional interpolation of the initial loss points according to the methodology given in [40]. In the case of the data format shown in Figure 3b, the traditional bilinear interpolation was used. The PDS efficiency $\eta_{PDS.i.m}$ was determined using Equation (6):

$$\eta_{PDS.i.m} = P_{mech.i} / (P_{mech.i} + P_{loss.i.m}). \tag{6}$$

4. PDS Energy Consumption

The mechanical power obtained according to Equation (4) and the interpolated PDS efficiency were used to calculate the electrical power consumed from the grid (7):

$$P_{1.i.m} = P_{mech.i} / \eta_{PDS.i.m}. \tag{7}$$

The annual electricity consumption of each PDS with the adopted flow-time diagram (Figure 2) of the pump unit was determined according to Equation (8):

$$E_{y.m} = 365 \cdot t_\Sigma \cdot \sum_{i=1}^{3} \left(P_{1.i.m} \cdot \frac{t_i}{t_\Sigma} \right). \tag{8}$$

where t_i is the time of running the pump at each point of the flow-time diagram; and t_Σ is the total duration of the daily operation of the pump (24 h).

The electricity cost was calculated using Equation (9):

$$C_{y.m} = E_{y.m} \cdot GT. \tag{9}$$

where $GT = $ EUR 0.1781/kWh is the electricity tariff for nonhousehold consumers in Germany in the first half of 2020, excluding VAT [46].

To compare the cost of energy consumed by different electric motors in a certain pump unit, the difference in the electricity cost relative to case $m = 3$ (PDS with the IE3 motor) was calculated as Equation (10):

$$S_{y.3m} = C_{y.3} - C_{y.m}. \tag{10}$$

It is known that the service life of a pump unit is about 15–20 years, and the most significant contribution to the lifetime expenses of a pump is made by the cost of consumed electricity, which can exceed 50–60% of the total cost [47,48]. In this study, we assumed that the design life of the pump unit was $n = 20$ years.

The net present value (*NPV*) for the lifetime energy cost was calculated according to Equation (11):

$$C_{LCCen.m} = C_{y.m}/(1 + (y - p))^n,$$ (11)

where $p = 0.02$ is the assumed annual inflation rate; and $y = 0.04$ is the interest rate [47,48].

The lifetime cost savings of PDSs with different IE4 motors ($m = 1, 2$) compared to a PDS with IE3 motors ($m = 3$) was evaluated as:

$$\Delta C_{LCCen.3m} = C_{LCCen.3} - C_{LCCen.m}.$$ (12)

5. Evaluation of CO_2 Emission Intensity

The yearly CO_2 emission intensity was evaluated according to Equation (13):

$$CDE_{y.m} = E_{y.m} \cdot EFE.$$ (13)

where $EFE = 418.8\ \text{g/kW·h}$ is the CO_2 emission factor for electricity consumption for Germany [49]. The yearly avoided CO_2 emissions were estimated as:

$$\Delta CDE_{y.3m} = CDE_{y.3} - CDE_{y.m}.$$ (14)

The emission intensity estimation using Equations (13) and (14) did not take into account the total volume of emissions. CO_2 emissions arise from the consumption of primary energy sources in power plants. Therefore, to assess the CO_2 emissions associated with the electricity calculated by Equation (8), it was necessary to use the primary energy factor (PEF) characterizing the average efficiency of conversion of primary energy into the final one [50,51]. Therefore, the annual emission including $PEF\ CDE^*_{y.m}$ and the corresponding avoided CO_2 emissions $\Delta CDE^*_{y.3m}$ were evaluated as the follows:

$$CDE^*_{y.m} = CDE_{y.m}\ PEF.$$ (15)

$$\Delta CDE^*_{y.3m} = CDE^*_{y.3} - CDE^*_{y.m}.$$ (16)

where $PEF = 2.2$ is the primary energy factor evaluated.

A default PEF value of 2.5 should be applied in accordance with directive 2012/27/EU [50]. However, recent studies indicated the need to revise this factor to 1.8–2.2 [52,53]. Based on this, $PEF = 2.2$ was adopted in this study.

6. Results and Discussions

Figure 4 shows the catalogue points of the pump head-flow characteristic calculated according to Equation (2) head-flow curves and reference control curve for the 2.2 kW pump as an example. Table 4 shows the calculation results for the pump performances during the operating cycle. Table 5 shows the interpolated PDS efficiency at the pump operating points indicated in Figure 2 for the four power ratings considered.

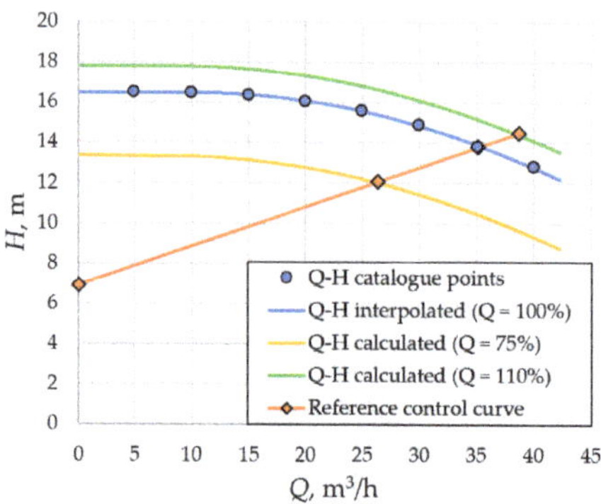

Figure 4. Calculated head-flow curves of the pump at 75, 100, and 110% flow and initial points of the 2.2 kW pump, and reference control curve.

Table 4. Performances of the pumps of different power ratings during the operating cycle.

No. of Pump	1 (NB 50-200/210, 2.2 kW)			2 (NB 80-315/305, 15 kW)			3 (NB 150-400/375, 75 kW)			4 (NB 250-500/445, 200 kW)		
No. of Load Point (i)	1	2	3	1	2	3	1	2	3	1	2	3
Q_i, %	75	100	110	75	100	110	75	100	110	75	100	110
Q_i, m^3/h	23.4	35.2	38.7	93.8	125.0	137.5	331.0	441.3	485.4	639.4	852.5	937.8
H_i, m	12.0	13.7	14.4	25.4	29.0	30.5	39.4	45.0	47.2	52.6	60.1	63.1
a_2		-0.0038			-5.5371×10^{-4}			-8.8848×10^{-5}			-3.7879×10^{-5}	
a_1		0.066			0.0398			0.0318			0.286	
a_0		16.27			33.18			48.83			63.00	
k		0.195			0.116			0.051			0.0352	
b		6.855			14.5			22.49			30.05	
$n_{pump.i}$, rpm	1310	1450	1508	1310	1450	1508	1310	1450	1508	1310	1450	1508
$n_{pump.i}$, %	90.3	100.0	104.0	90.3	100.0	104.0	90.3	100.0	104.0	90.3	100.0	104.0
$\eta_{pump \cdot i}$, %	68.9	69.8	69.5	75.9	76.8	76.3	82.9	84.4	84.1	79.6	81.5	81.2
$P_{mech.i}$, kW	1.25	1.88	2.18	8.54	12.86	14.95	42.82	64.09	74.29	115.1	171.3	198.6
$P_{mech \cdot i}$, %	56.9	85.5	99.2	56.9	85.7	99.7	57.1	85.5	99.1	57.6	85.7	99.3
$T_{pump.i}$, N·m	9.18	12.44	13.87	62.50	85.32	95.52	313.2	424.4	473.3	838.1	1126	1255

Table 5. Interpolated efficiencies of selected power drive systems.

m	Type of Motor, IE Class	Interpolated Motor Efficiency $\eta_{M.i.m}$, % in the Load Points											
		2.2 kW			15 kW			75 kW			200 kW		
		1	2	3	1	2	3	1	2	3	1	2	3
1	SynRM, IE4	86.8	88.5	89.1	91.6	92.2	92.4	93.1	93.5	93.6	94.2	94.5	94.4
2	IM, IE4	84.1	85.4	85.7	89.9	90.4	90.6	92.4	92.7	92.8	93.7	93.8	93.8
3	IM, IE3	82.1	82.4	82.6	88.5	89.1	89.3	91.8	91.9	92.0	93.1	93.2	93.2

Table 6 summarizes the results of calculations using Equations (8)–(12). The annual energy consumption is shown in Figure 5. According to these results, the PDS based on SynRM demonstrated the lowest annual electricity consumption and the highest cost savings. Based on the annual $S_{y.3m}$ and lifetime savings $\Delta C_{LCCen.3m}$, replacing the IE3 IM with the IE4 SynRM provided approximately 2 times more savings than replacing it with the IE4 IM for all power ratings considered. Table 7 summarizes the results for the CO_2 emissions according to Equations (13)–(16). The results for CO_2 emissions in Table 7 showed similar patterns; i.e., the PDS with IE4 SynRM allowed the highest avoided annual emissions: $\Delta CDE_{y.3m}$ and $\Delta CDE^*_{y.3m}$.

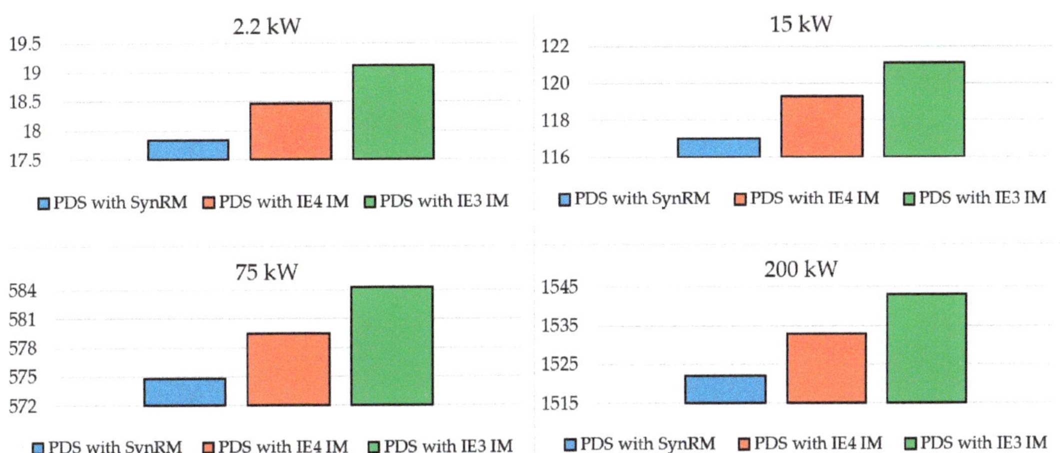

Figure 5. Annual energy consumption (MW·h).

Table 6. Results for the PDS electricity costs and cost savings.

m	Type of Motor, IE Class	$E_{y.m}$, MW·h	$C_{y.m}$, k€	$S_{y.3m}$, k€	$C_{LCCen.m}$, k€	$\Delta C_{LCCen.3m}$, k€
			2.2 kW			
1	SynRM, IE4	17.84	3.18	0.23	51.94	3.743
2	IM, IE4	18.47	3.29	0.12	53.80	1.884
3	IM, IE3	19.12	3.41	-	55.68	-
			15 kW			
1	SynRM, IE4	117.0	20.83	0.73	340.6	11.90
2	IM, IE4	119.2	21.24	0.32	347.3	5.216
3	IM, IE3	121.0	21.56	-	352.5	-
			75 kW			
1	SynRM, IE4	574.8	102.38	1.66	1674	27.22
2	IM, IE4	579.4	103.20	0.846	1688	13.83
3	IM, IE3	584.2	104.04	-	1702	-
			200 kW			
1	SynRM, IE4	1522	271.14	3.62	4437	59.12
2	IM, IE4	1533	272.97	1.79	4464	29.21
3	IM, IE3	1543	274.76	-	4493	-

Table 7. Results of calculating the annual CO_2 emissions.

m	Type of Motor, IE Class	Emissions Considering the Final Energy		Emissions Considering the Primary Energy	
		$CDE_{y.m}$, Tons	$\Delta CDE_{y.3m}$, Tons	$CDE^*_{y.m}$, Tons	$\Delta CDE^*_{y.3m}$, Tons
		2.2 kW			
1	SynRM, IE4	7.465	0.538	16.43	1.184
2	IM, IE4	7.737	0.271	17.02	0.596
3	IM, IE3	8.008	-	17.62	-
		15 kW			
1	SynRM, IE4	48.98	1.711	107.75	3.764
2	IM, IE4	49.94	0.750	109.87	1.650
3	IM, IE3	50.69	-	111.52	-
		75 kW			
1	SynRM, IE4	240.7	3.914	529.64	8.611
2	IM, IE4	242.7	1.989	533.87	4.376
3	IM, IE3	244.7	-	538.24	-
		200 kW			
1	SynRM, IE4	637.6	8.502	1402.7	18.70
2	IM, IE4	641.9	4.201	1412.2	9.242
3	IM, IE3	646.1	-	1426.6	-

The percentages shown in Table 8 and Figure 6 were the same for the cost savings and avoided CO_2 emissions.

Table 8. Cost savings (avoided CO_2 emissions, in %) relative to the PDS with an IE3 motor.

m	Type of Motor, IE Class	2.2 kW	15 kW	75 kW	200 kW
1	SynRM, IE4	6.72	3.38	1.60	1.32
2	IM, IE4	3.38	1.48	0.81	0.65

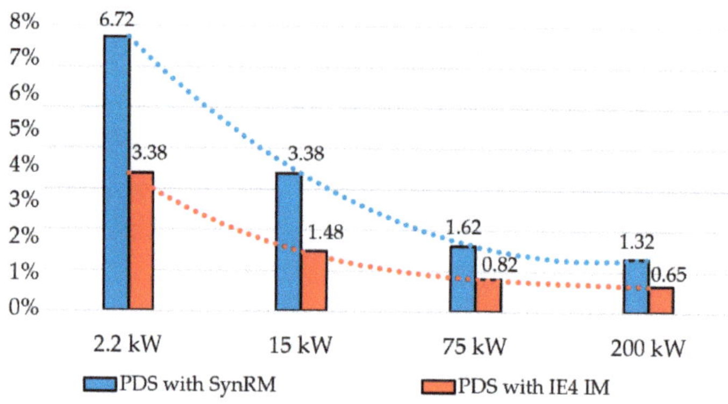

Figure 6. Cost savings/avoided CO_2 emissions (in %) for an output power range of 2.2–200 kW relative to the PDS with an IE3 motor.

As seen in Figure 6, the relative savings were about 5 times greater for low-power motors compared to high-power motors. Nevertheless, according to [54], for Europe, in the power range of 0.75–200 kW, the share of motors with rated power from 0.75 (1 hp) to 7.5 kW (10 hp) is 79.1%, and the share of motors with rated power from 7.5 (10 hp) to 75 kW (100 hp) is only 19.8%. Therefore, to fulfill the tasks of the European Green Deal, it is recommended to legislatively expand the mandatory power range for electric motors of the IE4 class to at least 7.5–200 kW, or even to 0.75–200 kW. However, the authors are aware that in the range of 0.75–7.5 kW, it is more difficult to achieve the IE4 class for induction motors [55].

Using the results shown in Table 8, the interpolation polynomials (17) for the PDS with SynRM and (18) for the PDS with IE4 IM were obtained, which can be used to calculate the cost savings, CS, % (avoided CO_2 emissions taking into account the final energy, ΔCDE, %; avoided CO_2 emissions taking into account the primary energy, ΔCDE^*, %) for intermediate power ratings not specified in Table 8, without carrying out a detailed calculation using Equations (1)–(16).

$$CS_{SynRM} = -0.02 \cdot P^3 + 0.91 \cdot P^2 - 5.93 \cdot P + 11.76, \tag{17}$$

$$CS_{IE4\ IM} = -0.125 \cdot P^3 + 1.37 \cdot P^2 - 5.135 \cdot P + 7.27. \tag{18}$$

7. Payback Period Assessment

Studies [56,57] described SynRMs of the IE4 class designed using the same stator as an IE3 IM. Based on this, the cost of the IE4 SynRM was taken to be equal to the cost of the IE3 IM. At the same time, the SynRM has a simpler rotor design (Figure 1), is more reliable, and requires fewer active materials for their production [58,59]. The SynRM was characterized by higher efficiencies at lower loads than IMs [55,59]. The IE4 energy efficiency class (and even IE5) for SynRM can be achieved relatively easily, without the use of expensive solutions typical for the class IE4 of IMs [55]: the use of copper conductors in the rotor, the use of higher-quality electrical steel, and an increase in the mass and volume of active materials, which, in turn, leads to an increase in the dimensions of the motor and, in some cases, the transition to the next frame size for the same rated power. Thus, the SynRM can be expected to cost no more than the IE3 class IM in mass production.

The disadvantages of SynRMs are that they have a slightly lower power factor than IMs [59], and therefore a slightly higher rated current. The current values of the considered electric motors, taken from the manufacturers' catalogues, are shown in Table 9. The rated currents of the IE3 IM and the IE4 IM of the same power rating were approximately the same. The rated current of the IE4 SynRM was 15–33% higher than the IE3 IM, which was higher than the overcurrent limit usually provided by frequency converter manufacturers. Due to this, it is required to use a more expensive converter with a higher power rating when using a SynRM, in comparison with an IM. The cost of the converter turned out to be 10–40% higher, which led to an increase in the cost of the PDS.

Let us calculate the payback period of using a PDS based on a SynRM and IE4 IM instead of PDS based on IE3 IM, for two cases, according to Formulas (19) and (20), where $C_{iic.m}$ is the turnkey costs of the m-th PDS, which are presented in Table 9. Formula (19) describes the calculation of the payback period for the case of a newly commissioned pump unit. Replacing the motor in a pump unit already put into operation is implied when calculating according to Formula (20):

$$T_{m1} = (C_{iic.m} - C_{iic.3})/S_{y.3m}, \tag{19}$$

$$T_{m2} = C_{iic.m}/S_{y.3m}. \tag{20}$$

To evaluate the cost of frequency converters and IE3 class IMs, their market prices were taken from manufacturers' websites [58,60]. It was assumed that the cost of the SynRM was the same as the cost of the IE3 IM. An analysis of recent studies [5,61–63] and available market prices showed that the difference in the turnkey price of motors of related energy efficiency classes was usually 15–30%. In this paper, it was assumed that the cost of the IE4 class IMs was 22.5% higher than the cost of the IE3 class IMs. Table 9 shows the assumed costs. For the considered power range, the cost of the frequency converter, as shown in Table 9, was 60–140% of the motor cost, and the difference in the price of the PDS based on SynRM and IM was in the range of −8–+5%. Based on the results in

Table 9, it can be concluded that the PDSs with SynRM and IE4 IM had similar initial costs. However, despite this, the payback period of the SynRM turned out to be significantly shorter (Figure 7) due to the greater energy savings, $S_{y.3m}$.

Table 9. Motor payback period.

m	Motor Type	Motor Current, A	Motor Price, €	Converter Price, €	PDS Price, $C_{iic.m}$, €	Payback Time 1, T_{m1}, Years	Payback Time 2, T_{m2}, Years
			2.2 kW				
1	SynRM, IE4	5.7	286	242	528	0.09	2.32
2	IM, IE4	4.5	351	222	573	0.56	4.95
3	IM, IE3	4.4	286	222	508	-	-
			15 kW				
1	SynRM, IE4	32.9	763	818	1581	0.34	2.22
2	IM, IE4	29	935	575	1510	0.54	4.71
3	IM, IE3	28.5	763	575	1338	-	-
			75 kW				
1	SynRM, IE4	173	2954	3085	6039	0.27	3.61
2	IM, IE4	133	3618	2637	6255	0.78	7.32
3	IM, IE3	130	2954	2637	5590	-	-
			200 kW				
1	SynRM, IE4	427	6883	9468	16,351	0.38	4.51
2	IM, IE4	345	8431	8092	16,524	0.87	9.28
3	IM, IE3	345	6883	8092	14,975	-	-

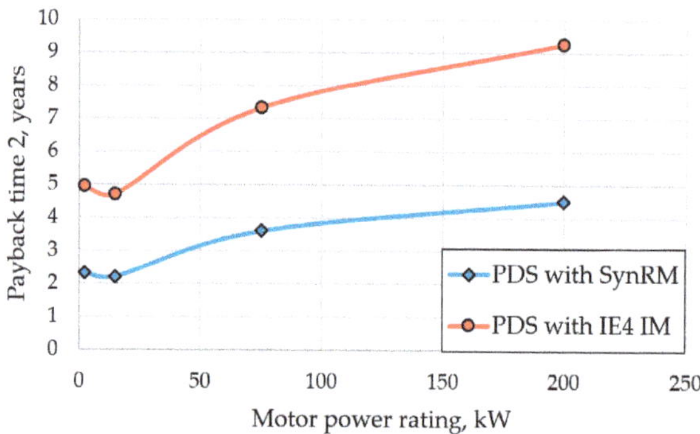

Figure 7. Payback time versus motor power rating.

8. Conclusions

This study provided a comparative assessment of metrics of energy consumption and CO_2 emission for various PDSs with a rated power in the range of 2.2–200 kW. The converter-fed PDSs with the IE3 IM, IE4 IM, and SynRM are considered. The analysis was based on extensive data from the technical specifications of the PDS [43–45] and pump manufacturers [27–30]. These data were processed using mathematical methods of polynomial and bilinear interpolation, regression, and a semianalytical model of a variable-speed pump unit. As a result of the analysis, the interpolating polynomials were constructed to evaluate the metrics of interest for any PDS from the entire considered power range.

This study showed that to achieve the goals of reducing greenhouse gas emissions, it is extremely important to use electric motors of higher energy-efficiency classes. The state of the art (IMs and SynRMs of the IE4 class) already allows effectively implementing the requirements of the European Commission [10], prescribing the mandatory use of IE4 class motors in the rated power range of 75–200 kW from 1 July 2023. The results showed that when replacing IE3 motors with IE4 motors, the largest reduction in CO_2 emissions was shown by the low-power motors, which make up the vast majority of electric motors in service [54]. This justifies the mandatory use of IE4 motors for lower power ratings.

In addition, it was shown that the SynRMs of the IE4 class demonstrated significantly better indicators of reduction of CO_2 emissions and energy savings than the IMs of the same energy-efficiency class. Moreover, SynRMs have several operational advantages, such as their simpler design and manufacturing technology, at a price comparable to the price of the IE3 IMs. As a result, when replacing an IE IM, the energy savings of an IE4 SynRM will be approximately 2 times higher, and the payback period will be approximately half that of the IE4 IM for the entire considered power range. At the same time, the influence of the SynRM's lower power factor compared to IMs and the associated need for frequency converters with a higher power rating does not significantly affect the payback period.

According to the results of this study, in order to achieve the goals of the European Green Deal [1,2], it is recommended to legislatively expand the mandatory power range of electric motors of IE4 class from 75–200 kW to at least 7.5–200 kW.

In future work, the potential for significant reductions in greenhouse gas emissions in other mass applications of electric machines will be investigated. For example, fans and compressors are in principle similar to pumps, but have different typical operating conditions. The subject of future work may be an assessment of the energy-saving potential, payback periods, and CO_2 emissions for various typical operating profiles of such equipment. Moreover, a fast-growing quantity of electric vehicles also consumes an increasing share of electricity, which makes analysis in this area relevant.

In addition, on the basis of the developed model, it is possible to optimize the operating points of multipump units, which include pumps operating in parallel on the same pipeline.

Author Contributions: Conceptual approach, V.D. and V.P.; data curation V.D. and V.G.; calculations and modeling, V.G. and V.K.; writing of original draft, V.D., V.G., V.K. and V.P.; visualization, V.G. and V.K.; review and editing, V.D., V.G., V.K. and V.P. All authors have read and agreed to the published version of the manuscript.

Funding: The work was partially supported by the Ministry of Science and Higher Education of the Russian Federation (through the basic part of the government mandate, Project No. FEUZ 2020-0060).

Institutional Review Board Statement: Not applicable.

Informed Consent Statement: Not applicable.

Data Availability Statement: In this article, for the analysis, the technical specifications of the manufacturers of electric motors and pumps were used [27–30,43–45].

Conflicts of Interest: The authors declare no conflict of interest. The funders had no role in the design of the study; in the collection, analyses, or interpretation of data; in the writing of the manuscript; or in the decision to publish the results.

References

1.   The European Green Deal. Available online: https://ec.europa.eu/info/strategy/priorities-2019-2024/european-green-deal_en (accessed on 23 January 2021).
2.   European Council Meeting (10 and 11 December 2020)—Conclusions. EUCO 22/20 CO EUR 17 CONCL 8. Available online: https://www.consilium.europa.eu/media/47296/1011-12-20-euco-conclusions-en.pdf (accessed on 23 January 2021).
3.   Chodakowska, E.; Nazarko, J. Assessing the performance of sustainable development goals of EU countries: Hard and soft data integration. *Energies* **2020**, *13*, 3439. [CrossRef]
4.   Siksnelyte, I.; Zavadskas, E.K. Achievements of the European Union countries in seeking a sustainable electricity sector. *Energies* **2019**, *12*, 2254. [CrossRef]
5.   Commission Staff Working Document Impact Assessment Accompanying the Document Commission Regulation (EU) 2019/1781 Laying Down Ecodesign Requirements for Electric Motors and Variable Speed Drives Pursuant to Directive 2009/125/EC of the European Parliament and of the Council and Repealing Commission Regulation (EC) No 640/2009. SWD/2019/0343 Final. Available online: https://ec.europa.eu/transparency/regdoc/rep/10102/2019/EN/SWD-2019-343-F1-EN-MAIN-PART-1.PDF (accessed on 23 January 2021).

6. Van Werkhoven, M.; Werle, R.; Brunner, C.U. 4E EMSA Policy guidelines for motor driven units: Pumps, fans and compressors. In Proceedings of the 10th International Conference on Energy Efficiency in Motor Driven System (EEMODS' 2017), Rome, Italy, 6–8 September 2017; Publications Office of the European Union: Luxembourg, 2018. Available online: http://publications.jrc.ec.europa.eu/repository/bitstream/JRC110714/eemods_2017_proceedings_v11(1).pdf (accessed on 23 January 2021). [CrossRef]
7. Rotating Electrical Machines—Part 30-1: Efficiency Classes of Line Operated AC Motors (IE Code). IEC 60034-30-1/ Ed. 1; IEC: 2014-03. Available online: https://webstore.iec.ch/publication/136 (accessed on 23 January 2021).
8. Rotating Electrical Machines—Part 30-2: Efficiency Classes of Variable Speed AC Motors (IE-Code) IEC 60034-30-2/ IEC: 2016-12. Available online: https://webstore.iec.ch/publication/30830 (accessed on 23 January 2021).
9. European Commission Regulation (EC), No. 640/2009 Implementing Directive 2005/32/ EC of the European Parliament and of the Council with Regard to Ecodesign Requirements for Electric Motors, (2009), Amended by Commission Regulation (EU) No 4/2014 of January 6, 2014. Document 32014R0004. Available online: https://eur-lex.europa.eu/legal-content/EN/TXT/?uri=CELEX%3A32014R0004 (accessed on 23 January 2021).
10. Commission Regulation (EU) 2019/1781 of 1 October 2019 Laying Down Ecodesign Requirements for Electric Motors and Variable Speed Drives Pursuant to Directive 2009/125/EC of the European Parliament and of the Council, Amending Regulation (EC) No 641/2009 with Regard to Ecodesign Requirements for Glandless Standalone Circulators and Glandless Circulators Integrated in Products and Repealing Commission Regulation (EC) No 640/2009. Available online: https://eur-lex.europa.eu/legal-content/EN/TXT/PDF/?uri=CELEX:32019R1781&from=EN (accessed on 23 January 2021).
11. Efficiency Regulations for Motors: International Norms. NORD DRIVESYSTEMS Group, S4700 Part. No. 6069202 / 4019. Available online: https://www.nord.com/cms/media/documents/bw/S4700_6069202_4019_Screen.pdf (accessed on 23 January 2021).
12. How International Standards for Electric Motor Systems Support Policies of Countries Using These in Their Regulations. Available online: https://www.iec.ch/government-regulators/electric-motors (accessed on 23 January 2021).
13. Shankar, V.K.A.; Umashankar, S.; Paramasivam, S.; Hanigovszki, N. A comprehensive review on energy efficiency enhancement initiatives in centrifugal pumping system. *Appl. Energy* **2016**, *181*, 495–513. [CrossRef]
14. Almeida, A.; Ferreira, F.; Duarte, A. Technical and economical considerations on super high-efficiency three-phase motors. *IEEE Trans. Ind. Appl.* **2014**, *50*, 1274–1285. [CrossRef]
15. Goman, V.; Oshurbekov, S.; Kazakbaev, V.; Prakht, V.; Dmitrievskii, V. Energy Efficiency analysis of fixed-speed pump drives with various types of motors. *Appl. Sci.* **2019**, *9*, 5295. [CrossRef]
16. Oshurbekov, S.; Kazakbaev, V.; Prakht, V.; Dmitrievskii, V. Comparative study of energy consumption of 15 kW induction motors of IE1 and IE2 efficiency classes in pump applications. In Proceedings of the XI International Conference on Electrical Power Drive Systems (ICEPDS), Saint-Petersburg, Russia, 4–7 October 2020; IEEE: Piscataway, NJ, USA, 2020; pp. 1–6. [CrossRef]
17. Ahonen, T.; Orozco, S.M.; Ahola, J.; Tolvanen, J. Effect of electric motor efficiency and sizing on the energy efficiency in pumping systems. In Proceedings of the 18th European Conference on Power Electronics and Applications (EPE'16 ECCE Europe), Karlsruhe, Germany, 6–8 September 2016; IEEE: Piscataway, NJ, USA, 2016; pp. 1–9. [CrossRef]
18. Van Rhyn, P.; Pretorius, J.H.C. Utilising high and premium efficiency three phase motors with VFDs in a public water supply system. In Proceedings of the IEEE 5th International Conference on Power Engineering, Energy and Electrical Drives (POWERENG), Riga, Latvia, 11–13 May 2015; IEEE: Piscataway, NJ, USA, 2015; pp. 497–502. [CrossRef]
19. Safin, N.; Kazakbaev, V.; Prakht, V.; Dmitrievskii, V. Calculation of the efficiency and power consumption of induction IE2 and synchronous reluctance IE5 electric drives in the pump application based on the passport specification according to the IEC 60034-30-2. In Proceedings of the 25th International Workshop on Electric Drives: Optimization in Control of Electric Drives (IWED), Moscow, Russia, 31 January–2 February 2018; IEEE: Piscataway, NJ, USA, 2018; pp. 1–5. [CrossRef]
20. Kazakbaev, V.; Prakht, V.; Dmitrievskii, V.; Ibrahim, M.N.; Oshurbekov, S.; Sarapulov, S. Efficiency analysis of low electric power drives employing induction and synchronous reluctance motors in pump applications. *Energies* **2019**, *12*, 1144. [CrossRef]
21. Rassolkin, A.; Heidari, H.; Kallaste, A.; Vaimann, T.; Acedo, J.P.; Romero-Cadaval, E. Efficiency map comparison of induction and synchronous reluctance motors. In Proceedings of the 26th International Workshop on Electric Drives: Improvement in Efficiency of Electric Drives (IWED), Moscow, Russia, 30 January–2 February 2019; IEEE: Piscataway, NJ, USA, 2019; pp. 1–4. [CrossRef]
22. Ferreira, F.J.T.E.; Baoming, G.; De Almeida, A.T. Reliability and operation of high-efficiency induction motors. *IEEE Trans. Ind. Appl.* **2016**, *52*, 4628–4637. [CrossRef]
23. Tabora, J.M.; De Lima Tostes, M.E.; De Matos, E.O.; Bezerra, U.H.; Soares, T.M.; De Albuquerque, B.S. Assessing voltage unbalance conditions in IE2, IE3 and IE4 classes induction motors. *IEEE Access* **2020**, *8*, 186725–186739. [CrossRef]
24. Ferreira, F.J.T.; Leprettre, B.; De Almeida, A.T. Comparison of protection requirements in IE2-IE3-and IE4-class motors. *IEEE Trans. Ind. Appl.* **2016**, *52*, 3603–3610. [CrossRef]
25. Wang, G. Data-driven energy models for existing VFD-motorpump systems. *Sci. Technol. Built Environ.* **2019**, *25*, 732–742. [CrossRef]
26. Hieninger, T.; Goppelt, F.; Schmidt-Vollus, R.; Schlucker, E. Energy-saving potential for centrifugal pump storage operation using optimized control schemes. *Energy Effic.* **2021**, *14*, 23. [CrossRef]
27. Grundfos Product Center: Technical Data and Curves NB 50-200/210 AF2ABAQE. Available online: https://product-selection.grundfos.com/products/nb-nbe-nbe-series-2000/nb/nb-50-200210-97837025?tab=variant-curves&pumpsystemid=1173717497 (accessed on 23 January 2021).

28. Grundfos Product Center: Technical Data and Curves NB 80-315/305 AF2ABAQE. Available online: https://product-selection.grundfos.com/products/nb-nbe-nbe-series-2000/nb/nb-80-315305-97839395?tab=variant-curves&pumpsystemid=1167816632 (accessed on 23 January 2021).

29. Grundfos Product Center: Technical Data and Curves NB 150-400/375 AF1ABAQE. Available online: https://product-selection.grundfos.com/products/nb-nbe-nbe-series-2000/nb/nb-150-400375-97837168?tab=variant-curves&pumpsystemid=1168447008 (accessed on 23 January 2021).

30. Grundfos Product Center: Technical Data and Curves NB 250-500/445 AF1ABAQE. Available online: https://product-selection.grundfos.com/products/nb-nbe-nbe-series-2000/nb/nb-250-500445-97921024?tab=variant-curves&pumpsystemid=1173934349 (accessed on 23 January 2021).

31. Tamminen, J.; Viholainen, J.; Ahonen, T.; Ahola, J.; Hammo, S.; Vakkilainen, E. Comparison of model-based flow rate estimation methods in frequency-converter-driven pumps and fans. *Energy Effic.* **2014**, *7*, 493–505. [CrossRef]

32. Liu, G.; Liu, M. Development of simplified in-situ fan curve measurement method using the manufacturers fan curve. *Build. Environ.* **2012**, *48*, 77–83. [CrossRef]

33. Nelik, L. *Centrifugal and Rotary Pumps. Fundamentals with Applications*; CRC Press: Boca Raton, FL, USA, 1999.

34. Wu, P.; Lai, Z.; Wu, D.; Wang, L. Optimization Research of Parallel Pump System for Improving Energy Efficiency. *J. Water Resour. Plan. Manag.* **2015**, *141*, 04014094. [CrossRef]

35. Arfaoui, J.; Rezk, H.; Al-Dhaifallah, M.; Elyes, F.; Abdelkader, M. Numerical Performance Evaluation of Solar Photovoltaic Water Pumping System under Partial Shading Condition using Modern Optimization. *Mathematics* **2019**, *7*, 1123. [CrossRef]

36. Goyal, N.; Ram, M.; Kumar, A.; Bisht, S.; Klochkov, Y. Reliability Measures and Profit Exploration of Windmill Water-Pumping Systems Incorporating Warranty and Two Types of Repair. *Mathematics* **2021**, *9*, 822. [CrossRef]

37. Belhaj Salem, M.; Fouladirad, M.; Deloux, E. Prognostic and Classification of Dynamic Degradation in a Mechanical System Using Variance Gamma Process. *Mathematics* **2021**, *9*, 254. [CrossRef]

38. Extended Product Approach for Pumps, Copyright © 2021 by Europump. Published by Europump. Available online: http://europump.net/uploads/Extended%20Product%20Approach%20for%20Pumps%20-%20A%20Europump%20guide%20(27OCT2014).pdf (accessed on 23 January 2021).

39. Stoffel, B. *Assessing the Energy Efficiency of Pumps and Pump Units. Background and Methodology*; Elsevier: Amsterdam, The Netherlands, 2015. [CrossRef]

40. *Adjustable Speed Electrical Power Drive Systems—Part 9-2: Ecodesign for Power Drive Systems, Motor Starters, Power Electronics and Their Driven Applications—Energy Efficiency Indicators for Power Drive Systems and Motor Starters*; IEC 61800-9-2/Ed1; IEC: Geneva, Switzerland, 2017.

41. Pellegrino, G.; Bojoi, R.; Guglielmi, P. Unified Direct-Flux Vector Control for AC Motor Drives. *IEEE Trans. Ind. Appl.* **2011**, *47*, 2093–2102. [CrossRef]

42. Li, Y.; Liu, M.; Lau, J.; Zhang, B. A novel method to determine the motor efficiency under variable speed operations and partial load conditions. *Appl. Energy* **2015**, *144*, 234–240. [CrossRef]

43. SinaSave Energy Saving and Amortization, Siemens Online Tool. Available online: https://www.sinasave.siemens.com (accessed on 23 January 2021).

44. Manufacturer's Statement ACS880-01 and IE4 SynRM Motor Package Efficiency. Drive: ACS880-01-202A-3, Motor: M3BL 280SMA, 3GBL282213-ADC, Pn 75 kW, 1500 rpm. Document No: FIVEN201506010267. Available online: https://library.e.abb.com/public/34f02c42d4b642b5888d22a429ef04c5/Manufacturers%20statement%20-%20IE4%20M3BL%20280SMA_ACS880_103A,%2075%20%20kW,%201500%20rpm.pdf (accessed on 23 January 2021).

45. Manufacturer's Statement ACS880-01 and IE4 SynRM Motor Package Efficiency. Drive: ACS880-01-427A-3, Motor: M3BL 315MLA, 3GBL312413-ADC, Pn 200 kW, 1500 rpm. Document No: FIVEN201506010275. Available online: https://library.e.abb.com/public/c00bb6eb084a40b3912317e345a73fe0/Manufacturers%20statement%20-%20IE4%20M3BL%20315MLA_ACS880_427A,%20200%20%20kW,%201500%20rpm.pdf (accessed on 23 January 2021).

46. Eurostat Data for the Industrial Consumers in Germany. Available online: http://ec.europa.eu/eurostat/statistics-explained/index.php/Electricity_price_statistics#Electricity_prices_for_industrial_consumers (accessed on 23 January 2021).

47. Pump Life Cycle Costs: A Guide to LCC Analysis for Pumping Systems, Executive Summary. (2001) Hydraulic Institute (Parsippany, NJ); Europump (Brussels, Belgium); Office of Industrial Technologies Energy Efficiency and Renewable Energy U.S. Department of Energy (Washington, DC). January 2001, pp. 1–19. Available online: https://searchworks.stanford.edu/view/4676735 (accessed on 23 January 2021).

48. Waghmode, L.; Sahasrabudhe, A. A comparative study of life cycle cost analysis of pumps. In Proceedings of the International Design Engineering Technical Conferences and Computers and Information in Engineering Conference (ASME 2010), Montreal, QC, Canada, 15–18 August 2010; Volume 6, pp. 491–500. [CrossRef]

49. CO_2 Intensity of Electricity Generation. European Environment Agency. Available online: https://www.eea.europa.eu/data-and-maps/data/co2-intensity-of-electricity-generation (accessed on 23 January 2021).

50. Directive 2012/27/EU of the European Parliament and of the Council of 25 October 2012 on Energy Efficiency, Amending Directives 2009/125/EC and 2010/30/EU and Repealing Directives 2004/8/EC and 2006/32/EC. Available online: https://eur-lex.europa.eu/legal-content/EN/TXT/?uri=celex%3A32012L0027 (accessed on 23 January 2021).

51. Wilby, M.R.; González, A.B.R.; Díaz, J.J.V. Empirical and dynamic primary energy factors. *Energy* **2014**, *73*, 771–779. [CrossRef]

52. Final Report. Evaluation of Primary Energy Factor Calculation Options for Electricity. Available online: https://ec.europa.eu/energy/sites/ener/files/documents/final_report_pef_eed.pdf (accessed on 23 January 2021).
53. Tucki, K.; Orynycz, O.; Mitoraj-Wojtanek, M. Perspectives for Mitigation of CO_2 Emission due to Development of Electromobility in Several Countries. *Energies* **2020**, *13*, 4127. [CrossRef]
54. de Castro Andrade, C.T.; Pontes, R.S.T. Economic analysis of Brazilian Policies for Energy Efficient Electric Motors. *Energy Policy* **2017**, *106*, 315–325. [CrossRef]
55. De Almeida, A.T.; Ferreira, F.J.; Baoming, G. Beyond Induction Motors—Technology Trends to Move Up Efficiency. *IEEE Trans. Ind. Appl.* **2014**, *50*, 2103–2114. [CrossRef]
56. Moghaddam, R.R.; Magnussen, F.; Sadarangani, C. Theoretical and Experimental Reevaluation of Synchronous Reluctance Machine. *IEEE Trans. Ind. Electron.* **2010**, *57*, 6–13. [CrossRef]
57. Dmitrievskii, V.; Prakht, V.; Kazakbaev, V.; Pozdeev, A.; Oshurbekov, S. Development of a high efficient electric drive with synchronous reluctance motor. In Proceedings of the 18th International Conference on Electrical Machines and Systems (ICEMS), Pattaya, Thailand, 25–28 October 2015; IEEE: Piscataway, NJ, USA, 2015; pp. 876–881. [CrossRef]
58. AC Electric Motor. Available online: https://www.acelectricmotor.co.uk/ (accessed on 23 January 2021).
59. Ozcelik, N.G.; Dogru, U.E.; Imeryuz, M.; Ergene, L.T. Synchronous Reluctance Motor vs. Induction Motor at Low-Power Industrial Applications: Design and Comparison. *Energies* **2019**, *12*, 2190. [CrossRef]
60. Three Phase VFD Pricelist. Available online: http://www.gohz.com/three-phase-vfd (accessed on 23 January 2021).
61. Report on Study on International Efficiency (IE) Efficiency Classes for Low Voltage AC Motors. Available online: https://www.emsd.gov.hk/filemanager/en/content_764/Report%20on%20International%20Efficiency%20Efficiency%20Classes%20for%20Low%20Voltage%20AC%20Motors.pdf (accessed on 23 January 2021).
62. Fact Sheet, No. 29–New Motor Technologies November 2018. Available online: https://www.topmotors.ch/sites/default/files/2018-11/E_MB_29_Motor_technologies.pdf (accessed on 23 January 2021).
63. Almeida, A. Motor Systems Technology Developments. In Proceedings of the 8th Motor Summit for Energy Efficient Motor Driven Systems Powered by Impact Energy, Zurich, Switzerland, 14–15 November 2018. Available online: https://motorsummit.ch/wp-content/uploads/2020/08/MS18_proceedings.pdf (accessed on 23 January 2021).

Article

Design of Constraints for Seeking Maximum Torque per Ampere Techniques in an Interior Permanent Magnet Synchronous Motor Control

Anton Dianov [1] and Alecksey Anuchin [2,*]

[1] This Is Engineering Inc., Seongnam 13449, Korea; anton.dianov@ieee.org
[2] Electric Drives Department, Moscow Power Engineering Institute, 111250 Moscow, Russia
* Correspondence: anuchinas@mpei.ru

Abstract: The efficient control of permanent magnet synchronous motors (PMSM) requires the development of a technique for loss optimization. The best approach is the implementation of power loss minimization algorithms, which are hard to model and design. Therefore, the developers typically involve maximum torque per ampere (MTPA) control, which optimizes Joule loss only. The conventional MTPA control requires knowledge of motor parameters and can only properly operate when these parameters are constant. However, motor parameters vary depending on operating conditions; thus, conventional techniques cannot be used. Furthermore, many industrial drives are designed for self-commissioning, and they do not have prior information on motor parameters. In order to solve this problem, various MTPA-seeking techniques, which track the minimum of motor current, have been developed. The dynamic performance between these seeking algorithms and maximum deviation from the true MTPA trajectory are defined by the constraints in most cases, in which proper design improves the dynamic behavior of MTPA-seeking algorithms. This paper considers a PMSM, which was designed to operate in the saturation area and whose MTPA trajectory significantly deviates from the same curve constructed for the initial unsaturated parameters. This paper considers existing approaches, explains their pros and cons, and demonstrates that these methods do not utilize full potential of the motor. A new constraint design was proposed and explained step by step. The experiment verifies the proposed technique and demonstrates improvements in efficiency and dynamic behavior of the seeking algorithm.

Keywords: synchronous motor; adaptive control; MTPA control; parameter variation; constraints design

Citation: Dianov, A.; Anuchin, A. Design of Constraints for Seeking Maximum Torque per Ampere Techniques in an Interior Permanent Magnet Synchronous Motor Control. *Mathematics* **2021**, *9*, 2785. https://doi.org/10.3390/math9212785

Academic Editor: Nicu Bizon

Received: 19 September 2021
Accepted: 29 October 2021
Published: 3 November 2021

Publisher's Note: MDPI stays neutral with regard to jurisdictional claims in published maps and institutional affiliations.

1. Introduction

Permanent magnet synchronous machines (PMSM) have a higher efficiency, torque-to-weight ratio, and power per volume value, which causes their popularity in motor drives, where efficiency and compactness are two of the main requirements. However, efficient control of PMSM requires full utilization of their potential and minimization of the consumed current at the commanded torque [1,2].

The feature of PMSMs is the presence of a reluctance torque component besides the main magnetic torque component. The reason for this feature is magnetic asymmetry along direct and quadrature axes, which is typically caused by the machine design. However, even-surface-mounted permanent magnet synchronous motors (SPMSM), which have equal direct and quadrature inductances at low-load conditions, typically demonstrate magnetic asymmetry under load. Therefore, efficient control systems have to consider this feature and utilize a reluctance torque component as well. Special control algorithms, which provide full utilization of motor torque, are called maximum torque per ampere (MTPA) techniques. These techniques differ by the type of operation (offline or online), usage of motor parameters (used or not required), etc., but an MTPA control became an

257

inevitable part of PMSM control systems, and it is hard to find a commercial motor drive without one of the MTPA algorithms.

The conventional approach to the calculation of the MTPA trajectory is differentiating the machine torque equation with respect to the amplitude of a stator current. For the purpose of simplification, the motor parameters are supposed to be stable and their dependencies on other variables are not considered [3,4]. However, motor parameters vary with the change of operating conditions: steel saturation impacts inductances, rotor temperature impacts flux-linkage, etc. Therefore, these variations have to be taken into account [5–7].

In order to adapt the MTPA control techniques to parameters variation, different online approaches were proposed. The authors of [8–15] proposed the usage of conventional MTPA equations together with the motor parameters online estimation techniques. The papers [8–10] considered motor inductance variation and proposed estimators for online inductance monitoring, whereas the papers [11,12] took into account only flux-linkage change and [13–15] considered simultaneous variation of these motor parameters. All these methods use the conventional MTPA equations obtained under the assumption that motor parameters are constant and do not vary; however, the motor parameters are updated using various estimation techniques. As a result, these approaches demonstrate the acceptable tolerance for motor drives where motor parameters vary slowly and their derivatives may be neglected, but they fail in the fast-dynamic systems, where motor parameter derivatives are significant. One more disadvantage of these methods is the necessity of motor parameter prior knowledge, which is used as a reference value for estimators and includes tuning of the adaptation mechanism.

In order to exclude these inconveniences, a group of techniques, which do not require motor parameters, have been proposed. These methods are called seeking algorithms, and they track minimum current consumption, which corresponds to the MTPA trajectory, using either injection of the additional signal [16–29] or perturbing the motor drive and analyzing response [30–38].

The algorithms proposed in [16–18] were designed for the direct torque control (DTC) and field-oriented control schemes, which use injections of high-frequency signals toward the motor, followed by an analysis of its response, which is used for extremum tracking. The main idea is to estimate the local torque derivative at a constant current and to find the position of a zero derivative, which corresponds to the MTPA condition. Similar methods adapted for open-loop v/f-constant control were studied in [21,22]. These techniques may operate without information of motor parameters. However, the knowledge of their approximate values may be used as a starting point for tuning and may significantly improve the dynamic.

The main disadvantage of injection-based techniques is noise and vibration caused by signal injection, which significantly restricts the application area of these methods; therefore, recently, a virtual signal injection control (VSIC) [23–29] was proposed, which eliminates the abovementioned drawback. The methods of this group inject a high-frequency signal into the system mathematically, without real modification of the signals applied to the motor. They use measured parameters: voltage current, speed, etc., and estimate the sign of the motor torque local derivative, which indicates the position of the operational point related to the torque extremum at the MTPA curve. At the next stage, this information is used to shift the operational point toward the torque extremum and is used to track that point. Since the stator resistance is neglected, when the authors developed VSIC, these algorithms worked better at higher speeds, but may fail in the low-speed range, where the applied voltage is comparable with the voltage drop across the stator. Furthermore, the sensitivity to stator resistance makes implementation of VSIC for motor drives impossible, which has a relatively high resistance (e.g., with aluminum windings) or which has to operate in a wide speed range.

Another approach for online tracking of MTPA conditions includes methods for real-time nonmodel-based optimization [39] and uses the perturb and observe (PandO) principle. The main idea of this approach is to perturb a system by modification of its input signal in the "test" purpose and observe the response to the applied perturbation. For this purpose, the authors of [31–34] suggested modification of the stator current phase with the following analysis of changes in its magnitude, while the authors of [36] adopted a similar idea to the stator voltage vector. The PandO techniques do not use either system models or parameters, which makes them convenient for a wide range of applications having unknown parameters or variables in a wide range. However, the main drawback is poorer dynamic and stability issues. If the operation of a seeking algorithm is not limited, it may incorrectly detect the gradient of the observing signal and move the system far from extremum, significantly worsening the performance and sometimes making the system unstable.

Most of the seeking techniques were designed under the assumption that the torque is constant or changes slowly. Therefore, they may fail in transients, where speed and torque are modified by the external factors. As a result, the system response to the injected signal or applied perturbation is mostly defined by that external factor, so seeking algorithms fail and move the operational point far away from the MTPA trajectory. As soon as the transient ends, the system operates as expected and tracking algorithms can return the operational point at the MTPA trajectory. In order to limit deviation from the MTPA trajectory, the seeking algorithms have to use constraints, which limit uncontrollable modification of the stator current and its phase. Furthermore, proper design of these constraints significantly decreases deviation from the correct trajectory and notably decrease reaction time of the system. In turn, lower deviations from the desired trajectory increase system efficiency in the dynamic modes. Therefore, proper selection of constrains is extremely important for the seeking techniques, where the dynamic response is one of the weakest points.

Despite numerous publications dedicated to the online techniques, the problem of constraints design was not studied in detail, and it is hard to find recommendations on their selection. In [37] it was recommended to limit the maximum phase of the stator current with a predefined constant; however, recommendations on its selection were not provided. The authors of [34] use the adjustment of the stator current phase in order to track the MTPA trajectory. They suggested the limitations of this angle with theoretical minimum and maximum values plus some gaps. This approach provides system stability, however, it results in poorer dynamic stability, because this limitation does not depend on the current operational point in the MTPA curve. In [17] it was suggested to use a theoretical MTPA curve calculated for rated and unsaturated parameters, as a reference value and slightly modify it with a small angle calculated by the MTPA tuning block. However, recommendations on the selection of the maximum modification angle were not provided.

As is clearly seen, the suggestions on the selection of constraints are general and quite simple; therefore, online techniques demonstrate poorer dynamic behavior and efficiency. Furthermore, the previous researchers did not consider the effects of motor parameters variations and the corresponding change of MTPA trajectory. Therefore, after analysis of the published research, it was decided to develop an algorithm for the design of constraints for the MTPA trajectory.

The contribution of this paper is the development of a constraint design algorithm for MTPA seeking techniques. The proposed algorithm was developed for operation with control systems of commercial motor drives; however, it can be easily extended to self-commissioning systems. The developed algorithm is compatible with all MTPA seeking techniques such as [30–32] and improves their dynamic behavior and efficiency. The proposed method was implemented and compared with existing algorithms. The experimental results proved superiority of the proposed technique and improvement of dynamic behavior and efficiency of the test motor drive.

2. Conventional MTPA Approach

2.1. Motor Equations

The simplified design of the interior permanent magnet synchronous motor (IPMSM) is shown in Figure 1. It has three pole pairs formed by the interior permanent magnets in the rotor and three-phase winding at the stator. The interaction of the flux formed by permanent magnets and the flux formed by the three-phase stator winding produces the torque.

Figure 1. Simplified design of IPMSM.

The basic electrical equations of IPMSM in the synchronous reference frame dq under assumption that hysteresis loss, eddy currents, etc., can be neglected, are:

$$\left.\begin{array}{l} u_d = i_d r_s + \frac{d\psi_d}{dt} - \omega\psi_q, \\ u_q = i_q r_s + \frac{d\psi_q}{dt} + \omega\psi_d; \end{array}\right\} \tag{1}$$

where:

u_d, u_q—d- and q-axis voltage components, respectively,
i_d, i_q—d- and q-axis currents components, respectively,
ψ_d, ψ_q—d- and q-axis flux linkages respectively,
r_s—stator resistance,
ω—electrical angular velocity.

The flux linkages cab be expressed as:

$$\left.\begin{array}{l} \psi_d = L_d i_d + \Psi_m, \\ \psi_q = L_q i_q; \end{array}\right\} \tag{2}$$

where:

L_d, L_q—full d- and q-axis inductances, respectively,
Ψ_m—permanent magnet flux linkage.

Combining (1) and (2) the motor equations in synchronous reference frame dq can be derived:

$$\left.\begin{array}{l} u_d = i_d r_s + L_d^{diff} \frac{di_d}{dt} - \omega L_q i_q, \\ u_q = i_q r_s + L_q^{diff} \frac{di_q}{dt} + \omega L_d i_d + \omega\Psi_m; \end{array}\right\} \tag{3}$$

where L_d^{diff} and L_q^{diff} are differential inductances of d- and q-axis, respectively.

The torque produced by the motor is defined as and contains full inductances:

$$T = \frac{3}{2}p i_q \big((L_d - L_q) i_d + \Psi_m \big) \tag{4}$$

where:

p—number of pole pairs.

Equations (3) and (4) are conventional equations of IPMSM, which are used for the design of an overwhelming majority of control systems.

2.2. MTPA Equations

As it could be found from the torque Equation (4), the total torque produced by machine includes magnetic component proportional to the rotor magnetic flux and reluctance component proportional to the difference between direct and quadrature inductances. The magnetic and reluctance torques depend on the phase of the stator current as sine and sine of doubled angle, respectively, therefore the resulting torque of a machine also has a maximum which has to be defined and used for efficient control.

In order to do this, the magnitude of the stator current vector is fixed at I_s and the relation between the current components, which corresponds to the maximum can be found. The quadrature component of the stator current is:

$$i_q = \sqrt{I_s^2 - i_d^2} \tag{5}$$

Combining (5) and (4) with the following differentiation, with respect to i_d, results in the expression, which is used for the calculation of the maximum of torque [16]:

$$i_d = -\frac{\Psi_m}{4(L_d - L_q)} - \sqrt{\frac{\Psi_m^2}{16(L_d - L_q)^2} + \frac{I_s^2}{2}} \tag{6}$$

Equation (6) can be rewritten in terms of the current components:

$$i_d = -\frac{\Psi_m}{2(L_d - L_q)} - \sqrt{\frac{\Psi_m^2}{4(L_d - L_q)^2} + i_q^2} \tag{7}$$

The last two equations define the MTPA condition of synchronous machines with constant parameters. They are used for the implementation of the conventional MTPA control and move stator current vector along the trajectory, the typical shape is depicted in Figure 2. Good examples of MTPA implementation according to this approach can be found in [40–42].

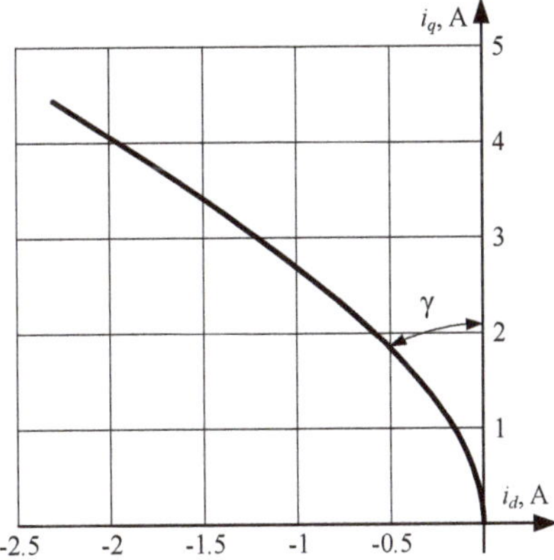

Figure 2. MTPA trajectory in $i_d i_q$ plane.

Since Equations (6) and (7) are calculation intensive, the MTPA trajectory is frequently written in terms of the stator current and its phase, which is a function of the stator current amplitude. In this case, the MTPA trajectory is described by the following equations:

$$\left.\begin{array}{l} i_d = -I_s \sin(\gamma), \\ i_q = I_s \cos(\gamma); \end{array}\right\} \tag{8}$$

$$\gamma(I_s) = \arccos\left(\frac{-\Psi_m + \sqrt{\Psi_m^2 + 8(L_d - L_q)^2 I_s^2}}{4(L_d - L_q)I_s}\right). \tag{9}$$

Despite the MTPA angle being hard for calculation, it is a smooth function, which can be easily approximated with a first- or second-order polynomial, which can be easily seen from Figure 3. Good examples of MTPA implementation according to this approach can be found in [43–45].

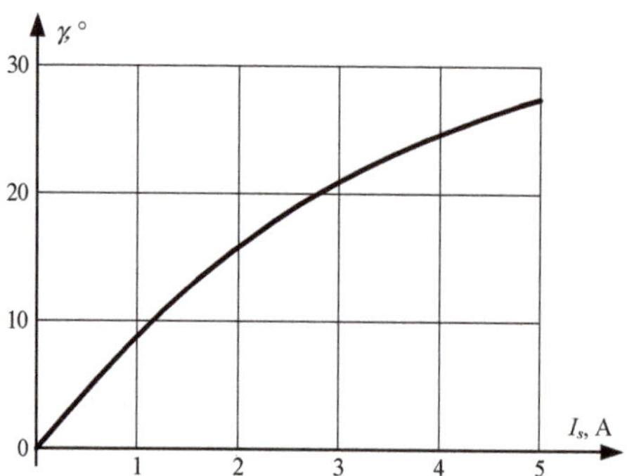

Figure 3. Dependence of the MTPA angle γ on the stator current amplitude.

3. Impact of the Motor Parameters Variation on the MTPA Trajectories

In order to design constraints properly, the impact of motor parameters on the MTPA trajectory must be analyzed. The analysis was performed for the motor used in the experimental verification. The rated parameters are demonstrated in Table 1.

Table 1. Motor rated parameters.

Parameter	Value
Number of pole pairs	$p = 3$
Rated power, kW	1.4
Rated torque, N·m	5.0
Phase resistance, Ω	2.05
d-axis inductance, mH	83
q-axis inductance, mH	115
Back-EMF constant, V·s/rad	0.2

3.1. Flux-Linkage Decrease

The rotor flux-linkage decreases with the rise of magnet temperature and may fall by 5–10%, depending on the magnet material. The degradation of magnets due to demagnetization may result in an additional 10%; therefore, the maximum decrease of flux-linkage can be considered as 20% [12]. A decrease of the rotor flux-linkages causes mitigation of the magnetic component of the motor torque. Thus, if the reluctance component is not changing, the MTPA angle increases, which is illustrated in Figures 4 and 5.

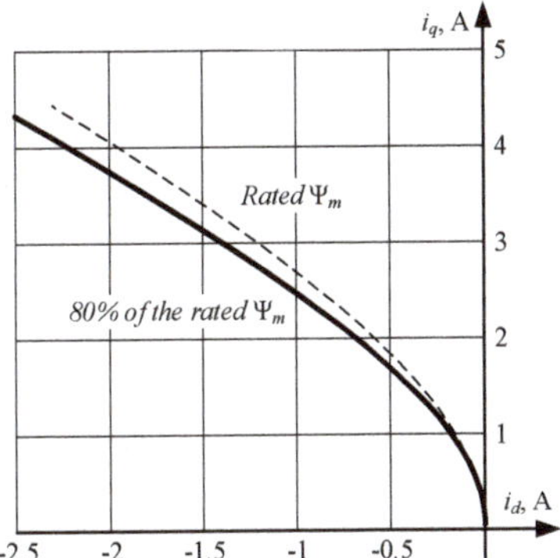

Figure 4. MTPA trajectories in $i_d i_q$ plane at different flux-linkages.

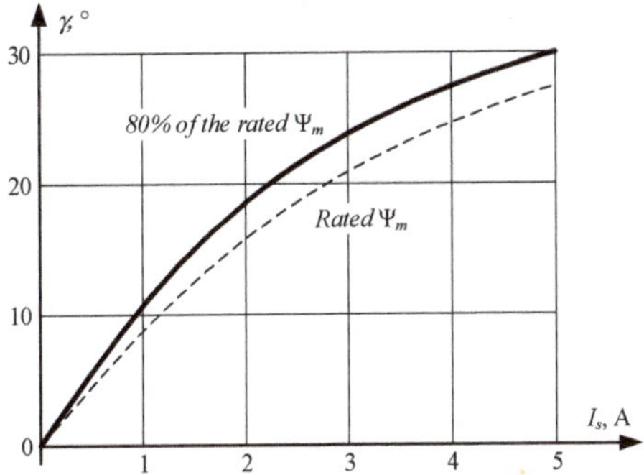

Figure 5. Dependence of the MTPA angles γ on the stator current amplitude at different flux-linkages.

3.2. Motor Inductances Decrease

The motor inductances decrease due to the steel saturation, which is caused by the amplitude of the applied stator current. Their variation significantly depends on the machine design and may be as high as 70–80% in synchronous reluctance motors (SynRM) and PM assisted synchronous reluctance motors (PMASynRM). In IPMSMs the maximum decrease of full inductance is typically about 50–60% [8].

At the same time, the reluctance torque of a synchronous motor depends on the difference between the direct and quadrature inductances; therefore, analysis of the inductance variation on the MTPA trajectory is more complicated, because both inductances vary simultaneously. Furthermore, the motor may saturate in the direct and quadrature directions in a different way, causing the inductance difference $\Delta L = L_d - L_q$ to rise, even if both inductances are falling. A good example of this phenomena is a motor under test, which is considered in this paper. The flux path of this electrical machine along the direct axis is saturated faster than the path along quadrature axis. As a result, the direct axis inductance decreases faster than the quadrature inductance and the inductance difference increases. However, with the following increase of the stator current, the quadrature inductance decreases, and the inductance difference reduces, reaching lower values than the initial values for the zero current. This phenomenon is illustrated in Figure 6, which demonstrates inductance variation with respect to the stator current.

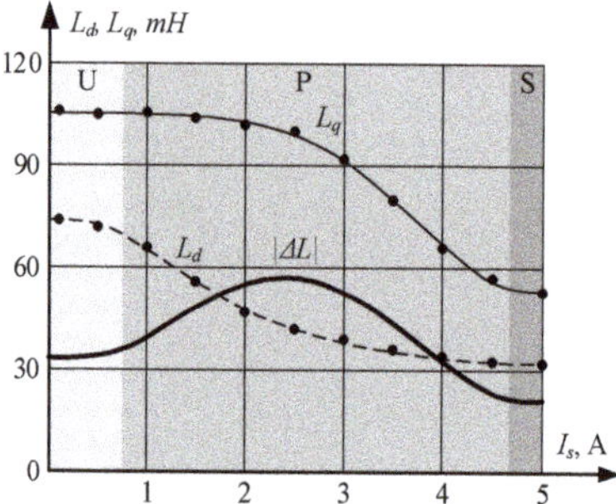

Figure 6. Motor inductances vs. current at the MTPA trajectory.

In this figure U, S and P denote unsaturated, saturated and partly saturated zones, respectively. It can be observed, that the inductance difference in the unsaturated (with rated parameters) zone ΔL_u, is less than the inductance difference in the partly saturated zone ΔL_p; however, it is higher than the inductance difference in the saturated zone ΔL_s. As a result, the true MTPA trajectory of this machine, which reflects complicated dependencies of motor inductances on the stator current, Figure 6, has complicated waveforms, which are demonstrated in Figures 7 and 8. In Figure 8, γ_u and γ_s correspond to the MTPA angle trajectories drawn for the unsaturated and saturated motor inductances, respectively, and γ stays for the true MTPA angle.

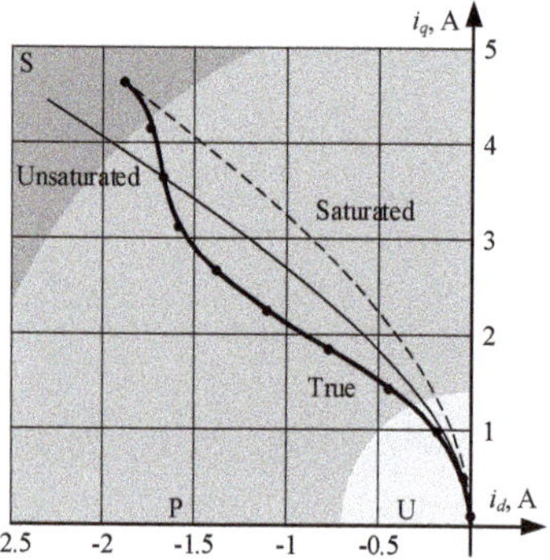

Figure 7. Test motor MTPA trajectories in $i_d i_q$ plane.

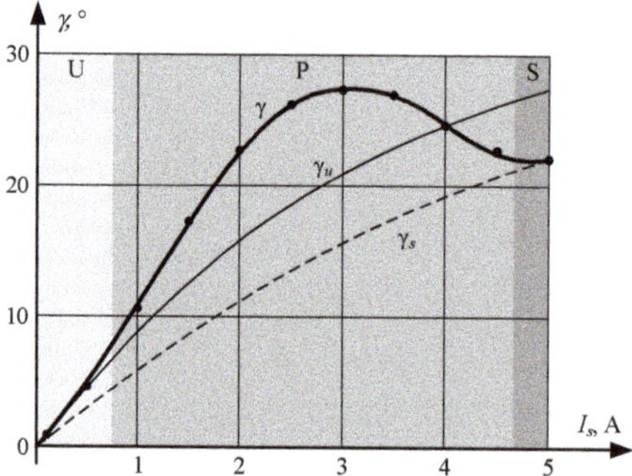

Figure 8. Dependence of the test motor MTPA angles γ with respect to the stator current amplitude.

It should be noted that some researchers consider unsaturated and saturated MTPA curves as boundary conditions for motor operation and design algorithm using this assumption [17]. However, the provided example demonstrates that such boundaries are incorrect for such motors. The MTPA trajectory may leave the area limited with curves for unsaturated and saturated conditions, and this phenomenon should be taken into account.

4. Proposed Method for Design of Constraints

In order to solve the abovementioned problems, a new algorithm for the design of constraints for MTPA seeking techniques has been developed and considered in this paper. This idea will be explained using a test motor as an example and the corresponding constraints for an MTPA trajectory seeking algorithm are constructed. For higher convenience, the MTPA trajectories are designed in the $i_d i_q$ plane and for an MTPA angle γ, simultaneously. Therefore, the proposed technique can be easily adapted for control schemes involving MTPA formulation based on the Equations (6)–(8).

The proposed method is demonstrated in Figure 9 and includes the following steps. Initially, the direct and quadrature inductances variations with the stator current increase are measured, and the corresponding dependence of the inductance difference ΔL is calculated. These measurements are repeated for several motor samples and the resulting data are averaged. After that, the MTPA trajectory for the measured inductances is plotted, i.e., curve "A" in Figures 10 and 11. At the next stage, the maximum possible decrease of the rotor flux-linkage is calculated, and the corresponding curve is plotted. For the test motor, which uses NdFeB magnets, the motor designers estimated the maximum temperature decrease is 3% and the maximum aging and demagnetization degradation is 5%, which results in a maximum reduction of 8%. The corresponding MTPA trajectories are depicted as curve "B" in Figures 10 and 11. After that, the maximum possible deviation of motor parameters from their rated values at the stage of production is taken into account. This deviation is not significant and is caused by assembling inaccuracies, materials from different lots, etc. The factory quality assurance engineers reported that the possible variation of the flux-linkage is less than 1%, while the inductance deviation may be as high as 12%. At the same time, the character of inductance dependencies, demonstrated in Figure 6, is almost unchanged and deviation results only in scaling of those curves. At the next stage, the previously constructed curves "A" and "B" are modified using obtained information on the maximum deviation of motor parameters, which results in curves "C" and "D" in Figures 10 and 11, respectively. These curves define the zone, limiting the possible location of the true MTPA trajectory for all motors in the series.

After the area containing the MTPA trajectory is defined, it must be extended with the gaps necessary for the proper operation of the MTPA seeking technique and stable detection of the torque extremum corresponding to the MTPA condition. The term "gap" relates to the disturbed parameter, and its dimension is the same; however, implementation of the constraints depends on the MTPA technique and typically is performed in the $i_d i_q$ or $I_s \gamma$ planes. These gaps depend on the exact MTPA tracking algorithm and have to enclose the area, which fully includes border trajectories and deviation from them caused by the injection of additional signals. The MTPA algorithm involved in the experiments, was a seeking technique, which permanently disturbs the phase angle of the stator current and measures its amplitude, trying to minimize it [33]. This algorithm can stably detect the maximum of the torque curve, when it lies in the middle of the disturbance step $\Delta \gamma$. Therefore, the gap, necessary for proper operation of this technique, which can guarantee the correct detection of extremum, is half of the disturbance angle, which was 2°. Thus, the curves "C" and "D" have to be moved at this gap angle, which results in "E" and "F" in Figures 10 and 11, respectively. These curves, "E" and "F", define our desired constraints, which can improve the performance of seeking an MTPA algorithm.

These curves could be implemented as a look-up table (LUT), approximating polynomial or a set of splines; however, the best way of approximation depends on the complexity of the constraint curves.

The proposed method of constraints design was developed for operation with a series of motors, where inductances could be measured in advance. However, it could be easily adapted for self-commissioning drives, which operate without prior knowledge of motor parameters. In this case, at the stage of parameter identification, the control system has to additionally identify dependence of the inductance difference on the stator current. After that, the curve "A" can be constructed. The construction of the curve "A" is made using data from Figure 6 by substituting inductances as functions from the current magnitude and the current magnitude itself into (6). Then evaluation of (5) allows to obtain both current components in $i_d i_q$ plane. If it is desired to evaluate MTPA angle γ with respect to the current magnitude, then Equation (9) is used.

At the next stage, the maximum variation of the flux-linkage has to be defined. Since this information is unavailable, the high-border estimation can be used: 12–15% for NdFeB magnets and 25–30% for ferrite magnets. Thus, the curve "B" can be constructed. The higher margin values increase response time of MTPA algorithms; therefore, it is desired to decrease the margins as much as possible. The recommended values of the flux linkage deviation came from the worst motor prototypes known to the authors and cover the overwhelming majority of motor drives. However, if a developer is confident that the maximum possible deviation of the flux linkage is less than the recommended values, the margins have to be decreased. Since self-commissioning routine tunes the drive for operation with the exact sample of motor, the curves "C" and "D", which define motor parameter variation at manufacturing, could be skipped. After that, the curves "E" and "F" can be constructed by moving the curves "A" and "B" at the gap angle, which is equal to the half of disturbance angle for the test algorithm.

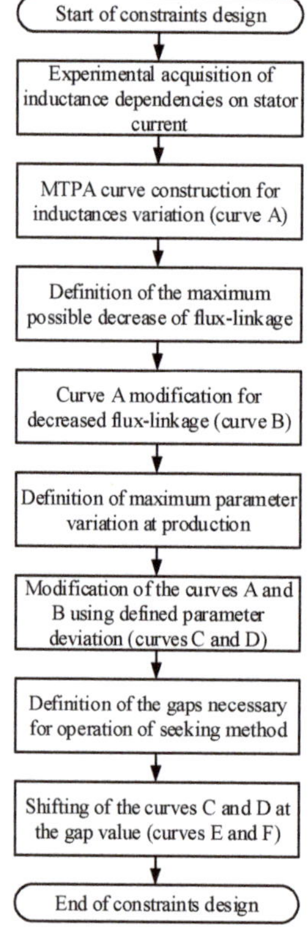

Figure 9. Flowchart of constraints design algorithm.

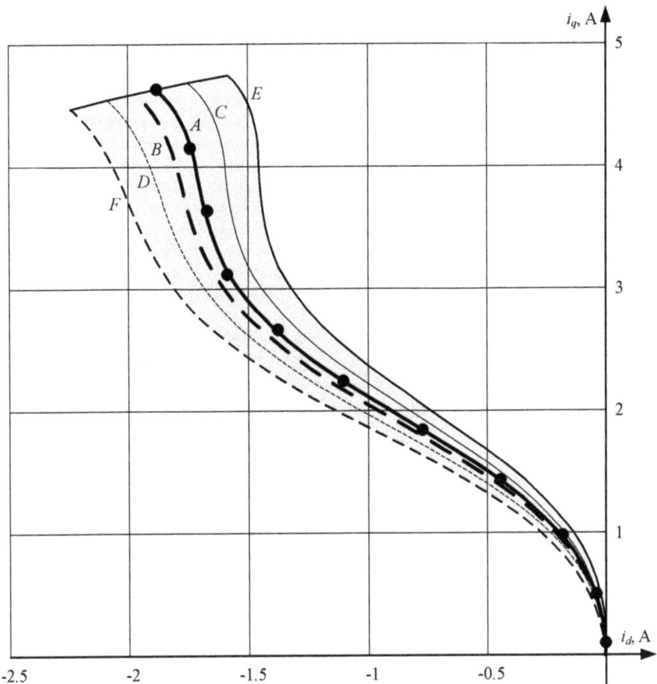

Figure 10. MTPA constraints design in $i_d i_q$ plane.

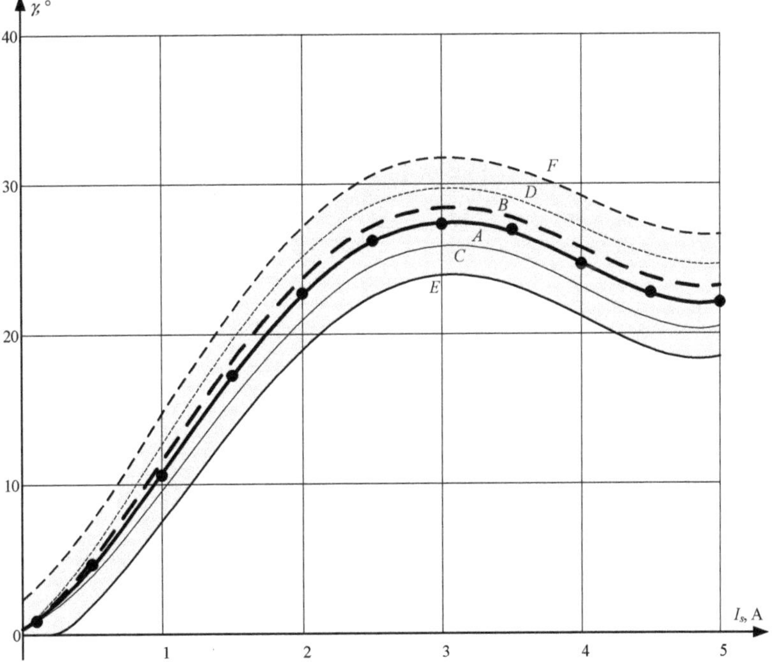

Figure 11. MTPA constraints design for MTPA angle γ.

5. Experimental Setup

The experimental verification of the proposed algorithm was performed using the test rig demonstrated in Figure 12. It included a prototype of the three phase PMSM developed for rotary and reciprocating compressors, the rated parameters are shown in Table 1. However, motor direct and quadrature inductances strongly depend on the stator current and vary as depicted in Figure 6.

Figure 12. Experimental setup.

The IPMSM control system implemented for verification of the proposed algorithm, was taken from a commercial motor drive discussed in [46] (Figure 13). It performs a sensorless control of IPM motors, using a back-EMF based speed and position estimation technique discussed in [47]. In order to exclude reverse rotation at the start, the estimator is enhanced with an initial rotor position estimation algorithm [48]. The performance of the estimation algorithm was verified with the help of an incremental position encoder [49,50], which demonstrated that it operates at speeds over 10 Hz with the maximum estimation error of 3~4 electrical degrees. Since the control system was designed for operation with compressors, it was enhanced with a silent stoppage algorithm considered in [51].

The control scheme contains an outer speed loop and inner current loops in the synchronous reference frame, where the rotor speed and position are provided by the estimation algorithm. The information on the motor electrical signals is provided by the DC-link voltage and two shunt-based current sensors, placed at the bottom legs of the inverter.

The IPMSM control system includes a field-weakening algorithm, which increases the maximum operation speed by up to +50% of the rated one by demagnetizing the rotor field with a negative i_d current. The motor drive includes an adaptive MTPA control algorithm, required to increase efficiency. This algorithm uses a seeking technique, which was considered in [31]; therefore, this method for MTPA control can be used for the experimental validation of the proposed idea.

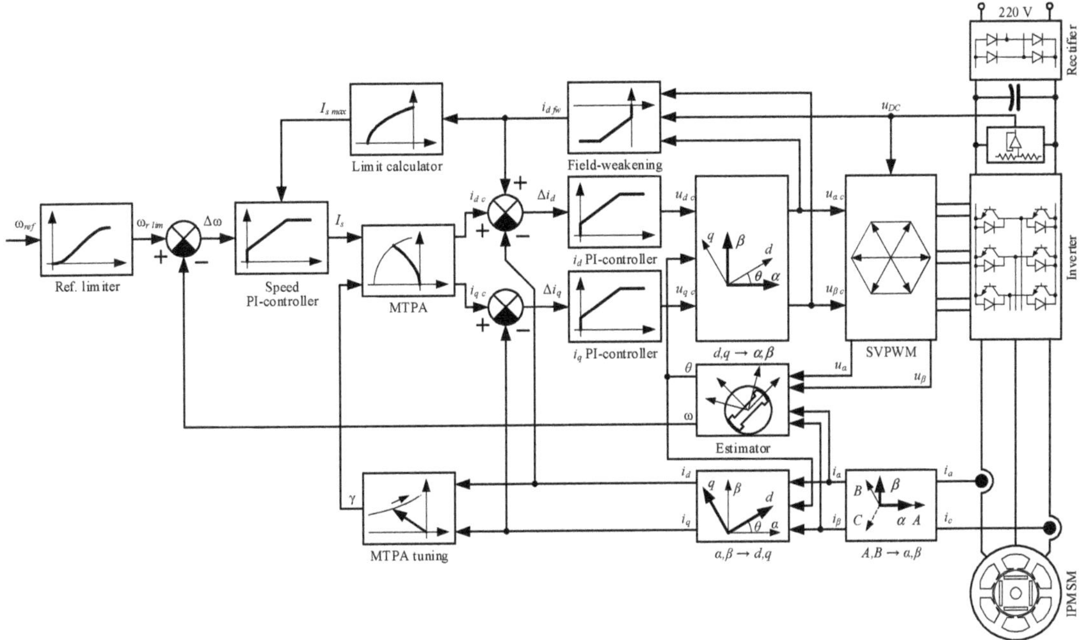

Figure 13. Control block diagram of the IPM motor drive with seeking MTPA algorithm.

The motor under test was connected with a Magtrol water cooling power brake 1-PB-65 capable of providing load torque up to 10 N·m. This power brake was controlled by the Magtrol DSP6001 dynamometer controller, which received commands from the computer and reported speed and torque. This load equipment provided low-ripple and precise torque application, which is necessary for the accurate measurements of speed-torque characteristics of the motor drive. The electrical signals were measured by the Yokogawa DL 850 oscilloscope, capable of operating with raw data and oscillograms and a Yokogawa WT-1800E power analyzer used for measurements of power and efficiency.

The inverter involved in these experiments was a commercial device designed for the control of rotary compressors of air conditioners. It was based on the FSBB15CH60F (15 A/600 V) Fairchild smart power module and designated for operation in 220 V (50/60) Hz standard grids. This inverter had a conventional structure for low-cost applications, which involved two shunt-based current sensors in the inverter legs and one voltage sensor in the DC-link. The signals from the sensors were pre-amplified by internal operational amplifiers of the microcontroller and then processed by a 12-bit ADC. The control system of the inverter used in the experiments included the iHart i910 Cortex M3 core microcontroller operating at 80 MHz and controlled the inverter switches at 10 kHz. The experimental sample of the inverter was extended with an RS-232 communication interface used for the connection with the computer in order to send commands and monitor the internal data.

6. Experimental Results

In order to check the performance of the proposed method, a number of experiments have been carried out. The seeking MTPA algorithm used for testing, involves the PandO principle and permanently disturbs the phase of the stator current, analyzing changes of its magnitude. Thus, the constraints for this technique were set as curves "E" and "F" from Figure 11 and implemented using LUT, containing 33 points for each curve with linear interpolation between them. The disturbance angle $\Delta\gamma$ used by the seeking algorithm was $4°$ and its calculation step N was selected as 20 electrical revolutions. The maximum calculation time $T_{\max}$ of the MTPA tuning algorithm was set to 0.5 s, which results in a minimum motor speed, at which an MTPA tuning algorithm can operate:

$$n_{\min} = \frac{60 \cdot N}{p \cdot T_{\max}} = \frac{60 \cdot 20}{3 \cdot 0.5} = 800 \text{ rpm} \tag{10}$$

This speed perfectly fits the requirements of the control system of compressors, where the operational area requires speeds over 900~1200 rpm.

In order to check the dynamic behavior of the seeking algorithm with different constraints, the load torque contained equal intervals of constant load, and positive and negative ramps. For the performance evaluation of the proposed idea, the measurements were performed at the lowest allowed operational speed, which is the most difficult condition for the seeking algorithms. Despite the minimum operational speed of the selected seeking technique being 800 rpm, the experiments were conducted at 1200 rpm, which is the minimum permitted speed for reciprocating the compressors utilizing the considering drive. After obtaining the data on estimated MTPA angle, the efficiency of the motor was measured in operation of the MTPA-seeking algorithm with the considered sets of constraints. In this test, the data were averaged at the interval of ten load periods (200 s), which minimizes the impact of random perturbations.

For comparison of the developed method with the prior art, the constraints were designed according to their recommendations.

6.1. Fixed Limits

In this method the constraints for the stator current phase were selected to consider the case where motor parameters are unknown and self-commissioning is not performed. The fixed predefined numbers equal to the theoretical minimum and maximum values where selected. As can be observed from Figure 11, the minimum value of the MTPA angle γ is 0°, whereas the maximum angle is defined by curve "F" (about 32°). For a pure reluctance motor the maximum angle cannot exceed 45° if it is desired to cover 100% of all motor designs. These angles were selected as the limits for the variation of the seeking algorithm disturbance factor. These limits are depicted in Figure 14 together with the MTPA trajectories and curves used for the design.

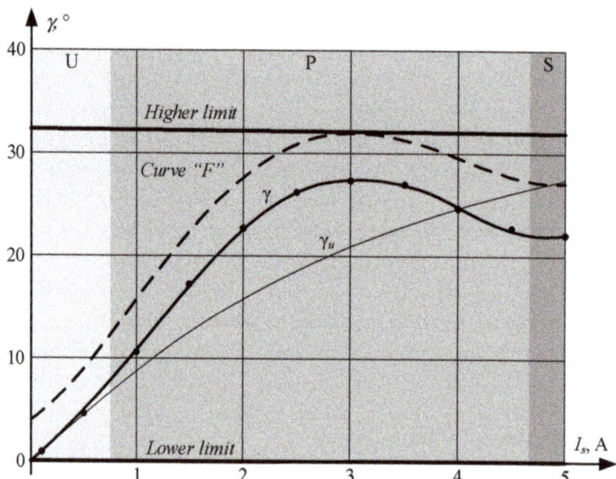

Figure 14. Constraints design for MTPA angle γ using fixed limits.

6.2. Rated MTPA Curve with Gaps

According to this method, the constraints are designed using the MTPA curve for rated motor parameters, which is shifted to higher and lower directions at the fixed gap. In this experiment, the gap was selected as the seeking algorithm disturbance angle $\Delta\gamma$, which is equal to 4°. At the same time, the MTPA angle for the test motor may not be negative; therefore, the lower limit is set to zero in this area, where parallel shift of the rated MTPA curve resulted in negative values. These limits are depicted in Figure 15 together with the MTPA trajectories and curves used for design.

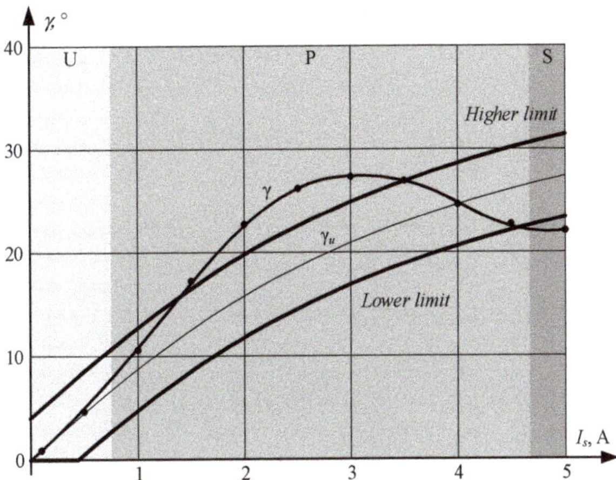

Figure 15. Constraints design for MTPA angle γ using rated MTPA curve with gaps.

6.3. Operation at Low Load

In this experiment, the performance of the proposed technique was evaluated and compared with existing algorithms at low loads, where the motor was not saturated (zone "U"). Two cycles of the load torque used in this experiment are depicted in Figure 16a. The performance of seeking the algorithm operation with "Fixed limits", "Rated MTPA curve with gaps" and proposed constraints are demonstrated in Figure 16b–d, respectively. The measured efficiency for these constraint sets was: 47.4%, 51.9% and 52.1%, respectively.

6.4. Operation at Medium Load

In this experiment, the performance of the proposed technique was evaluated and compared with existing algorithms at low loads, where the motor was partly saturated (zone "P"). Two cycles of the load torque used in this experiment are depicted in Figure 17a. The performance of the seeking algorithm operation with "Fixed limits", "Rated MTPA curve with gaps" and proposed constraints are demonstrated in Figure 17b–d, respectively. The measured efficiency for these constraint sets was: 85.2%, 87.1% and 88.4%, respectively.

6.5. Operation at High Load

In this experiment, the performance of the proposed technique was evaluated and compared with existing algorithms at low loads, where the motor was highly saturated (zone "S"). Two cycles of the load torque used in this experiment are depicted in Figure 18a. The performance of the seeking algorithm operation with "Fixed limits", "Rated MTPA curve with gaps" and proposed constraints are demonstrated in Figure 18b–d, respectively. The measured efficiency for these constraint sets was: 85.6%, 86.1% and 86.3%, respectively.

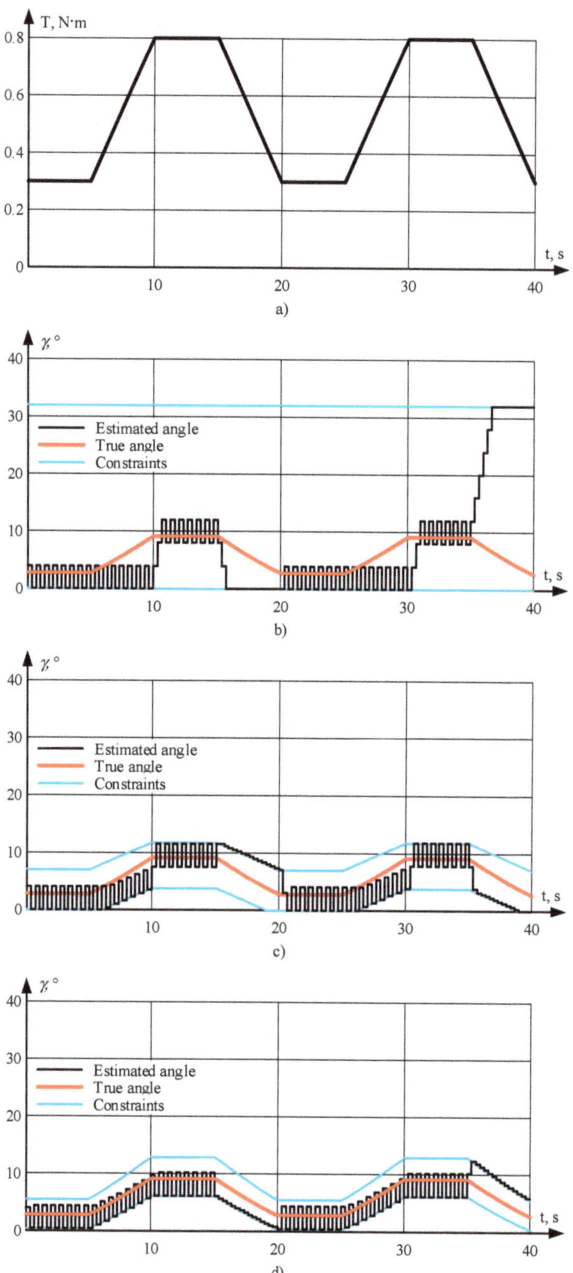

Figure 16. Operation of seeking algorithm with different constraints at low load: (**a**) Torque profile; (**b**) Fixed limits; (**c**) Rated MTPA curve with gaps; and (**d**) Proposed technique.

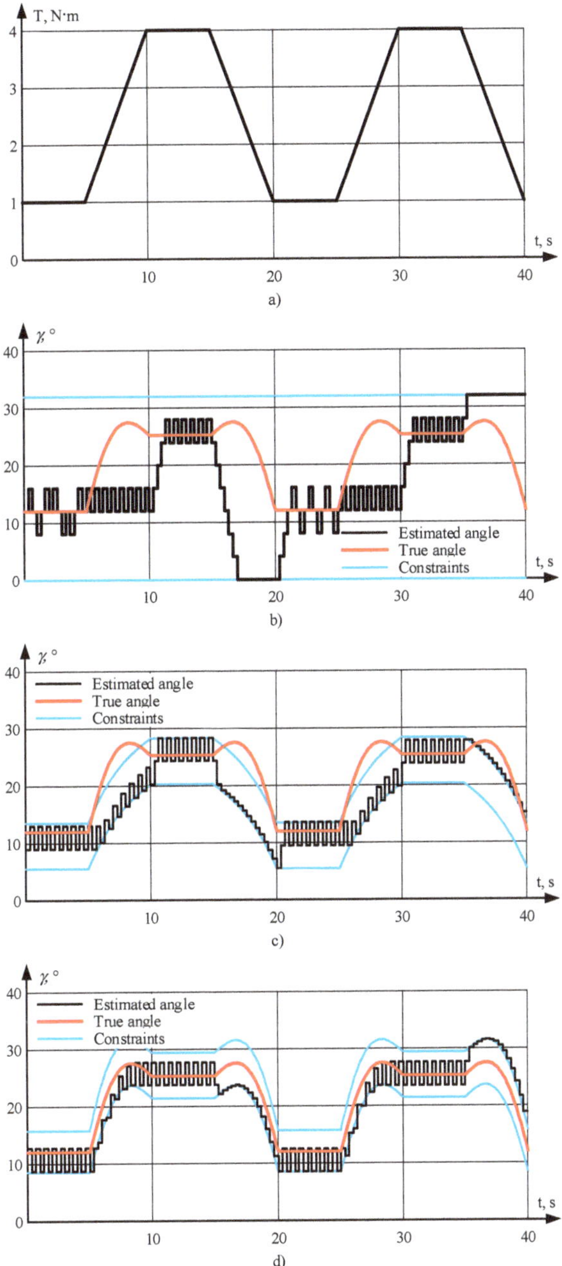

Figure 17. Operation of the seeking algorithm with different constraints at medium load. (**a**) Torque profile; (**b**) Fixed limits; (**c**) Rated MTPA curve with gaps; and (**d**) Proposed technique.

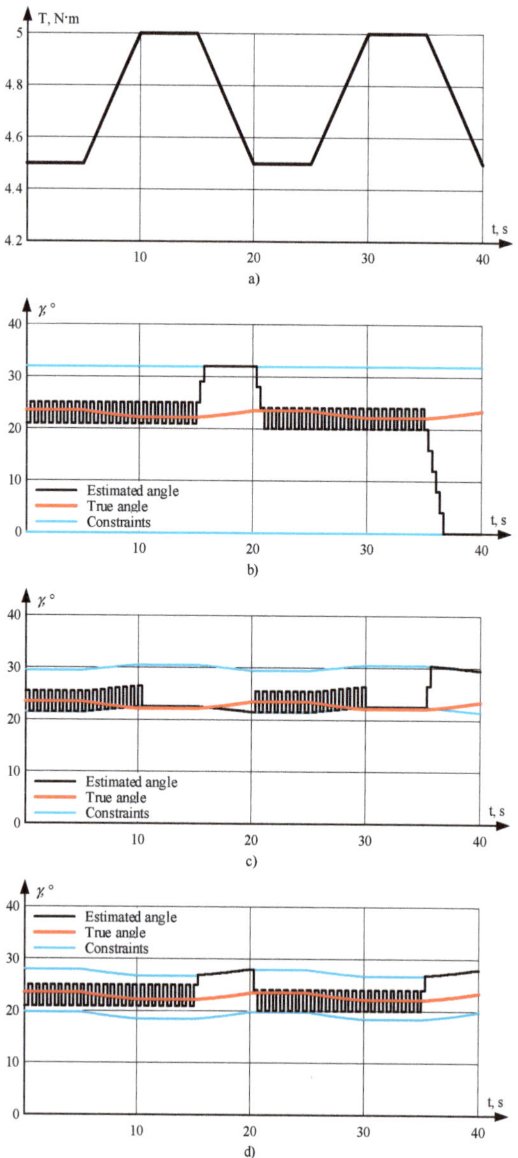

Figure 18. Operation of the seeking algorithm with different constraints at high load. (**a**) Torque profile; (**b**) Fixed limits; (**c**) Rated MTPA curve with gaps; and (**d**) Proposed technique.

7. Discussion

It is clearly observable from the provided experimental results, that the seeking algorithm does not operate properly at transients. Furthermore, its behavior depends on the sign of the torque derivative: if it is positive, the estimated angle fluctuates near its previous value, if the derivative is negative, the estimated angle goes to a lower or higher limit, depending on the direction at the previous stage. As a result, the constraints design plays an important role in the dynamic performance by limiting this uncontrollable change.

The experimental results showed that the operation of the seeking algorithm with fixed constraints results in a poorer dynamic and significantly decreases motor efficiency in dynamic modes. Therefore, this method may be recommended only for simple systems, which mainly operate under constant load torque.

The seeking algorithm with constraints designed using a rated MTPA curve with gaps, demonstrates better dynamic and efficiency than the same algorithm with the fixed constraints. It demonstrates acceptable results in the regions, where the true MTPA curve is close to the rated curve. However, its operation is not optimal at the loads, where the true MTPA trajectory significantly deviates from the rated MTPA curve and leaves the area enclosed by the constraints. The experimental motor is a good example of this case, which is illustrated by Figure 17. As a result, this method of constraints design can be suggested for motors, where the MTPA trajectory does not significantly deviate from the rated MTPA trajectory.

It can be observed that the operation of the seeking algorithm with constraints, which is designed according to the proposed method, demonstrated better efficiency and dynamic behavior. Therefore, it is recommended for the motor drives, where the motor was designed for operation with considerable saturation and where the MTPA curve significantly deviates from the curve constructed for the unsaturated parameters.

At the same time, it should be noted that the operation of the seeking algorithm with a different set of constraints at a constant or slowly changing load torque is similar and they differ only in dynamic.

8. Conclusions

This paper proposes a new method for constraints design for MTPA seeking techniques. This algorithm takes into account possible motor parameter variation due to operational conditions and deviation at the stage of production. The authors considered existing approaches to constraints design and experimentally compared performance of the MTPA-seeking algorithm with constraints designed according to existing and proposed algorithms. It was shown that the proposed method improves efficiency and dynamic operation of MTPA seeking techniques, especially for the motors, which operate in the saturation zone and whose MTPA curve significantly deviates from the MTPA trajectory calculated for the unsaturated parameters.

Author Contributions: Conceptualization A.D.; methodology A.A.; software A.D.; validation A.D. and A.A.; formal analysis A.D. and A.A.; investigation A.D.; resources A.A.; data curation A.A.; writing—original draft preparation A.D.; writing—review and editing A.D. and A.A.; visualization A.D.; supervision A.D.; project administration A.D.; funding acquisition A.A. All authors have read and agreed to the published version of the manuscript.

Funding: This research is supported by the Russian Science Foundation grant (Project №21-19-00696).

Institutional Review Board Statement: Not applicable.

Informed Consent Statement: Not applicable.

Data Availability Statement: Not applicable.

Conflicts of Interest: The authors declare no conflict of interest.

References

1. Sun, J.; Lin, C.; Xing, J.; Jiang, X. Online MTPA Trajectory Tracking of IPMSM Based on a Novel Torque Control Strategy. *Energies* **2019**, *12*, 3261. [CrossRef]
2. Tinazzi, F.; Zigliotto, M. Torque Estimation in High-Efficency IPM Synchronous Motor Drives. *IEEE Trans. Energy Convers.* **2015**, *30*, 983–990. [CrossRef]
3. Preindl, M.; Bolognani, S. Optimal State Reference Computation with Constrained MTPA Criterion for PM Motor Drives. *IEEE Trans. Power Electron.* **2015**, *30*, 4524–4535. [CrossRef]
4. Inoue, T.; Inoue, Y.; Morimoto, S.; Sanada, M. Maximum Torque Per Ampere Control of a Direct Torque-Controlled PMSM in a Stator Flux Linkage Synchronous Frame. *IEEE Trans. Ind. Appl.* **2016**, *52*, 2360–2367. [CrossRef]

5. Decker, S.; Brodatzki, M.; Bachowsky, B.; Schmitz-Rode, B.; Liske, A.; Braun, M.; Hiller, M. Predictive Trajectory Control with Online MTPA Calculation and Minimization of the Inner Torque Ripple for Permanent-Magnet Synchronous Machines. *Energies* **2020**, *13*, 5327. [CrossRef]
6. Li, Z.; Li, H. MTPA control of PMSM system considering saturation and cross-coupling. In Proceedings of the 15th International Conference on Electrical Machines and Systems, Sapporo, Japan, 11–24 October 2012; pp. 1–6.
7. Miao, Y.; Miao, Y.; Ge, H.; Preindl, M.; Ye, J.; Cheng, B.; Emadi, A. MTPA Fitting and Torque Estimation Technique Based on a New Flux-Linkage Model for Interior-Permanent-Magnet Synchronous Machines. *IEEE Trans. Ind. Appl.* **2017**, *53*, 5451–5460. [CrossRef]
8. Carraro, M.; Tinazzi, F.; Zigliotto, M. Estimation of the direct-axis inductance in PM synchronous motor drives at standstill. In Proceedings of the 2013 IEEE International Conference on Industrial Technology, Cape Town, South Africa, 25–28 February 2013; pp. 313–318.
9. Kim, H.B.; Hartwig, J.; Lorenz, R.D. Using on-line parameter estimation to improve efficiency of IPM machine drives. In Proceedings of the IEEE 33rd Power Electronics Specialists Conference, Cairns, Australia, 23–27 June 2002; pp. 815–820.
10. Phowanna, P.; Boonto, S.; Konghirun, M. Online parameter identification method for IPMSM drive with MTPA. In Proceedings of the 18th International Conference on Electrical Machines and Systems, Pattaya City, Thailand, 25–28 October 2015; pp. 1–6.
11. Feng, G.; Lai, C.; Mukherjee, K.; Kar, N.C. Online PMSM Magnet Flux-Linkage Estimation for Rotor Magnet Condition Monitoring Using Measured Speed Harmonics. *IEEE Trans. Ind. Appl.* **2017**, *53*, 2786–2794. [CrossRef]
12. Xiao, S.; Griffo, A. PWM-based flux linkage and rotor temperature estimations for permanent magnet synchronous machines. *IEEE Trans. Energy Convers.* **2020**, *35*, 6061–6069. [CrossRef]
13. Liu, Q.; Hameyer, K. High-Performance Adaptive Torque Control for an IPMSM With Real-Time MTPA Operation. *IEEE Trans. Energy Convers.* **2017**, *32*, 571–581. [CrossRef]
14. Shinohara, A.; Inoue, Y.; Morimoto, S.; Sanada, M. Maximum Torque Per Ampere Control in Stator Flux Linkage Synchronous Frame for DTC-Based PMSM Drives Without Using q-Axis Inductance. *IEEE Trans. Ind. Appl.* **2017**, *53*, 3663–3671. [CrossRef]
15. Mohamed, Y.A.R.I.; Lee, T.K. Adaptive self-tuning MTPA vector controller for IPMSM drive system. *IEEE Trans. Energy Convers.* **2006**, *21*, 636–644. [CrossRef]
16. Bolognani, S.; Sgarbossa, L.; Zordan, M. Self-tuning of MTPA current vector generation scheme in IPM synchronous motor drives. In Proceedings of the European Conference on Power Electronics and Applications, Aalborg, Denmark, 2–5 September 2007; pp. 1–10.
17. Bolognani, S.; Petrella, R.; Prearo, A.; Sgarbossa, L. Automatic Tracking of MTPA Trajectory in IPM Motor Drives Based on AC Current Injection. *IEEE Trans. Ind. Appl.* **2010**, *47*, 105–114. [CrossRef]
18. Antonello, R.; Carraro, M.; Zigliotto, M. Maximum-torque-per-ampere operation of anisotropic synchronous permanent-magnet motors based on extremum seeking control. *IEEE Trans. Ind. Electron.* **2014**, *61*, 5086–5093. [CrossRef]
19. Chen, Q.; Gu, L.; Lin, Z.; Liu, G. Extension of Space-Vector-Signal-Injection Based MTPA Control into SVPWM Fault-Tolerant Operation for Five-Phase IPMSM. *IEEE Trans. Ind. Electron.* **2019**, *66*, 6089–6101. [CrossRef]
20. Yousefi-Talouki, A.; Pescetto, P.; Pellegrino, G.-M.L.; Boldea, I. Combined Active Flux and High-Frequency Injection Methods for Sensorless Direct-Flux Vector Control of Synchronous Reluctance Machines. *IEEE Trans. Power Electron.* **2017**, *33*, 2447–2457. [CrossRef]
21. Lee, K.; Han, Y. MTPA control strategy based on signal injection for V/f scalar-controlled surface permanent magnet synchronous machine drives. *IEEE Access* **2020**, *8*, 96036–96044. [CrossRef]
22. Sue, S.M.; Hung, T.W.; Liaw, J.H.; Li, Y.F.; Sun, C.Y. A new MTPA control strategy for sensorless V/f controlled PMSM drives. In Proceedings of the 6th IEEE Conference on Industrial Electronics and Applications, Beijing, China, 21–23 June 2011; pp. 1–5.
23. Sun, T.; Koc, M.; Wang, J. MTPA Control of IPMSM Drives Based on Virtual Signal Injection Considering Machine Parameter Variations. *IEEE Trans. Ind. Electron.* **2018**, *65*, 6089–6098. [CrossRef]
24. Sun, T.; Wang, J.; Chen, X. Maximum torque per ampere (MTPA) control for interior permanent magnet synchronous machine drives based on virtual signal injection. *IEEE Trans. Power Electron.* **2015**, *30*, 5036–5045. [CrossRef]
25. Sun, T.; Wang, J.; Koc, M. On Accuracy of Virtual Signal Injection based MTPA Operation of Interior Permanent Magnet Synchronous Machine Drives. *IEEE Trans. Power Electron.* **2017**, *32*, 7405–7408. [CrossRef]
26. Tang, Q.; Shen, A.; Luo, P.; Shen, H.; Li, W.; He, X. IPMSMs Sensorless MTPA Control Based on Virtual q-Axis Inductance by Using Virtual High-Frequency Signal Injection. *IEEE Trans. Ind. Electron.* **2019**, *67*, 136–146. [CrossRef]
27. Wang, J.; Huang, X.Y.; Fang, Y.T.; Niu, F.; Wu, L.J. Power Perturbation Based Virtual Signal Injection Control of MTPA for IPMSM Drive System. In Proceedings of the 13th International Conference on Electrical Machines, Alexandroupoli, Greece, 3–6 September 2018; pp. 1–6.
28. Chen, Q.; Zhao, W.; Liu, G.; Lin, Z. Extension of Virtual-Signal-Injection-Based MTPA Control for Five-Phase IPMSM Into Fault-Tolerant Operation. *IEEE Trans. Ind. Electron.* **2018**, *66*, 944–955. [CrossRef]
29. Sun, T.; Long, L.; Yang, R.; Li, K.; Liang, J. Extended Virtual Signal Injection Control for MTPA Operation of IPMSM Drives with Online Derivative Term Estimation. *IEEE Trans. Power Electron.* **2020**, *35*, 6061–6069.
30. Niazi, P.; Toliyat, H.A.; Goodarzi, A. Robust Maximum Torque per Ampere (MTPA) Control of PM-Assisted SynRM for Traction Applications. *IEEE Trans. Veh. Technol.* **2007**, *56*, 1538–1545. [CrossRef]

31. Dianov, A.; Anuchin, A. Adaptive maximum torque per ampere control of sensorless permanent magnet motor drives. *Energies* **2020**, *13*, 5071. [CrossRef]
32. Dianov, A.; Anuchin, A. Adaptive maximum torque per ampere control for IPMSM drives with load varying over mechanical revolution. *IEEE J. Emerg. Sel. Top. Power Electron.* **2020**. [CrossRef]
33. Dianov, A.; Anuchin, A.; Bodrov, A. Robust MTPA Control for Steady State Operation of Low-Cost IPMSM Drives. *IEEE J. Emerg. Sel. Top. Ind. Electron.* **2021**. [CrossRef]
34. Dianov, A.; Young-Kwan, K.; Sang-Joon, L.; Sang-Taek, L. Robust self-tuning MTPA algorithm for IPMSM drives. In Proceedings of the 34th Annual Conference of IEEE Industrial Electronics, Orlando, FL, USA, 10–13 November 2008; pp. 1355–1360.
35. Wang, G.; Li, Z.; Zhang, G.; Yu, Y.; Xu, D. Quadrature PLL-Based High-Order Sliding-Mode Observer for IPMSM Sensorless Control with Online MTPA Control Strategy. *IEEE Trans. Energy Convers.* **2012**, *28*, 214–224. [CrossRef]
36. Chaoui, H.; Okoye, O.; Khayamy, M. Current sensorless MTPA for IPMSM drives. *IEEE/ASME Trans. Mechatron.* **2017**, *22*, 1585–1593. [CrossRef]
37. Lin, F.-J.; Huang, M.-S.; Chen, S.-G.; Hsu, C.-W. Intelligent Maximum Torque per Ampere Tracking Control of Synchronous Reluctance Motor Using Recurrent Legendre Fuzzy Neural Network. *IEEE Trans. Power Electron.* **2019**, *34*, 12080–12094. [CrossRef]
38. Daryabeigi, E.; Zarchi, H.A.; Markadeh, G.A.; Soltani, J.; Blaabjerg, F. Online MTPA Control Approach for Synchronous Reluctance Motor Drives Based on Emotional Controller. *IEEE Trans. Power Electron.* **2014**, *30*, 2157–2166. [CrossRef]
39. Krstic, M. Extremum seeking control. In *Encyclopedia of Systems and Control*; Baillieul, J., Samad, T., Eds.; Springer: London, UK, 2014.
40. Joo, K.J.; Park, J.S.; Lee, J. Study on Reduced Cost of Non-Salient Machine System Using MTPA Angle Pre-Compensation Method Based on EEMF Sensorless Control. *Energies* **2018**, *11*, 1425. [CrossRef]
41. Guo, Q.; Zhang, C.; Li, L.; Zhang, J.; Wang, M. Maximum Efficiency per Torque Control of Permanent-Magnet Synchronous Machines. *Appl. Sci.* **2016**, *6*, 425. [CrossRef]
42. Baek, S.K.; Oh, H.K.; Park, J.H.; Shin, Y.J.; Kim, S.W. Evaluation of Efficient Operation for Electromechanical Brake Using Maximum Torque per Ampere Control. *Energies* **2019**, *12*, 1869. [CrossRef]
43. Ye, M.; Shi, T.; Wang, H.; Li, X.; Xia, C. Sensorless-MTPA Control of Permanent Magnet Synchronous Motor Based on an Adaptive Sliding Mode Observer. *Energies* **2019**, *12*, 3773. [CrossRef]
44. Dianov, A.; Su, K.N.; Kwan, K.Y. Future Drives of Home Appliances: Elimination of the Electrolytic DC-Link Capacitor in Electrical Drives for Home Appliances. *IEEE Ind. Electron. Mag.* **2015**, *9*, 10–18. [CrossRef]
45. Li, K.; Wang, Y. Maximum Torque Per Ampere (MTPA) Control for IPMSM Drives Based on a Variable-Equivalent-Parameter MTPA Control Law. *IEEE Trans. Power Electron.* **2019**, *34*, 7092–7102. [CrossRef]
46. Dianov, A. Estimation of the mechanical position of reciprocating compressor for silent stoppage. *IEEE Open J. Power Electron.* **2020**, *1*, 64–73. [CrossRef]
47. Dianov, A.; Kim, N.S.; Lim, S.M. Sensorless starting of direct drive horizontal axis washing machines. *J. Int. Conf. Electr. Mach. Syst.* **2014**, *3*, 148–154. [CrossRef]
48. Dianov, A.; Anuchin, A.S.; Kozachenko, V.F. Initial Rotor Position Detection of PM Motors. In Proceedings of the EPE Power Electronics and Motion Control Conference, Riga, Latvia, 2–4 September 2004; pp. 1–6.
49. Anuchin, A.; Dianov, A.; Briz, F. Synchronous Constant Elapsed Time Speed Estimation Using Incremental Encoders. *IEEE/ASME Trans. Mechatron.* **2019**, *24*, 1893–1901. [CrossRef]
50. Anuchin, A.; Dianov, A.; Shpak, D.; Astakhova, V.; Fedorova, K. Speed estimation algorithm with specified bandwidth for incremental position encoder. In Proceedings of the 17th International Conference on Mechatronics—Mechatronika, Prague, Czech Republic, 7–9 December 2016; pp. 1–6.
51. Dianov, A. Stoppage noise reduction of reciprocating compressors. *IEEE Trans. Ind. Appl.* **2021**, *57*, 4376–4384. [CrossRef]

 *mathematics*

MDPI

Article

Inverter Volt-Ampere Capacity Reduction by Optimization of the Traction Synchronous Homopolar Motor

Vladimir Prakht [1,*], Vladimir Dmitrievskii [1], Alecksey Anuchin [2] and Vadim Kazakbaev [1]

1   Department of Electrical Engineering, Ural Federal University, 620002 Yekaterinburg, Russia; vladimir.dmitrievsky@urfu.ru (V.D.); vadim.kazakbaev@urfu.ru (V.K.)
2   Department of Electric Drives, Moscow Power Engineering Institute, 111250 Moscow, Russia; anuchinas@mpei.ru
*   Correspondence: va.prakht@urfu.ru; Tel.: +7-343-375-45-07

Abstract: The synchronous homopolar motor (SHM) with an excitation winding on the stator and a toothed rotor is a good alternative to traction induction motors for hybrid mining trucks. The main problem in the design of the SHM electric drives is that the magnetic flux forms three-dimensional loops and, as a result, the lack of high-quality optimization methods, which leads to the need to overrate the installed power of the inverter. This article discusses the procedure and results of optimization of a commercially available 370 kW traction SHM using the Nelder–Mead method. The objective function is composed to mainly improve the following characteristics of the traction SHM: total motor power loss and maximum armature winding current. In addition, terms are introduced into the objective function to make it possible to limit the voltage, the loss in the excitation winding, and the maximum magnetic flux density in the non-laminated sections of the magnetic core. As a result of the optimization, the motor losses and the maximum current required by the motor from the inverter were significantly reduced. The achieved reduction in the maximum current allows the cost of the IGBT modules of the inverter to be reduced by 1.4 times (by $ 2295), and also allows the AC component of the DC-link current to be reduced by the same amount.

Keywords: Nelder–Mead method; mining dump truck; optimal design; synchronous homopolar motor; traction drive

Citation: Prakht, V.; Dmitrievskii, V.; Anuchin, A.; Kazakbaev, V. Inverter Volt-Ampere Capacity Reduction by Optimization of the Traction Synchronous Homopolar Motor. *Mathematics* **2021**, *9*, 2859. https://doi.org/10.3390/math9222859

Academic Editor: Alessandro Niccolai

Received: 10 October 2021
Accepted: 9 November 2021
Published: 11 November 2021

Publisher's Note: MDPI stays neutral with regard to jurisdictional claims in published maps and institutional affiliations.

1. Introduction

Synchronous homopolar machines (SHMs) with an excitation winding on the stator are used in a number of applications, such as aircraft and ground vehicle generators, welding generators, and flywheel energy storage devices [1–3]. The main advantages of the SHM are the structural simplicity of the toothed rotor and the high reliability of the machine as a whole due to the absence of an excitation winding or a squirrel cage on the rotor. A number of studies have proposed the use of SHM in traction applications, due to the disadvantages of induction motors commonly used in these applications, such as the low reliability of the welded rotor cage, high rotor losses, difficulties in sensorless control, and difficulty of employing pure electric brakes at zero rotational speed due to the thermal cycling of the inverter semiconducting devices [3]. The complex design of the magnetic core causes difficulties in using traditional 2D FEM models to assess the performances of the SHM. For this reason, a number of original calculation methods have been proposed for the SHM, including 3D FEM, 2D FEM, one-dimensional magnetic circuits, and their various combinations [4–7]. In [3], a method for mathematical optimization of the traction SHM was proposed. It has been shown that by applying optimization, it is possible to significantly reduce the losses and torque ripple of the SHM. However, in [3], the current of the armature winding was not reduced sufficiently during the optimization to allow a reduction of the power rating and the cost of the traction inverter.

In this study, compared with [3], the objective optimization function is modified to significantly reduce the armature winding current and, as a result, to use cheaper IGBT modules (650 A) compared to those in the non-optimized SHM (1000 A). At the optimization, the restrictions imposed by the standard cross-sections of rectangular winding wires are taken into account. Additionally, the optimization was aimed at the total power loss reduction.

The one-criterion Nelder–Mead method is applied in this work to optimize the SHM design. An important advantage of the Nelder–Mead method over other methods that are often used to optimize electrical machines [8,9] is the significant savings in computational time, which makes it possible to increase the number of parameters to optimize, as well as to apply more complex optimization criteria, for the calculation of which it is necessary to calculate several load points of the machine [10]. This advantage is important for optimizing traction machines with a wide speed control range.

2. Geometry of the Traction SHM

Figure 1 shows the sketches of the nine-phase traction SHM. The machine has three sets of stacks on the stator and on the rotor. A nine-phase six-pole armature winding is placed on the stator. An excitation winding consisting of two coils is placed between the stator stacks. The stator has 54 slots. The rotor has no windings. Each rotor stack has 6 teeth, and the teeth of adjacent rotor stacks are offset by 30 mechanical degrees. In Figure 2a, the dependence of the maximal torque on the rotational speed of the electric drive of the BELAZ 75570 mining dump truck is shown. Figure 2b demonstrates the circuit of the nine-phase traction SHM inverter. The SHM traction inverter consists of three individual three-bridge inverters and a single-phase breaker for the supply of the excitation winding [4]. A more detailed description of the nine-phase traction SHM and the inverter is given in [4,11].

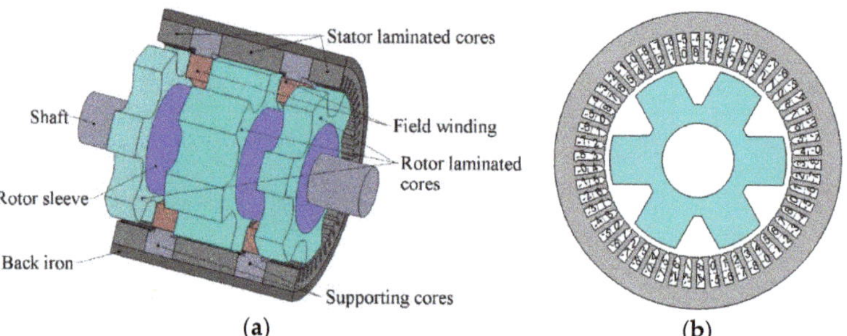

Figure 1. SHM design features: (**a**) general view of the SHM. The armature winding on the stator is not depicted so as not to obstruct; (**b**) SHM cross-section.

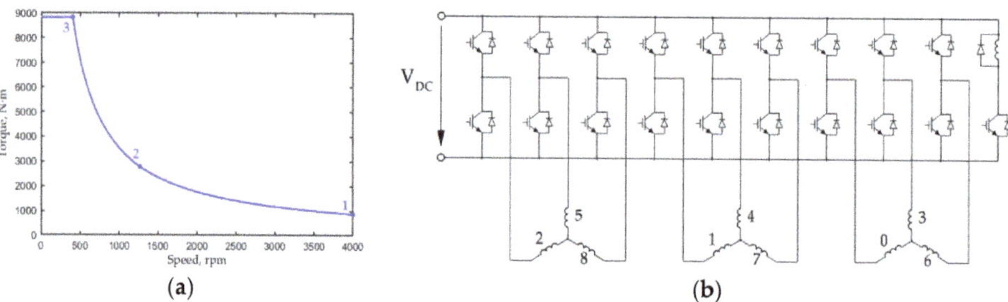

Figure 2. SHM diagrams: (**a**) traction characteristic of the motor; (**b**) inverter schematic; the numbers from 1 to 9 are the numbers of the phases of the SHM.

3. Objective Function for Optimizing the Traction SHM

Figure 2b shows the required torque-speed characteristic of the rear wheel drive of the considered truck [4]. The traction drive must provide the maximum torque of 8833 N·m from zero (standstill) to the rated speed of 400 rpm. In the speed range of 400 rpm and above, the drive must be capable of producing power of at least 370 kW. The maximum required rotation speed while maintaining the mechanical power of 370 kW is 4000 rpm. Table 1 indicates the motor operating points considered in the optimization procedure: maximum torque at rated speed, rated power at maximum speed, and an intermediate point with geometric mean values of the torque and speed.

Table 1. Operating points of the SHM taken into account in the optimization procedure.

Mode Number	Torque, N·m	Rotational Speed, rpm	Mechanical Power, kW
1	883	4000	370
2	2793	1265	370
3	8833	400	370

In [3], during the optimization of the traction SHM, it was possible to significantly reduce the motor losses, and the torque ripple in comparison with the non-optimized SHM [4]. The line voltage is reduced from 940 to 772 V. However, the peak armature current was reduced only slightly (from 886 to 816 A). Therefore, it is necessary to use the inverter with IGBT modules FF1000R17IE4 [12] with a rated current of 1000 A for both the non-optimized SHM [4] and the optimized SHM.

The aim of this paper is to demonstrate the possibility of using the cheaper and less powerful power module FF650R17IE4 [13] for the considered powertrain. For this purpose, it is necessary to increase the number of turns of the stator winding and change the geometry of the slot. At the same time, the maximum voltage limitation constraint must be set.

Therefore, when optimizing the traction SHM in this study, the following main goals were set: (1) the armature winding current must be significantly reduced; this will make it possible to switch to cheaper power IGBT modules FF650R17IE4 [13] in the inverter; (2) the amplitude of the voltage must not exceed 1000 V; and (3) it is necessary to reduce the total motor losses in comparison with the non-optimized design of the SHM [3].

When estimating the motor losses, ranges 1–2 and 2–3 on the motor torque-speed curve (Figure 2a) are considered. It is assumed that average losses in the ranges are equal to the arithmetic mean of the losses at their boundaries (points 1 and 2 and points 2 and 3, respectively) and that the motor will be equally likely to run between the points 1 and 2 and between points 2 and 3 on its torque–speed curve. For this reason, the weighted average losses were chosen as the first optimization objective:

$$<P_{losses}> = (P_{losses1} + 2 \cdot P_{losses2} + P_{losses3})/4. \tag{1}$$

The motor torque ripples were also chosen as an optimization objective. Both non-symmetrized (created by one combination of rotor and stator stacks) TR and symmetrized (created by a whole motor consisting of three combinations of stator and rotor stacks) torque ripple $TRsym$ are considered [3]. The last optimization objective is the maximum armature winding current I_3 that is achieved at the maximum torque (operating point 3). Taking into account all the objectives, the motor optimization function is formulated as:

$$F_0 = \ln(<P_{losses}>) + 0.7 \ln(I_3) + 0.05 \cdot \ln[\max(TR_{sym})] + 0.025 \cdot \ln[\max(TR)]. \tag{2}$$

Formula (2) suggests that $<P_{losses}>$ is the most valuable objective. The second most important objective is I_3. A 1% decrease in I_3 is as valuable as a 0.7% decrease in $<P_{losses}>$. The decrease in $\max(TR_{sym})$ and $\max(TR)$ is not so significant. Decreasing $\max(TR_{sym})$

and max(TR) by 1% is just as important as decreasing <P_{losses}> by 0.05% and 0.025%, respectively. However, including the max(TR_{sym}) and max(TR) terms allows the increase in motor torque ripple during optimization to be limited.

The optimization procedure also takes into account the following constraints:

$$U_{DC1} < 1000 \text{ V}; B_3 < 1.65 \text{ T}; P_{exc} < 12{,}000 \text{ W}, \tag{3}$$

where U_{DC1} is the maximum voltage reached at the maximum speed (operating point 1), P_{exc} is the power loss in the excitation winding, and B_3 is the maximum flux density in the non-laminated sections of the magnetic circuit (the rotor sleeve and the motor housing).

The optimization procedure is based on the one-criterion unconstrained Nelder–Mead method. There are various ways to implement constraints when using the Nelder–Mead method. For example, an objective function can take an infinite value when constraints are not met. However, this approach leads to a rapid decrease in the volume of the simplex. Additionally, constraints (3) should be satisfied in the initial design and along the optimization. To avoid these drawbacks, the 'soft constraints' are applied to the objective function (2). The constraining terms begin to increase rapidly if constraints are not met:

$$F = F_0 + k_1 \times f(U_{DC1}/1000[V] - 1) + k_2 \times f(B_3/1.65[T] - 1) + k_3 \times f(P_{exc}/12{,}000[W] - 1),$$
$$\text{where } f(x) = \begin{cases} x, x > 0 \\ 0 \end{cases}. \tag{4}$$

As a consequence, the objective function allows constraints to be violated in order to prevent a rapid decrease in the volume of the simplex. However, as it will be shown below, the optimized design will still satisfy constraints (3), if the factors k_1, k_2, and k_3 are large enough (exceed the corresponding Lagrange multipliers). In this study, it is assumed that $k_1 = k_2 = k_3 = 1.5$. Due to the choice of the objective function (4), the choice of the initial approximation is not limited by constraints (3). As will be seen below, constraints (3) will be violated in the initial approximation.

4. Initial Design Parameters and Variable Parameters Used for Optimizing the Traction SHM

Figure 3 shows the main geometric parameters of the non-optimized traction SHM [4]. The parameters that are fixed and varied during optimization are shown in Tables 2 and 3, correspondingly. The outer dimensions of the motor (the length of the motor without winding end parts L = 545 mm and the stator housing outer radius $R_{housing}$ = 367 mm) did not change during optimization. The rotor yoke thickness and the stator stack height were also not varied. Due to the fact that the outer radius of the stator housing $R_{housing}$ remains constant, the inner radius of the stator changes as the thickness h of the stator housing changes. The outer radius of the rotor also depends on the width of the air gap δ. To ensure equal conditions of flow of the excitation magnetic flux in the axial direction through the stator housing and the rotor sleeve, the areas of their cross-sections are assumed to be equal.

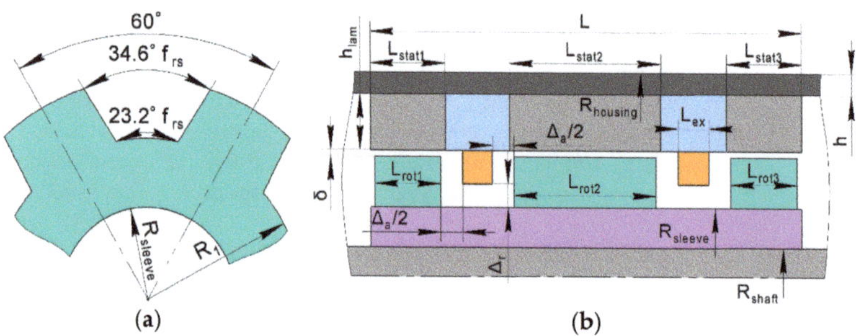

Figure 3. Geometric parameters of the SHM: (**a**) rotor radial dimensions; (**b**) stator and rotor axial and radial dimensions.

Table 2. Some geometric parameters of the SHM that were not varied during the optimization.

Parameter	Value [4]
Machine length without end winding parts L, mm	545
Lengths of the stator stacks, L_{stat1}; L_{stat2}; L_{stat3}, mm	101; 197; 101
The lengths of the rotor stacks, L_{rot1}; L_{rot2}; L_{rot3}, mm	92; 184; 92
Axial clearance between excitation winding and rotor, Δ_a, mm	30
Radial clearance between field winding and rotor Δ_r, mm	27
Rotor yoke thickness R_1-R_{sleeve}, mm	22.8
Shaft radius R_{shaft}, mm	70
Stator lamination height h_{lam}, mm	65
External radius of the stator housing $R_{housing}$, mm	367

Table 3. Parameters that were varied during the optimization.

Parameter	Initial Value before the Optimization [4]
Housing thickness h, mm	36
Total stator stacks length L_{stator}, mm	399
Airgap width δ, mm	2.3
Rotor slot factor f_{rs}	1
Angles of field weakening at operating points 1,2,3, electrical radians	0.61; 0.3; 0.25
Magnetic monopole densities at operating points 1,2,3, Wb/m	0.48; 0.63; 1.2

As a result, the outer radius of the rotor sleeve is determined by formula:

$$R_{sleeve} = \sqrt{(R_{shaft}^2 + R_{housing}^2 - [R_{housing} - h]^2)}. \tag{5}$$

In [4], R_{sleeve} equals 161 mm, while (5) provides R_{sleeve} = 167 mm. Since the thickness of the rotor yoke $R_1 - R_{sleeve}$ is not varied, the depth of the rotor slot changes not only with a change in the outer diameter of the rotor, but also with R_{sleeve}.

Variation of the angular dimensions between the rotor teeth along the rotor inner radius and along the rotor outer radius was carried out in concert by multiplying both dimensions by the coefficient f_{rs}. The excitation winding resistance is 10.2 Ohms in [4]. The longitudinal and radial dimensions occupied by the excitation winding between two stator stacks are L_{ex} = 43 mm and h_{exc} = 78 mm [3]. During the optimization, $L_{ex} = (L - L_{stat})/2 - \Delta_a$ changed along with the L_{stat} variation. $h_{exc} = R_{housing} - h - h_{lam} - R_{sleeve} - \Delta_r$ changed due to the variation in h, as well as due to the variation in R_{sleeve} which is a function of h (5). The resistance of the excitation winding changes with the dimensions of the field winding as 10.2 Ohm $\times$ 43 mm $\times$ 78 mm/h_{exc}/L_{ex}. The number of turns of the excitation winding is equal to 340 and assumed to be unchanged. In this study, the number of turns of the armature winding increased for better utilization of the supply voltage, which results in an increased height of the stator slots and decreased thickness of the stator lamination yoke. To restrict the growth of the stator slots, the thinner winding was chosen. The winding details are provided in Table 4. The main characteristics of the motor prototype described in [4] are shown on the left side of Table 5. The main characteristics of the motor obtained after the above changes (initial design) are shown on the right side of Table 5.

Table 4. Winding parameters that were non-varied during the optimization.

Parameter	SHM Prototype Described in [4]	New Initial Design
Number of turns per stator armature layer	5	7
Number of parallel strands per turn of the stator armature coil	2	2
Dimensions of armature wire winding, mm^2	3.15×4.5	2.5×4.5
The height of the stator slot part filled with the wire, mm	36.4	41.1
Excitation winding resistance, Ohm	10.2	16.8

Table 5. Comparison of the characteristics of the SHM prototype described in [4] and the characteristics of the new initial design used as the starting point for the optimization.

Value	SHM Prototype Described in [4]				New Initial Design (before the Optimization)			
Operating point	1	2	3	Brake mode	1	2	3	Brake mode
Speed, rpm	4000	1265	400	1100	4000	1265	400	1100
Current, A ampl	197	408	886	643	142	296	669	485
Mechanical power, kW	370	370	370	−540	370	370	370	−540
Active power, kW	412	387	404	−508	412	387	405	−509
Efficiency, %	89.8	95.4	90.0	93.8	89.8	95.3	89.8	94.0
Total losses, kW	41.9	18.0	41.0	32.2	42.2	18.1	42.1	32.4
Power factor	0.99	0.82	0.91	−0.65	0.99	0.81	0.88	−0.62
Line voltage, V ampl	940	472	196	462	1303	661	272	642
Not symmetrized torque ripple, N·m	71.9	61.5	24.1	42.1	71.9	62.1	24.1	43.5
Symmetrized torque ripple, N·m	21.0	12.4	2.8	8.4	20.8	12.3	2.6	8.1
Excitation current, A	5.6	8.1	26.3	10.7	5.5	8.1	24.8	10.8
Flux density in non-laminated parts of the magnetic core, T	0.59	0.77	1.46	0.77	0.59	1.04	1.65	0.98

Therefore, according to (5), the drop of the excitation magnetomotive force (MMF) on the rotor sleeve is reduced due to the increase in R_{sleeve}, in comparison with [3]. On the other hand, the depth of the rotor slots and the rotor saliency decrease. The resistance of the excitation winding also changes. In addition, the initial design used as a starting point for optimization differs from [3] in the parameters of the armature winding as Table 4 indicates. As can be seen from Table 5, the change in the parameters of the armature winding shown in Table 4 led to a significant increase in the line voltage up to 1303 V at the first operating point. This voltage value significantly exceeds the maximum allowable voltage in the DC-link of the mining dump truck power supply and cannot be implemented in practice. However, in the next section it will be shown that, using the objective function (4) and the Nelder–Mead method, it is possible to significantly improve all the main characteristics of the SHM without exceeding the voltage limit of 1000 V.

5. Optimization of Traction HSM Using the Nelder–Mead Method

The traditional Nelder–Mead algorithm [14], the 2-D FEM based mathematical model of the SHM, according to [4], and the objective function (4) were used in the SHM optimization process. This optimization was applied to the new initial design with a larger number of turns and a modified stator slot shape (see Table 4) to further reduce the armature winding current and comply with the maximum voltage limitation simultaneously.

The optimization procedure varied the 10 SHM parameters listed in Table 2. Figure 4 shows the cross-section of the motor and the magnitude of the magnetic flux density before and after optimization at the operating point 3 (see Table 1) with the maximum torque and the most saturated conditions. Regions of the cross-section with an extreme saturation level over 2 T are highlighted with black outlines. It can be seen that after the optimization, the area of the regions with maximum saturation decreased. Table 6 shows the modified design parameters of the SHM after optimization. Figure 5 demonstrates the change during optimization of such values as the total losses $<P_{losses}>$, the armature current amplitude I_3 at operating point 3, the maximum line voltage, and the value of the objective function F (4). Table 7 compares the main characteristics of the motor before and after optimization. As Table 6 shows, at operating points 1, 2, and 3 (motor mode operation), the total losses are reduced by 1.09, 1.19, and 1.04 times, respectively. After the optimization, the torque ripple only slightly decreased at operating point 1. The voltage at operating point 1 decreased 1.3 times; therefore, the maximum voltage is 988 V and does not exceed the constraint of 1000 V indicated in (3). Additionally, at operating point 3, the amplitude value of the armature winding current decreased from 669 to 601 A. Although in the initial design, due to the increase in the number of turns, the amplitude value of the armature winding current is much less than in [4], the initial approximation is not feasible due to the line voltage constraint. The use of optimization made it possible not only to reduce the line voltage in operating point 1 to an allowable level, but also to further reduce the armature winding current.

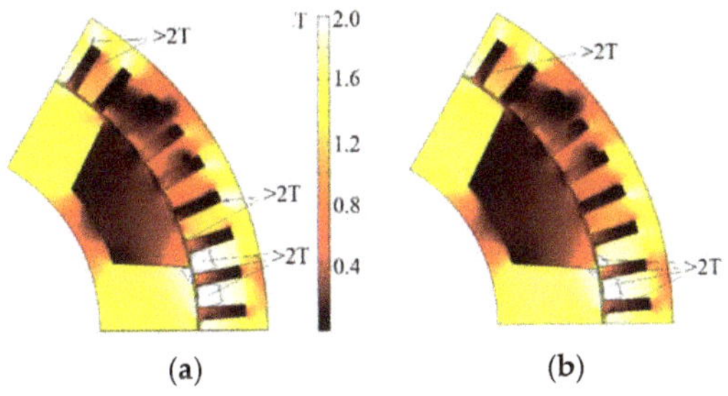

Figure 4. The motor cross-section and the magnitude plot of flux density; black outlines mark the extreme saturation level (>2 T): (**a**) before optimization; (**b**) after optimization; it can be observed that after optimization, the area of regions with flux density >2 T noticeably decreased.

Table 6. Varied design parameters of the traction SHM after the optimization.

Parameter	Optimal Design
Housing thickness h, mm	32.8
Total stator stacks length L_{stator}, mm	431
Airgap width δ, mm	2.41
Rotor slot factor f_{rs}	1.10
Angles of field weakening at operating points 1,2,3, electrical radians	0.762; 0.400; 0.364
Magnetic monopole densities at operating points 1,2,3, Wb/m	0.331; 0.678; 1.139

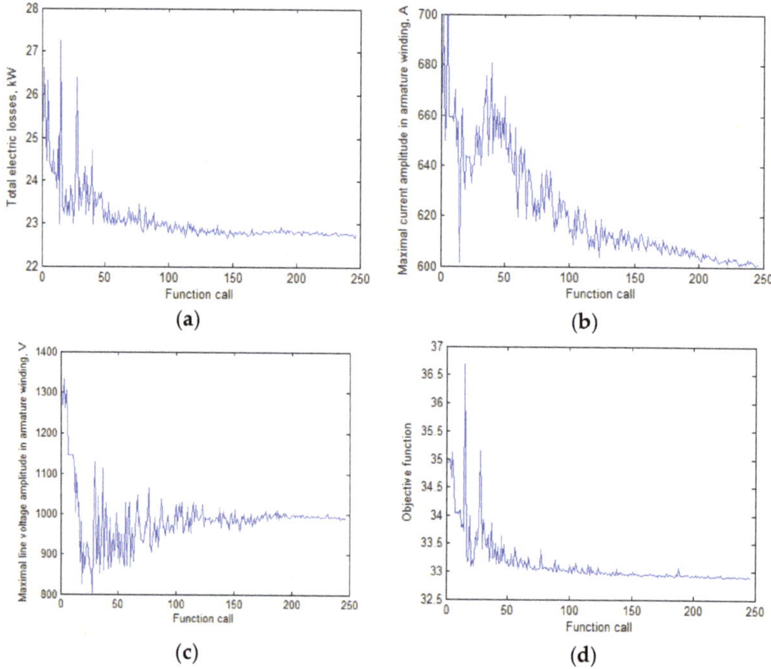

Figure 5. Variations in the objectives during the optimization: (**a**) total losses $<P_{losses}>$; (**b**) maximum current amplitude in the armature winding; (**c**) maximum voltage amplitude in the armature winding; (**d**) objective function F.

Although Table 7 demonstrates the SHM performances at only 3 points in the motor mode indicated in Figure 2a, the calculation shows that the optimized drive can deliver the target mechanical power of 370 kW over the entire speed range from 400 to 4000 rpm.

Since the maximum amplitude current value for the optimized SHM is less than 650 A, then the FF650R17IE4 IGBT modules can be used in the inverter, instead of the FF1000R17IE4 modules, which are used in the commercially available prototype of the traction SHM drive [4]. The cost of the FF1000R17IE4 module is $ 840 while the cost of the FF650R17IE4 module is 1.4 times less and is $ 585. The prices are from the website of the IGBT module manufacturer [15]. Thus, for the 9-phase inverter, the savings on IGBT modules alone are $(840 - 585) \times 9 = \$ 2295$. In addition, the AC current component in the capacitor bank of the DC link will decrease by 1.4 times, which will make it possible to reduce the capacitance of the capacitor bank of the DC link, and will also lead to a decrease in the cost of the inverter.

Table 7. Optimization results.

Value	New Initial Design (Before the Optimization)				After Optimization			
Operating point	1	2	3	Brake mode	1	2	3	Brake mode
Speed, rpm	4000	1265	400	1100	4000	1265	400	1100
Current, A ampl	142	296	669	485	176	255	601	429
Mechanical power, kW	370	370	370	−540	370	370	370	−540
Mechanical losses, kW	17.57	0.65	0.05	0.45	17.57	0.65	0.05	0.45
Conductive winding losses, kW	1.4	6.1	30.9	16.2	2.1	4.5	25.0	12.7

Table 7. *Cont.*

Value	New Initial Design (Before the Optimization)				After Optimization			
Eddy-current winding losses, kW	5.2	2.5	1.1	4.3	6.0	2.1	1.0	3.8
Stator core losses, kW	15.1	7.5	2.4	9.4	10.9	5.9	2.5	8.2
Rotor core losses, kW	2.5	0.6	0.1	0.6	1.5	0.6	0.1	0.5
Excitation losses, kW	0.4	0.8	7.5	1.4	0.5	1.5	11.9	2.0
Active power, kW	412	387	405	−509	408	384	399	−514
Efficiency, %	89.8	95.3	89.8	94.0	90.5	96.1	90.1	94.9
Total losses (motor), kW	42.2	18.1	42.1	32.4	38.7	15.2	40.5	27.6
Line voltage, V ampl	1303	661	272	642	988	632	275	621
Symmetrized torque ripple, N·m	20.8	12.3	2.6	8.1	18.9	12.5	3.0	9.4
Excitation current, A	5.5	8.1	24.8	10.8	5.5	9.5	26.6	10.8
Flux density in non-laminated parts of the magnetic core, T	0.59	0.77	1.46	0.77	0.48	0.98	1.65	0.91
Power factor	0.99	0.81	0.88	−0.62	0.97	0.96	0.96	−0.74

6. Conclusions

Inadequate design methods for synchronous homopolar machines (SHMs) can result in the need to significantly overrate the installed power of the traction inverter in applications requiring operation over a wide constant power speed range. This article discusses the novel procedure and results of optimization of the commercially available 370 kW traction SHM using the Nelder–Mead method. The objective function was composed to improve/minimize the basic characteristics of the traction SHM, such as the total motor power loss and maximum armature winding current. To obtain the feasible optimized design, necessary constraints were imposed. As a result of the optimization, the motor losses and the maximum current required by the motor from the inverter were significantly reduced. The achieved reduction in the maximum current allowed the cost of the IGBT modules of the inverter to be reduced by 1.4 times (by $2295), and also allowed a reduction of the AC component of the DC link current.

Author Contributions: Conceptual approach, A.A., V.D. and V.P.; data duration, V.D. and V.K.; software, V.D. and V.P.; calculations and modeling, A.A., V.D., V.K. and V.P.; writing—original draft, A.A., V.D., V.K. and V.P.; visualization, V.D. and V.K.; review and editing, A.A., V.D., V.K. and V.P. All authors have read and agreed to the published version of the manuscript.

Funding: The research was performed with the support of the Russian Science Foundation grant (Project № 21-19-00696).

Institutional Review Board Statement: Not applicable.

Informed Consent Statement: Not applicable.

Data Availability Statement: All data are contained within the article.

Acknowledgments: The authors thank the editors and reviewers for their careful reading and constructive comments.

Conflicts of Interest: The authors declare no conflict of interest.

References

1. Kalsi, S.; Hamilton, K.; Buckley, R.G.; Badcock, R.A. Superconducting AC Homopolar Machines for High-Speed Applications. *Energies* **2019**, *12*, 86. [CrossRef]
2. Bianchini, C.; Immovilli, F.; Bellini, A.; Lorenzani, E.; Concari, C.; Scolari, M. Homopolar generators: An overview. In Proceedings of the 2011 IEEE Energy Conversion Congress and Exposition, Phoenix, AZ, USA, 17–22 September 2011; pp. 1523–1527. [CrossRef]
3. Dmitrievskii, V.; Prakht, V.; Anuchin, A.; Kazakbaev, V. Design Optimization of a Traction Synchronous Homopolar Motor. *Mathematics* **2021**, *9*, 1352. [CrossRef]
4. Dmitrievskii, V.; Prakht, V.; Anuchin, A.; Kazakbaev, V. Traction Synchronous Homopolar Motor: Simplified Computation Technique and Experimental Validation. *IEEE Access* **2020**, *8*, 185112–185120. [CrossRef]
5. Ye, C.; Yang, J.; Xiong, F.; Zhu, Z.Q. Relationship between homopolar inductor machine and wound-field synchronous machine. *IEEE Trans. Ind. Electron.* **2020**, *67*, 919–930. [CrossRef]
6. Yang, J.; Ye, C.; Liang, X.; Xu, W.; Xiong, F.; Xiang, Y.; Li, W. Investigation of a Two-Dimensional Analytical Model of the Homopolar Inductor Alternator. *IEEE Trans. Appl. Supercond.* **2018**, *28*, 5205205. [CrossRef]
7. Belalahy, C.; Rasoanarivo, I.; Sargos, F. Using 3D reluctance network for design a three phase synchronous homopolar machine. In Proceedings of the 2008 34th Annual Conference of IEEE Industrial Electronics, Orlando, FL, USA, 10–13 November 2008; pp. 2067–2072. [CrossRef]
8. Cupertino, F.; Pellegrino, G.; Gerada, C. Design of synchronous reluctance machines with multiobjective optimization algorithms. *IEEE Trans. Ind. Appl.* **2014**, *50*, 3617–3627. [CrossRef]
9. Krasopoulos, C.T.; Beniakar, M.E.; Kladas, A.G. Robust Optimization of High-Speed PM Motor Design. *IEEE Trans. Magn.* **2017**, *53*, 1–4. [CrossRef]
10. Prakht, V.; Dmitrievskii, V.; Kazakbaev, V. Optimal Design of Gearless Flux-Switching Generator with Ferrite Permanent Magnets. *Mathematics* **2020**, *8*, 206. [CrossRef]
11. Anuchin, A. Development of Digital Systems for Efficient Control of Traction Electric Equipment for Hybrid Electric Vehicles. Ph.D. Thesis, Moscow Power Engineering Institute, Moscow, Russia, 2018; pp. 1–445. Available online: https://mpei.ru/diss/Lists/FilesDissertations/365-%D0%94%D0%B8%D1%81%D1%81%D0%B5%D1%80%D1%82%D0%B0%D1%86%D0%B8%D1%8F.pdf (accessed on 10 October 2021). (In Russian)
12. FF1000R17IE4, IGBT-Modules, Technical Information, Revision 3.2, Infineon. November 2013. Available online: https://www.infineon.com/dgdl/Infineon-FF1000R17IE4-DS-v03_02-EN.pdf?fileId=db3a30431ff9881501201dc994a34980 (accessed on 10 October 2021).
13. FF650R17IE4, IGBT-Modules, Technical Information, Revision 3.3, Infineon. November 2013. Available online: https://www.infineon.com/dgdl/Infineon-FF650R17IE4-DS-v03_03-EN.pdf?fileId=db3a30431ff9881501201dcfe2a54986 (accessed on 10 October 2021).
14. Nelder, J.A.; Mead, R. A Simplex Method for Function Minimization. *Comput. J.* **1965**, *7*, 308–313. [CrossRef]
15. IGBT Modules. Product Description. Available online: https://www.infineon.com/cms/en/product/power/igbt/igbt-modules/ (accessed on 10 October 2021).

MDPI

St. Alban-Anlage 66

4052 Basel

Switzerland

Tel. +41 61 683 77 34

Fax +41 61 302 89 18

www.mdpi.com

Mathematics Editorial Office

E-mail: mathematics@mdpi.com

www.mdpi.com/journal/mathematics

www.ingramcontent.com/pod-product-compliance
Lightning Source LLC
LaVergne TN
LVHW071522170726
843515LV00008B/2043